Nonlinear Dynamics and Entropy of Complex Systems with Hidden and Self-Excited Attractors

Nonlinear Dynamics and Entropy of Complex Systems with Hidden and Self-Excited Attractors

Special Issue Editors

Christos Volos
Sajad Jafari
Jacques Kengne
Jesus M. Munoz-Pacheco
Karthikeyan Rajagopal

MDPI • Basel • Beijing • Wuhan • Barcelona • Belgrade

MDPI

Special Issue Editors

Christos Volos
Aristotle University of Thessaloniki
Greece

Sajad Jafari
Amirkabir University of Technology
Iran

Jacques Kengne
University of Dschang
Cameroon

Jesus M. Munoz-Pacheco
Autonomous University of Puebla
Mexico

Karthikeyan Rajagopal
Defence University
Ethiopia

Editorial Office
MDPI
St. Alban-Anlage 66
4052 Basel, Switzerland

This is a reprint of articles from the Special Issue published online in the open access journal *Entropy* (ISSN 1099-4300) from 2018 to 2019 (available at: https://www.mdpi.com/journal/entropy/special_issues/Nonlinear_Entropy).

For citation purposes, cite each article independently as indicated on the article page online and as indicated below:

LastName, A.A.; LastName, B.B.; LastName, C.C. Article Title. *Journal Name* **Year**, *Article Number*, Page Range.

ISBN 978-3-03897-898-5 (Pbk)
ISBN 978-3-03897-899-2 (PDF)

Contents

About the Special Issue Editors

Christos Volos, Dr., received his Physics Diploma degree (1999), M.Sc. degree in electronics (2002), and Ph.D. degree (2008) in chaotic electronics from the Physics Department, Aristotle University of Thessaloniki, Greece. He also currently serves here as an Assistant Professor and a member of the Laboratory of Nonlinear Systems—Circuits & Complexity (LaNSCom).

His current research interests include chaos, nonlinear systems, mem-elements, design of analog and mixed signal electronic circuits, neural networks, chaotic electronics and their applications (secure communications, cryptography, robotics), and chaotic synchronization and control.

Dr. Volos has published 125 papers in international journals, 40 book chapters, and 58 international conference papers, in addition to editing 7 books on the topic of nonlinear circuits and systems. Furthermore, he is an Editorial Board Member of 8 international journals and has been a Guest Editor of 15 Special Issues in international journals. Dr. Volos is also closely associated with several international journals, where he serves as a reviewer.

Sajad Jafari, Dr., was born in Kermanshah, Iran, in 1983. He received his B.Sc. (2005), M.Sc. (2008), and Ph.D. (2013) degrees in biomedical engineering from the Biomedical Engineering Department, Amirkabir University of Technology, Tehran, Iran. He is also currently serving here as an Assistant Professor, a position he has held since 2013.

His research interests include artificial intelligence, optimization, pattern recognition, and especially nonlinear and chaotic signals and systems. Dr. Jafari serves as Editor in the journals *International Journal of Bifurcation and Chaos* (World Scientific) and *AEÜ—International Journal of Electronics and Communications* (Elsevier).

Jacques Kengne, Dr., was born in Cameroon in 1971. He received his M.Sc. (2007) and Ph.D. (2011) degrees in electronics from the Faculty of Sciences, University of Dschang, Cameroon. From 2010 to 2012, he worked as a Lecturer at the Department of Electrical Engineering, IUT-FV, at the University of Dschang. He has served as Senior Lecturer from 2012 until 2016, when he was appointed in his current position as Associate Professor.

Prof. Kengne has authored or co-authored more than 60 journal papers. He also serves as a reviewer for renowned international journals, including *IJBC*; *Commun Nonlinear Sci. Numer Simulat.*; *Int. Journal of Dynamics and Control*; *Nonlinear Dynamics*; *Chaos, Solitons & Fractals*; and *IEEE Trans Circuits Systems*. His research interests include nonlinear systems and circuits, chaos, multistability, and chaos synchronization with applications.

Jesus M. Munoz-Pacheco, Dr., received his M.Sc. (2005) and Ph.D. (2009) degrees from the National Institute for Astrophysics, Optics and Electronics (INAOE), Mexico. He has served as Associate Professor–Researcher at Polytechnic University of Puebla from 2008 until 2012, when he was appointed to his current position as Full Professor–Researcher at Autonomous University of Puebla (BUAP).

His research interests include the design of chaotic systems using electronic circuits (integrated and discrete), chaos control and synchronization, fractional order chaotic systems, and chaos-based applications. Within these areas, he has authored and co-authored 1 book, 7 book chapters, 6 patents (filed in Mexico), 41 journal articles, and around 30 IEEE conference papers. Dr. Munoz-Pacheco is a regular member (Level I) of the National System of Researchers, an outstanding program founded by the Mexican Government through the National Council for Science and Technology (CONACYT). He is Associate Editor of the *International Journal of Nonlinear Dynamics and Control*, and Guest Editor for the JCR-journals *Mathematical Problems in Engineering*, *Complexity*, and *Entropy*. Moreover, Dr. Munoz-Pacheco regularly serves as a reviewer for high impact-factor journals (including *TCAS-I* and *TCAS-II*; *Nonlinear Dynamics*; and *Chaos, Solitons & Fractals*).

Karthikeyan Rajagopal, Dr., is presently a Senior Researcher and Professor in the Center for Nonlinear Dynamics of Defence University, Ethiopia. He completed his Ph.D. in electronics and communication engineering, specializing in chaos-based secure communication engineering.

His post-graduate work has focused on embedded system technologies with an emphasis on programming of real-time targets. He has over 100 international journal papers indexed in SCI and his present research areas include fractional order nonlinear systems and control, time delay systems, FPGA, and LabVIEW implementations of fractional order systems. He has several projects funded by the Ministry of National Defence, Ethiopia, and Institute of Research and Development, Defence University, Ethiopia.

entropy

MDPI

Editorial

Nonlinear Dynamics and Entropy of Complex Systems with Hidden and Self-Excited Attractors

Christos K. Volos [1,*], Sajad Jafari [2], Jacques Kengne [3], Jesus M. Munoz-Pacheco [4] and Karthikeyan Rajagopal [5]

[1] Laboratory of Nonlinear Systems, Circuits & Complexity (LaNSCom), Department of Physics, Aristotle University of Thessaloniki, Thessaloniki 54124, Greece
[2] Nonlinear Systems and Applications, Faculty of Electrical and Electronics Engineering, Ton Duc Thang University, Ho Chi Minh City 700000, Vietnam; sajad.jafari@tdtu.edu.vn
[3] Department of Electrical Engineering, University of Dschang, P.O. Box 134 Dschang, Cameroon; kengnemozart@yahoo.fr
[4] Faculty of Electronics Sciences, Autonomous University of Puebla, Puebla 72000, Mexico; jesusm.pacheco@correo.buap.mx
[5] Center for Nonlinear Dynamics, Institute of Research and Development, Defence University, P.O. Box 1041 Bishoftu, Ethiopia; rkarthiekeyan@gmail.com
* Correspondence: volos@physics.auth.gr

Received: 1 April 2019; Accepted: 3 April 2019; Published: 5 April 2019

Keywords: hidden attractor; complex systems; fractional-order; entropy; chaotic maps; chaos

In the last few years, entropy has been a fundamental and essential concept in information theory. It is also often used as a measure of the degree of chaos in systems; e.g., Lyapunov exponents, fractal dimension, and entropy are usually used to describe the complexity of chaotic systems. Thus, it will be important to study entropy in nonlinear systems. Additionally, there has been an increasing interest in a new classification of nonlinear dynamical systems including two kinds of attractors: self-excited attractors and hidden attractors. Self-excited attractors can be localized straightforwardly by applying a standard computational procedure. Some interesting examples of systems with self-excited attractors are chaotic systems with different kinds of symmetry, with multi-scroll attractors, with multiple attractors, and with extreme multistability.

In systems with hidden attractors, we have to develop a specific computational procedure to identify the hidden attractors because the equilibrium points do not help in their localization. Some examples of this kind of system are chaotic dynamical systems with no equilibrium points, with only stable equilibria, with curves of equilibria, with surfaces of equilibria, and with non-hyperbolic equilibria. There is evidence that hidden attractors play a vital role in various fields ranging from phase-locked loops, oscillators, describing convective fluid motion, a model of the drilling system, information theory and cryptography to multilevel DC/DC converters. Furthermore, hidden attractors may lead to unexpected and disastrous responses.

The overall purpose of this Special Issue lies in gathering the latest scientific trends on the advanced topics of dynamics, entropy, fractional-order calculus, and applications in complex systems with hidden attractors and self-excited attractors.

In the paper "A New Chaotic System with Multiple Attractors: Dynamic Analysis, Circuit Realization and S-Box Design", Qiang Lai, Akif Akgul, Chunbiao Li, Guanghui Xu, and Ünal Çavuşoğlu report a novel three-dimensional chaotic system with three nonlinearities. The system has one stable equilibrium, two stable equilibria, and one saddle-node, two saddle foci and one saddle-node for different parameters. Also, an electronic circuit is given for implementing the chaotic attractors of the system, and an S-Box is developed for cryptographic operations [1].

In the paper "A New Chaotic System with a Self-Excited Attractor: Entropy Measurement, Signal Encryption, and Parameter Estimation", Guanghui Xu, Yasser Shekofteh, Akif Akgül, Chunbiao Li and Shirin Panahi introduce a new chaotic system with an engineering application for signal encryption. The implementation and manufacturing are performed via a real circuit as a random number generator. Also, the authors provide a parameter estimation method to extract chaotic model parameters from the real data of the chaotic circuit using a Gaussian mixture model (GMM) and two optimization algorithms: WOA (Whale Optimization Algorithm), and MVO (Multi-Verse Optimizer) [2].

In the paper "A Novel Algorithm to Improve Digital Chaotic Sequence Complexity through CCEMD and PE", Chunlei Fan, Zhigang Xie, and Qun Ding introduce a three-dimensional chaotic system with a hidden attractor. The complex dynamic behaviors of the system are analyzed by Poincaré cross-sections, equilibria, and initial values. Further, they have designed a new algorithm based on complementary ensemble empirical mode decomposition (CEEMD) and permutation entropy (PE) that can effectively enhance digital chaotic sequence complexity [3].

In the paper "A New Two-Dimensional Map with Hidden Attractors", Chuanfu Wang and Qun Ding investigate the hidden dynamics of a new two-dimensional map inspired by Arnold's cat map and study the existence of fixed points and their stabilities in detail [4].

In the paper "Stochastic Entropy Solutions for Stochastic Nonlinear Transport Equations", Rongrong Tian and Yanbin Tang analyze the existence and uniqueness of the stochastic entropy solution for a nonlinear transport equation with a stochastic perturbation. They prove the continuous dependence of stochastic robust entropy solutions on the coefficients and the nonlinear functions [5].

In the paper "Multivariate Multiscale Complexity Analysis of Self-Reproducing Chaotic Systems", Shaobo He, Chunbiao Li, Kehui Sun and Sajad Jafari propose a chaotic system with infinitely many attractors. Multiscale multivariate permutation entropy (MMPE) and multiscale multivariate Lempel–Ziv complexity (MMLZC) are employed to analyze the complexity of these self-reproducing chaotic systems with infinitely many chaotic attractors [6].

In the paper "A New Fractional-Order Chaotic System with Different Families of Hidden and Self-Excited Attractors", Jesus M. Munoz-Pacheco, Ernesto Zambrano-Serrano, Christos Volos, Sajad Jafari, Jacques Kengne and Karthikeyan Rajagopal introduce a new fractional-order chaotic system with a single parameter and four nonlinearities. One striking feature is that by varying the system parameter, the fractional-order system generates several complex dynamics: self-excited attractors, hidden attractors, the coexistence of hidden attractors, and multistability. Moreover, the complexity of the system is analyzed by computing its spectral entropy and Brownian-like motions [7].

In the paper "A New Chaotic System with Stable Equilibrium: Entropy Analysis, Parameter Estimation, and Circuit Design", Tomasz Kapitaniak, S. Alireza Mohammadi, Saad Mekhilef, Fawaz E. Alsaadi, Tasawar Hayat and Viet-Thanh Pham present a new three-dimensional chaotic system with one stable equilibrium. This system is a multistable dynamic system in which the strange attractor is hidden. To show the feasibility and ability in engineering applications of the proposed system, an entropy analysis, parameter estimation, and circuit design are performed [8].

In the paper "Optimization of Thurston's Core Entropy Algorithm for Polynomials with a Critical Point of Maximal Order", Gamaliel Blé and Domingo González discuss some properties of the topological entropy systems generated by polynomials of degree d in their Hubbard tree. An optimization of Thurston's core entropy algorithm is developed for a family of polynomials of degree d [9].

In the paper "Strange Attractors Generated by Multiple-Valued Static Memory Cell with Polynomial Approximation of Resonant Tunneling Diodes", Jiri Petrzela studies the multiple-valued memory system (MVMS) composed by a pair of resonant tunneling diodes (RTD). For specific values of system parameters, such a tunnel shows a double-spiral chaotic attractor. The existence of these types of strange attractors is proved using the largest Lyapunov exponents (LLE) and computer-aided simulation of the designed lumped circuit using only commercially available active elements [10].

In the paper "The Co-existence of Different Synchronization Types in Fractional-order Discrete-time Chaotic Systems with Non–identical Dimensions and Orders", Samir Bendoukha, Adel Ouannas, Xiong Wang, Amina-Aicha Khennaoui, Viet-Thanh Pham, Giuseppe Grassi, and Van Van Huynh analyze the co-existence of different synchronization types for fractional-order discrete-time chaotic systems with different dimensions. They show that through appropriate nonlinear control, projective synchronization (PS), full state hybrid projective synchronization and inverse full state hybrid projective synchronization (IFSHPS), generalized synchronization is achieved [11].

In the paper "The Complexity and Entropy Analysis for Service Game Model Based on Different Expectations and Optimal Pricing", Yimin Huang, Xingli Chen, Qiuxiang Li and Xiaogang Ma propose a multichannel dynamic service game model to analyze the relations between the manufacturer and the retailer under optimal pricing. Theoretical analysis of the model and numerical simulations from the perspective of entropy theory, game theory, and chaotic dynamics are conducted. Chaotic and complex behaviors are observed causing the system's entropy to increase when the manufacturer adjusts the service decision quickly [12].

In the paper "Dynamics and Complexity of a New 4D Chaotic Laser System", Hayder Natiq, Mohamad Rushdan Md Said, Nadia M. G. Al-Saidi and Adem Kilicman introduce a new 4D chaotic laser system with three equilibria and only two quadratic nonlinearities. Dynamics analysis, including stability of symmetric equilibria and the existence of coexisting multiple Hopf bifurcations on these equilibria, are investigated. Moreover, the complexity of the laser system reveals that system time series can locate and determine the parameters and initial values that show coexisting attractors [13].

In the paper "Entropy Analysis and Neural Network-Based Adaptive Control of a Non-Equilibrium Four-Dimensional Chaotic System with Hidden Attractors", Hadi Jahanshahi, Maryam Shahriari-Kahkeshi, Raúl Alcaraz, Xiong Wang, Vijay P. Singh, and Viet-Thanh Pham present a non-equilibrium four-dimensional chaotic system with hidden attractors. Its dynamical behavior is investigated using a bifurcation diagram, as well as three well-known entropy measures: approximate entropy, sample entropy, and Fuzzy entropy. Additionally, an adaptive radial-basis-function neural network (RBF-NN)-based control method is proposed [14].

In the paper "Adaptive Synchronization of Fractional-Order Complex Chaotic system with Unknown Complex Parameters", Ruoxun Zhang, Yongli Liu and Shiping Yang investigate the problem of synchronization of fractional-order complex-variable chaotic systems (FOCCS) with unknown complex parameters. Based on the complex-variable inequality and stability theory for fractional-order complex-valued systems, a new scheme is presented for adaptive synchronization of FOCCS with unknown complex parameters [15].

In the paper "Chaotic Map with No Fixed Points: Entropy, Implementation and Control", Van Van Huynh, Adel Ouannas, Xiong Wang, Viet-Thanh Pham, Xuan Quynh Nguyen, and Fawaz E. Alsaadi propose a map without equilibrium. The map has no fixed point but exhibits chaos. The entropy of this new map has been calculated. Also, experimental observations of the map using an open micro-controller platform are given [16].

In the paper "Dynamics and Entropy Analysis for a New 4-D Hyperchaotic System with Coexisting Hidden Attractors", Licai Liu, Chuanhong Du, Xiefu Zhang, Jian Li, and Shuaishuai Shi report a new no-equilibrium 4-D hyperchaotic multistable system with coexisting hidden attractors. One prominent feature is that by varying a system parameter or initial value, the system can generate several nonlinear complex attractors: periodic, quasiperiodic, multiple topologies chaotic, and hyperchaotic [17].

The Guest Editors hope that you will enjoy reading this Special Issue devoted to this exciting and fast-evolving field, and that it will motivate researchers to pursue further advances in the emerging areas of complex systems with hidden and self-excited attractors.

Acknowledgments: We express our thanks to the authors of the above contributions, and to the journal Entropy and MDPI for their support during this work.

Conflicts of Interest: The authors declare no conflict of interest.

References

1. Lai, Q.; Akgul, A.; Li, C.; Xu, G.; Çavuşoğlu, Ü. A New Chaotic System with Multiple Attractors: Dynamic Analysis, Circuit Realization and S-Box Design. *Entropy* **2018**, *20*, 12. [CrossRef]
2. Xu, G.; Shekofteh, Y.; Akgül, A.; Li, C.; Panahi, S. A New Chaotic System with a Self-Excited Attractor: Entropy Measurement, Signal Encryption, and Parameter Estimation. *Entropy* **2018**, *20*, 86. [CrossRef]
3. Fan, C.; Xie, Z.; Ding, Q. A Novel Algorithm to Improve Digital Chaotic Sequence Complexity through CCEMD and PE. *Entropy* **2018**, *20*, 295. [CrossRef]
4. Wang, C.; Ding, Q. A New Two-Dimensional Map with Hidden Attractors. *Entropy* **2018**, *20*, 322. [CrossRef]
5. Tian, R.; Tang, Y. Stochastic Entropy Solutions for Stochastic Nonlinear Transport Equations. *Entropy* **2018**, *20*, 395. [CrossRef]
6. He, S.; Li, C.; Sun, K.; Jafari, S. Multivariate Multiscale Complexity Analysis of Self-Reproducing Chaotic Systems. *Entropy* **2018**, *20*, 556. [CrossRef]
7. Munoz-Pacheco, J.M.; Zambrano-Serrano, E.; Volos, C.; Jafari, S.; Kengne, J.; Rajagopal, K. A New Fractional-Order Chaotic System with Different Families of Hidden and Self-Excited Attractors. *Entropy* **2018**, *20*, 564. [CrossRef]
8. Kapitaniak, T.; Mohammadi, S.A.; Mekhilef, S.; Alsaadi, F.E.; Hayat, T.; Pham, V.-T. A New Chaotic System with Stable Equilibrium: Entropy Analysis, Parameter Estimation, and Circuit Design. *Entropy* **2018**, *20*, 670. [CrossRef]
9. Blé, G.; González, D. Optimization of Thurston's Core Entropy Algorithm for Polynomials with a Critical Point of Maximal Order. *Entropy* **2018**, *20*, 695. [CrossRef]
10. Petrzela, J. Strange Attractors Generated by Multiple-Valued Static Memory Cell with Polynomial Approximation of Resonant Tunneling Diodes. *Entropy* **2018**, *20*, 697. [CrossRef]
11. Bendoukha, S.; Ouannas, A.; Wang, X.; Khennaoui, A.-A.; Pham, V.-T.; Grassi, G.; Huynh, V.V. The Co-existence of Different Synchronization Types in Fractional-order Discrete-time Chaotic Systems with Non–identical Dimensions and Orders. *Entropy* **2018**, *20*, 710. [CrossRef]
12. Huang, Y.; Chen, X.; Li, Q.; Ma, X. The Complexity and Entropy Analysis for Service Game Model Based on Different Expectations and Optimal Pricing. *Entropy* **2018**, *20*, 858. [CrossRef]
13. Natiq, H.; Said, M.R.M.; Al-Saidi, N.M.G.; Kilicman, A. Dynamics and Complexity of a New 4D Chaotic Laser System. *Entropy* **2019**, *21*, 34. [CrossRef]
14. Jahanshahi, H.; Shahriari-Kahkeshi, M.; Alcaraz, R.; Wang, X.; Singh, V.P.; Pham, V.-T. Entropy Analysis and Neural Network-Based Adaptive Control of a Non-Equilibrium Four-Dimensional Chaotic System with Hidden Attractors. *Entropy* **2019**, *21*, 156. [CrossRef]
15. Zhang, R.; Liu, Y.; Yang, S. Adaptive Synchronization of Fractional-Order Complex Chaotic system with Unknown Complex Parameters. *Entropy* **2019**, *21*, 207. [CrossRef]
16. Huynh, V.V.; Ouannas, A.; Wang, X.; Pham, V.-T.; Nguyen, X.Q.; Alsaadi, F.E. Chaotic Map with No Fixed Points: Entropy, Implementation and Control. *Entropy* **2019**, *21*, 279. [CrossRef]
17. Liu, L.; Du, C.; Zhang, X.; Li, J.; Shi, S. Dynamics and Entropy Analysis for a New 4-D Hyperchaotic System with Coexisting Hidden Attractors. *Entropy* **2019**, *21*, 287. [CrossRef]

entropy

MDPI

Article

A New Chaotic System with Multiple Attractors: Dynamic Analysis, Circuit Realization and S-Box Design

Qiang Lai [1,*] , Akif Akgul [2], Chunbiao Li [3], Guanghui Xu [4] and Ünal Çavuşoğlu [5]

[1] School of Electrical and Automation Engineering, East China Jiaotong University, Nanchang 330013, China
[2] Department of Electrical and Electronics Engineering, Faculty of Technology, Sakarya University, Serdivan 54187, Turkey; aakgul@sakarya.edu.tr
[3] School of Electronic and Information Engineering, Nanjing University of Information Science and Technology, Nanjing 210044, China; chunbiaolee@nuist.edu.cn
[4] School of Electrical and Electronic Engineering, Hubei University of Technology, Wuhan 430068, China; xgh@hbut.edu.cn
[5] Department of Computer Engineering, Faculty of Computer and Information Sciences, Sakarya University, Serdivan 54187, Turkey; unalc@sakarya.edu.tr
* Correspondence: chaos1963@ecjtu.jx.cn or laiqiang87@126.com; Tel.: +86-0791-8704-6216

Received: 17 November 2017; Accepted: 25 December 2017; Published: 27 December 2017

Abstract: This paper reports about a novel three-dimensional chaotic system with three nonlinearities. The system has one stable equilibrium, two stable equilibria and one saddle node, two saddle foci and one saddle node for different parameters. One salient feature of this novel system is its multiple attractors caused by different initial values. With the change of parameters, the system experiences mono-stability, bi-stability, mono-periodicity, bi-periodicity, one strange attractor, and two coexisting strange attractors. The complex dynamic behaviors of the system are revealed by analyzing the corresponding equilibria and using the numerical simulation method. In addition, an electronic circuit is given for implementing the chaotic attractors of the system. Using the new chaotic system, an S-Box is developed for cryptographic operations. Moreover, we test the performance of this produced S-Box and compare it to the existing S-Box studies.

Keywords: new chaotic system; multiple attractors; electronic circuit realization; S-Box algorithm

1. Introduction

The discovery of the well-known Lorenz attractor [1] in 1963 opened the upsurge of chaos research. In the decades thereafter, a large number of meaningful achievements on chaos control, chaotification, synchronization and chaos application have emerged continuously. Great changes have also been made to the understanding of chaos. Scholars began to think more about a way to produce chaos rather than blindly suppress chaos. The generation of chaotic attractors in three-dimensional autonomous ordinary differential systems has been of particular interest. As we all know, a multitude of typical systems with chaotic attractors were found, including Rössler system, Chen system, Sprott system, Lü system, etc. [2–8].

With the further research of chaos, scientists found that some nonlinear dynamic systems not only have a chaotic attractor but also coexist with multiple attractors for a set of fixed parameter values. The coexisting attractors may be fixed points, limit cycles, strange attractors, etc. The number and type of attractors are usually associated with parameters and initial conditions of the system. Each attractor has its own basin of attraction which is composed of the initial conditions leading to long-term behavior that settles onto the attractor. The phenomenon of multiple attractors can be seen in many biological systems and physical systems [9–11]. In recent years, the low-dimensional autonomous chaotic systems with multiple attractors have aroused scholars' research enthusiasm.

Li and Sprott found multiple attractors in chaotic systems by numerical analysis and introduced the offset boosting method and conditional symmetry method for producing multiple attractors in differential systems [12–16]. Kengne et al. analyzed the multiple attractors of simple chaotic circuits, which can be described by differential equations [17,18]. Bao and Xu put forward some memristor-based circuit systems with multiple chaotic attractors [19,20]. Lai et al. proposed some three-dimensional and four-dimensional continuous chaotic systems with multiple attractors [21–23]. Wei et al. attempted to reveal the intrinsic mechanism of the multiple attractors by analyzing the bifurcation of the system [24]. The investigation of chaos and multiple coexisting attractors is indeed a very interesting research issue in academia. It helps to recognize the dynamic evolution of the actual system and promote the study of complexity science.

Chaotic systems have been found to be used in many areas. The most valuable application is cryptology. Chaotic system, in view of its rich dynamic behaviors and initial sensitivity, provides the mixing and spreading properties, which are the general requirements of encryption [25,26]. The S-Box is known as the most basic unit with scrambling function in block encryption algorithms. A good S-Box can make the encryption algorithm have higher security and better ability to withstand attacks. Although there have been many works on chaotic S-Box design, it is still important to generate S-Box according to some unique chaotic systems. Before applying the chaotic system to engineering fields, it is necessary to realize it through electronic circuits in order to prove its real existence. Based on circuit theory and simple circuit elements, chaotic signals can be generated in oscilloscopes. So far, the electronic circuit has become an important tool for the analysis of chaotic systems [27–30].

This present paper considers a special polynomial chaotic system with the following features: (i) it has three nonlinearities xz, yz, xyz and the invariance of transformation $(x, y, z) \mapsto (-x, -y, z)$; (ii) it performs a butterfly attractor; and (iii) it has a stable equilibrium, an unstable equilibrium and two stable equilibria, three unstable equilibria for different parameter conditions, and experiences mono-stability, bi-stability, mono-periodicity, bi-periodicity, one strange attractor and two coexisting strange attractors. After investigating the dynamic behavior of the system, an electronic circuit and an S-Box are designed according to the system.

The paper is arranged as follows: Section 2 describes the chaotic system and shows its butterfly attractor. Section 3 analyzes the stability of the equilibria. Section 4 studies the dynamic behavior of the system. Section 5 considers the electronic circuit realization of the system. Section 6 establishes the S-Box according to the system, and Section 7 summarizes the conclusions of this paper.

2. The Description of a Chaotic System

The chaotic system proposed in this paper can be expressed as the following set of differential equations:

$$\begin{cases} \dot{x} = ax - yz, \\ \dot{y} = -by + xz, \\ \dot{z} = -cz + xyz + k, \end{cases} \tag{1}$$

with state vector $(x, y, z) \in R^3$ and parameter vector $(a, b, c, k) \in R^4$. A butterfly attractor can be observed by numerical simulation on Matlab software (Matlab 8.0, MathWorks, Natick, MA, USA). The phase portraits of system (1) under parameters $(a, b, c, k) = (4, 9, 4, 4)$ and initial condition $(1, 1, 1)$ are shown in Figure 1. It visually demonstrates that system (1) displays an attractor as the system trajectories will eventually move to a bounded region. The Lyapunov exponents of the system are calculated as $l_1 = 1.7729$, $l_2 = 0.0000$, $l_3 = -7.5949$. The Lyapunov dimension is $D_l = 2 - l_1/l_3 = 2.2334$, so it can be determined that the attractor is a chaotic attractor. The time series of z generated from two very close initial conditions $(1, 1, 1)$ and $(1, 1, 1.001)$ are plotted in Figure 2. At the beginning, they are almost the same, but their differences are increasing after a number of iterations. That is to say, system (1) is sensitive dependence on initial conditions and its future behavior is unpredictable in the long term. The Poincaré map of system (1) is obtained via selecting the sections

$\Delta_1 = \{(x,y) \in R^2 \,|z = 10\}$ and $\Delta_2 = \{(y,z) \in R^2 \,|x = 0\}$. As shown in Figure 3, the Poincaré map is a sheet of point set. It is consistent with the nature of chaos.

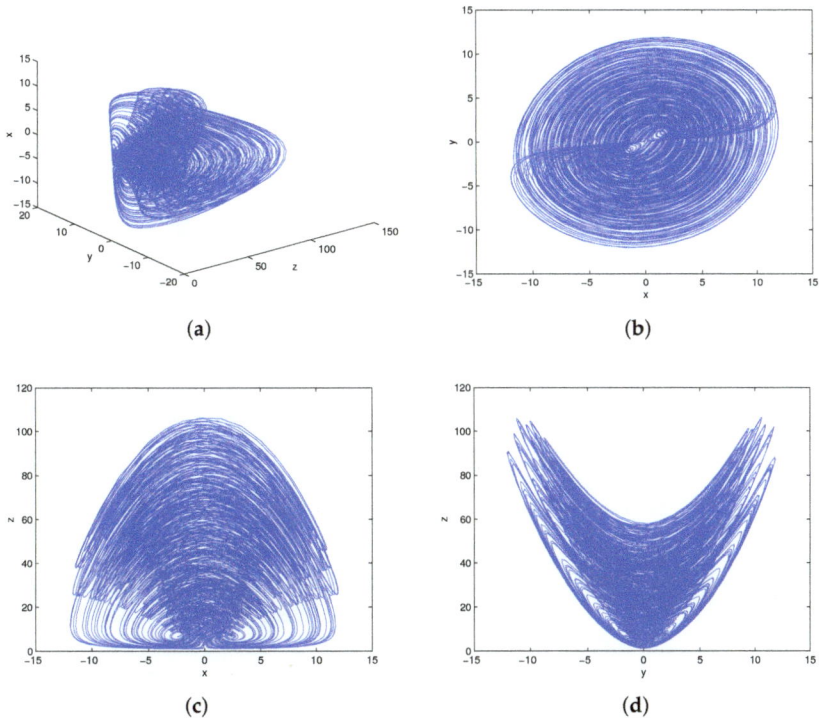

(a)

(b)

(c)

(d)

Figure 1. The butterfly attractor of system (1): (a) $x - y - z$; (b) $x - y$; (c) $x - z$; (d) $y - z$.

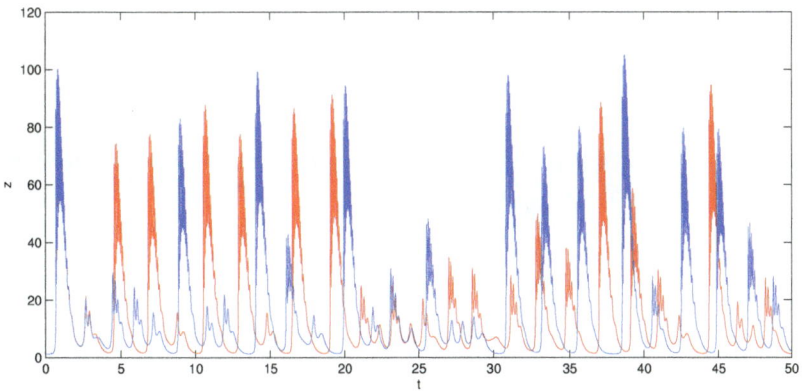

Figure 2. The time series of variable z generated from initial conditions $(1, 1, 1)$ (red color) and $(1, 1, 1.001)$ (blue color).

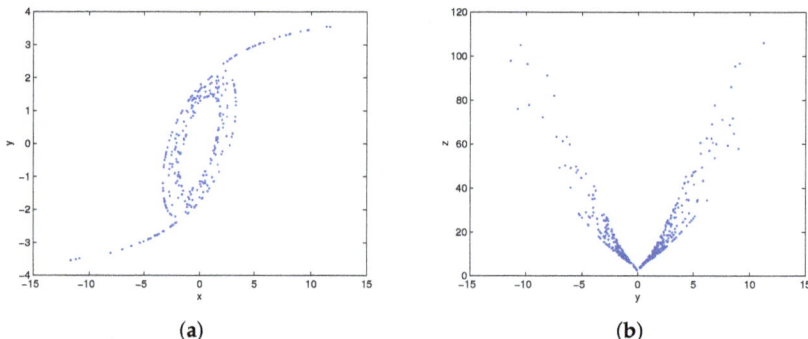

(a)   (b)

Figure 3. The Poincaré maps of system (1) with crossing sections: (a) Δ_1; (b) Δ_2.

3. The Stability of Equilibria

Suppose that parameters a, b, c, k are all positive real numbers. The equilibria of system (1) can be obtained by solving $\dot{x} = \dot{y} = \dot{z} = 0$. If $k \geq c\sqrt{ab}$, system (1) has only one equilibrium $O(0, 0, k/c)$. If $k < c\sqrt{ab}$, system (1) has three equilibria as follows:

$$O(0, 0, k/c),$$

$$O_1(\sqrt{(c\sqrt{ab} - k)/a}, \sqrt{(c\sqrt{ab} - k)/b}, \sqrt{ab}),$$

$$O_2(-\sqrt{(c\sqrt{ab} - k)/a}, -\sqrt{(c\sqrt{ab} - k)/b}, \sqrt{ab}).$$

Proposition 1. *Suppose that $b > a > 0, k > 0$, and the parameter c satisfies the following condition:*

$$\frac{k}{\sqrt{ab}} < c < \frac{2k[(a^2 + b^2)\sqrt{ab} + k(b - a)]}{\sqrt{ab}[(a + b)^2\sqrt{ab} + k(b - a)]}, \tag{2}$$

then the equilibria O_1 and O_2 of system (1) are asymptotically stable.

Proof. By linearizing the system (1) at the equilibrium, the Jacobian matrix is given by

$$H = \begin{pmatrix} a & -z & -y \\ z & -b & x \\ yz & xz & xy - c \end{pmatrix}. \tag{3}$$

By using $|\lambda I - H| = 0$, the corresponding characteristic equation evaluated at O_1, O_2 is obtained as

$$\lambda^3 + w_1\lambda^2 + w_2\lambda + w_3 = 0, \tag{4}$$

where

$$w_1 = (b - a + \frac{k}{\sqrt{ab}}),$$

$$w_2 = (a - b)(c - \frac{2k}{\sqrt{ab}}),$$

$$w_3 = 4ab(c - \frac{k}{\sqrt{ab}}).$$

□

According to the Routh–Hurwitz criterion, the equilibria O_1, O_2 are stable if all the roots of Equation (4) have negative real parts. This requires that $w_1 > 0$, $w_2 > 0$, $w_3 > 0$ and $w_1w_2 > w_3$. It is easy to verify that $w_1 > 0$, $w_2 > 0$, $w_3 > 0$ if $b > a > 0, k > 0$ and

$$\frac{k}{\sqrt{ab}} < c < \frac{2k}{\sqrt{ab}}. \tag{5}$$

To make $w_1w_2 > w_3$, the parameter c should meet

$$c < c_0 = \frac{2k[(a^2 + b^2)\sqrt{ab} + k(b - a)]}{\sqrt{ab}[(a + b)^2\sqrt{ab} + k(b - a)]}. \tag{6}$$

Since $c_0 < \frac{2k}{\sqrt{ab}}$, then O_1, O_2 are asymptotically stable if $b > a > 0, k > 0$, $\frac{k}{\sqrt{ab}} < c < c_0$. When the parameter c passes through the critical value c_0, then double Hopf bifurcation occur with two limit cycles branched from O_1, O_2 and system (1) loses its stability.

Proposition 2. *Suppose that $b > a > 0, c > 0, k > 0$, then: (i) the equilibrium O is unstable for $c > \frac{k}{\sqrt{ab}}$; and (ii) the equilibrium O is asymptotically stable for $c \leq \frac{k}{\sqrt{ab}}$.*

Proof. The characteristic equation evaluated at O is given by

$$(\lambda + c)[c^2\lambda^2 + (b - a)c^2\lambda + k^2 - abc^2] = 0. \tag{7}$$

If $c > \frac{k}{\sqrt{ab}}$, then Equation (7) has a root with a positive real part. Thus, O is unstable. If $c < \frac{k}{\sqrt{ab}}$, all the roots of Equation (7) have negative real parts, which implies that O is stable. When $c = \frac{k}{\sqrt{ab}}$, Equation (7) has three roots $\lambda_1 = 0, \lambda_2 = a - b, \lambda_3 = -c$. Therefore, O is non-hyperbolic equilibrium. It can be verified that O is asymptotically stable by applying the center manifold theorem. □

4. The Evolution of Multiple Attractors

Detailed investigation of the complex dynamic behaviors of system (1) is presented in this section. Simulation experiments including bifurcation diagrams, phase portraits, Lyapunov exponents, and Poincaré maps give a close and intuitive look at system (1). There is a wealth of chaotic dynamics associated with the fractal properties of the attractor in system (1). With the change of parameters, system (1) experiences stable state, periodic state and chaotic state. For different initial values, system (1) performs different types of attractors with independent domains of attraction.

4.1. Dynamic Evolution with Parameter c

Consider the dynamic evolution of system (1) with respect to parameter c under the given parameter conditions $a = 2, b = 8, k = 4$. The bifurcation diagrams of system (1) versus $c \in (0, 6)$ are shown in Figure 4a, where the red color branch and blue color branch are yielded from initial values $x_{01} = (1, 1, 1)$, $x_{02} = (-1, -1, 1)$, respectively. The overlapped regions of the red color and blue color branches indicate that the trajectories of x_{01}, x_{02} eventually tend to the same attractor, while the separated regions indicate that the trajectories of x_{01}, x_{02} tend to different attractors. Figure 4b is the Lyapunov exponents of system (1) with initial value x_{01}. It shows that the system (1) experiences stable state, periodic state, chaotic state with the variation of c. When $c \in (0, 1)$, system (1) is mono-stable as it has only one stable equilibrium. When $c \in (1, 1.396)$, system (1) performs bi-stability with respect to the existence of two stable equilibria. As c increases across the critical value $c_0 = 1.396$, system (1) occurs double Hopf bifurcation at the equilibria. When $c \in (1.396, 1.516)$, system (1) performs bi-periodicity. When $c \in (1.516, 2.257)$, system (1) changes into mono-periodic state. When $c \in (2.821, 3.096)$, system (1) yields two strange attractors from initial values x_{01}, x_{02}. When $c \in (3.310, 4.167) \cup (4.370, 5.324) \cup (5.480, 6)$, system (1) has only one chaotic attractor. Table 1 describes the attractors of system (1) with different values of c. The phase portraits in Figure 5 illustrate the existence of different types of attractors in system (1).

Table 1. Attractors of system (1) with different values of c.

Value of c	Equilibrium Point	Type of Attractor	Figure
$c = 0.8$	Stable point: $(0,0,5)$	A point attractor	Figure 5a
$c = 1.2$	Saddle node: $(0,0,3.3333)$ Stable point: $(\pm0.6325, \pm0.3162, 4)$	A pair of point attractors	Figure 5b
$c = 1.4$	Saddle node: $(0,0,2.8571)$ Saddle focus: $(\pm0.8944, \pm0.4472, 4)$	A pair of limit cycles	Figure 5c
$c = 1.6$	Saddle node: $(0,0,2.5)$ Saddle focus: $(\pm1.0954, \pm0.5477, 4)$	A symmetric limit cycle	Figure 5d
$c = 2.9$	Saddle node: $(0,0,1.3793)$ Saddle focus: $(\pm1.9494, \pm0.9748, 4)$	A pair of strange attractors	Figure 5e
$c = 3.1$	Saddle node: $(0,0,1.2903)$ Saddle focus: $(\pm2.0494, \pm1.0247, 4)$	A pair of limit cycles	Figure 5f
$c = 3.2$	Saddle node: $(0,0,1.25)$ Saddle focus: $(\pm2.0976, \pm1.0488, 4)$	A symmetric limit cycle	Figure 5g
$c = 3.6$	Saddle node: $(0,0,1.1111)$ Saddle focus: $(\pm2.2804, \pm1.1402, 4)$	A butterfly strange attractor	Figure 5h

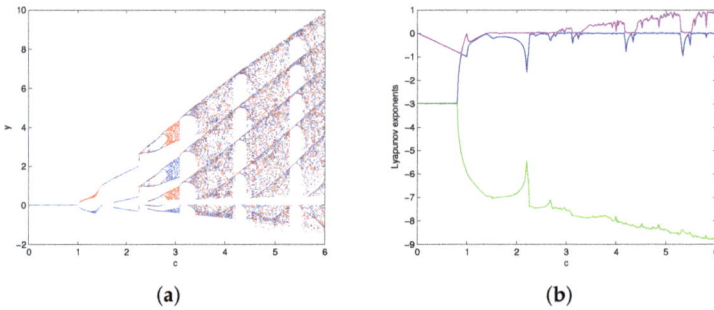

(a)

(b)

Figure 4. The bifurcation diagrams (**a**) and Lyapunov exponents (**b**) of system (1) versus $c \in (0,6)$.

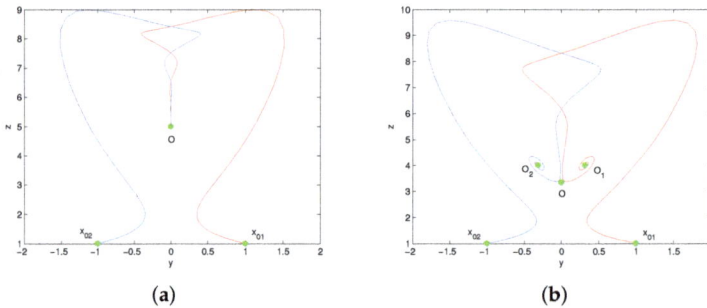

(a)

(b)

Figure 5. *Cont.*

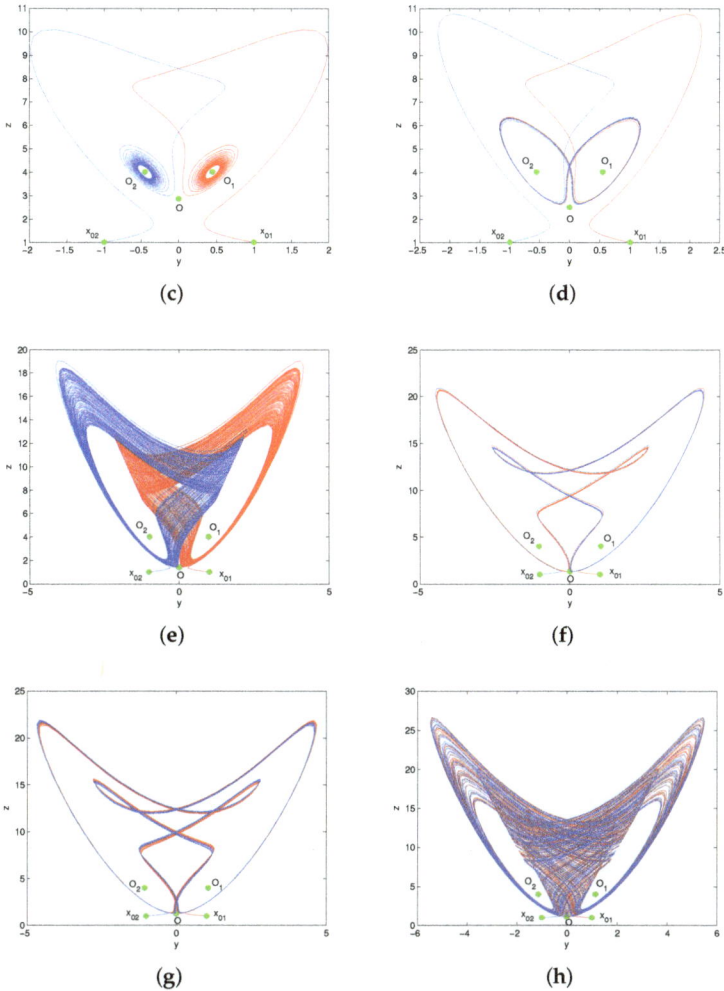

Figure 5. The phase portraits of system (1) with: (**a**) $c = 0.8$; (**b**) $c = 1.2$; (**c**) $c = 1.4$; (**d**) $c = 1.6$; (**e**) $c = 2.9$; (**f**) $c = 3.1$; (**g**) $c = 3.2$; (**h**) $c = 3.6$.

4.2. Dynamic Evolution with Parameter k

The bifurcation diagrams of system (1) for parameters $(a, b, c) = (4, 9, 4)$, $k \in (5, 25)$ are shown in Figure 6a, where the red color branch and blue color branch are yielded from initial values x_{01}, x_{02}, respectively. Obviously, the state of system (1) changes from chaos to period and then to stable when parameter k increases from 5 to 25. It also can be illustrated by the Lyapunov exponents in Figure 6b. The maximum Lyapunov exponent is positive with $c \in (5, 13.6) \cup (13.9, 14.8) \cup (15.4, 15.9)$, negative with $c \in (19.8, 25)$, and equal to zero with $c \in (13.7, 13.8) \cup (14.9, 15.3) \cup (16, 19.7)$. For $c = 5, 15, 18, 25$, we can observe a strange attractor, a limit cycle, and a stable point of system (1), with their phase portraits are shown in Figure 7. For $c = 19, 20$, we can observe two coexisting periodic attractors and two coexisting point attractors of system (1), as shown in Figure 8.

(a)

(b)

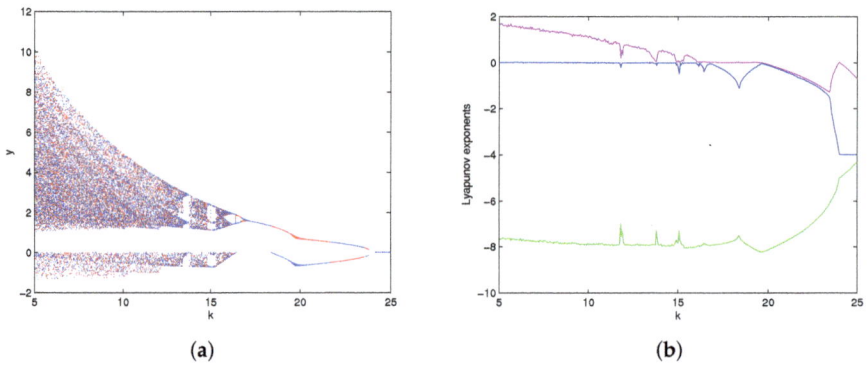

Figure 6. The bifurcation diagrams (**a**) and Lyapunov exponents (**b**) of system (1) versus $k \in (5, 25)$.

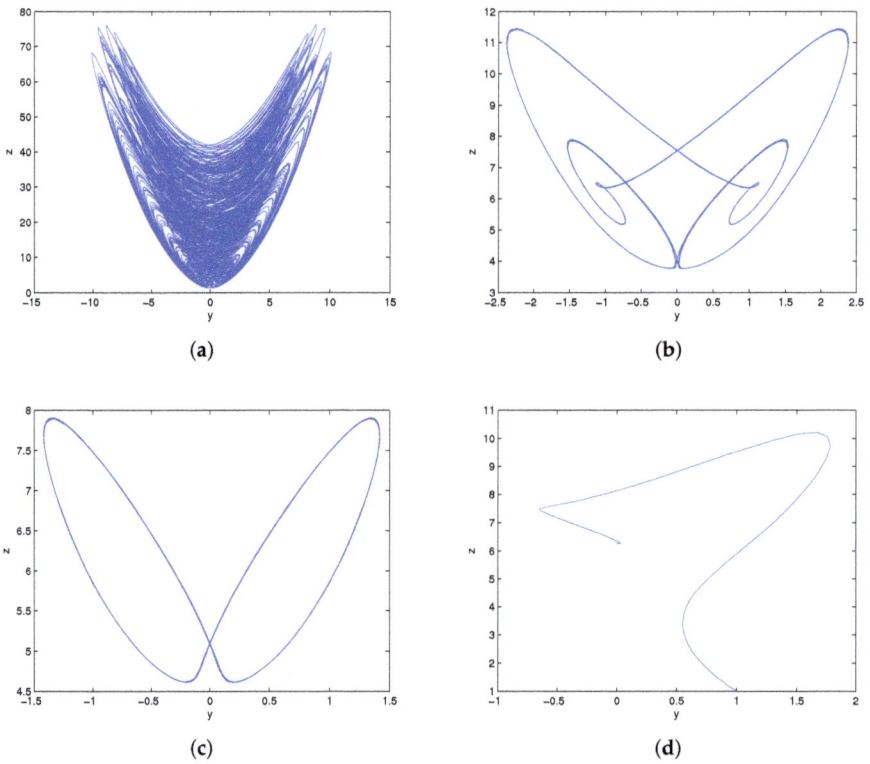

(a)

(b)

(c)

(d)

Figure 7. The phase portraits of system (1) with: (**a**) $k = 5$; (**b**) $k = 15$; (**c**) $k = 18$; (**d**) $k = 25$.

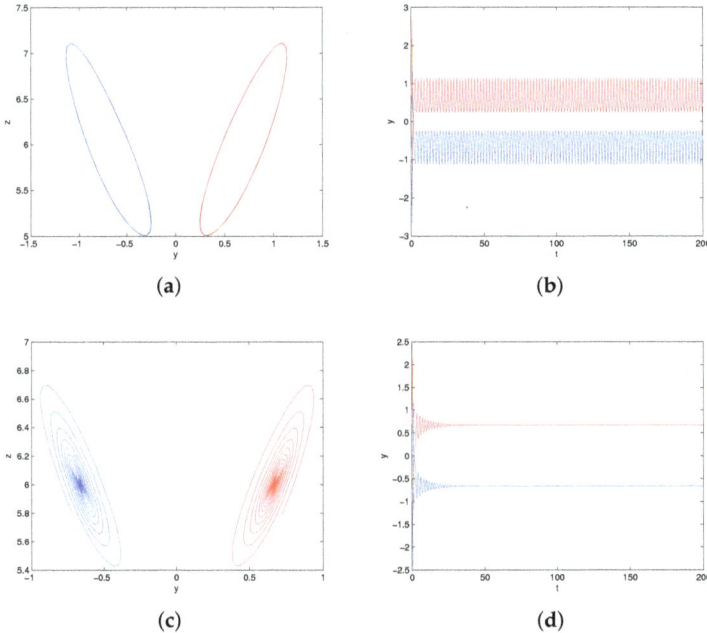

Figure 8. The coexisting attractors of system (1): (**a**) projections on $x - y$ with $k = 19$; (**b**) time series of y with $k = 19$; (**c**) projections on $x - y$ with $k = 20$; (**d**) time series of y with $k = 20$.

5. Electronic Circuit Realization

There are many works that are related to chaos based applications in the literature [31–36]. Here, we will present the circuit realization of system (1) for realistically obtaining its chaotic attractors. The numerical simulation in Figure 5e displays two coexisting strange attractors in system (1) for $(a, b, c, k) = (2, 8, 2.9, 4)$ and initial conditions $(\pm 1, \pm 1, 1)$. For circuit realization of this state of system (1), we need to refrain from saturation of circuit elements, and the effective way to achieve this goal is to reduce the voltage values of the circuit via scaling the variables of system (1). In the process of scaling, we assume $(X, Y, Z) = (x, y, z/2)$ and then the scaled system is obtained as

$$\begin{cases} \dot{X} = aX - 2YZ, \\ \dot{Y} = -bY + 2XZ, \\ \dot{Z} = -cZ + XYZ + \frac{k}{2}. \end{cases} \quad (8)$$

Figure 9 gives the new phase portraits of the scaled system (8) for $(a, b, c, k) = (2, 8, 2.9, 4)$. Evidently, the scaling process does not cause fundamental changes to the system (1), but just limits the variables to a smaller region $\Omega = \{(x, y, z) \, | x, y \in (-5, 5), z \in (0, 10)\}$.

The circuit diagram of system (8) raised by the OrCAD-PSpice programme (OrCAD 16.6, OrCAD company, Hillsboro, OR, USA) is presented in Figure 10. It has three input (or output) signals with respect to the variables X, Y, Z, and the operations between signals realized via the basic electronic materials including resistors, capacitor, TL081 operational amplifiers (op-amps), and AD633 multipliers. By fixing $R1 = R3 = 20 \text{ K}\Omega$, $R2 = 200 \text{ K}\Omega$, $R4 = 50 \text{ K}\Omega$, $R5 = R6 = 100 \text{ K}\Omega$, $R7 = 138$ KΩ, $R8 = 3000 \text{ K}\Omega$, $R9 = 4 \text{ K}\Omega$, $C1 = C2 = C3 = 1 \text{ nF}$, $Vn = -15 \text{ V}$, $Vp = 15 \text{ V}$ and executing the circuit on electronic card shown in Figure 11, we can obtain the outputs of circuit in the oscilloscope. The oscilloscope graphics in Figure 12 show good consistency with the numerical simulations in

Figure 9. Hence, we can come to a conclusion that the coexisting chaotic attractors in system (1) are physically obtained.

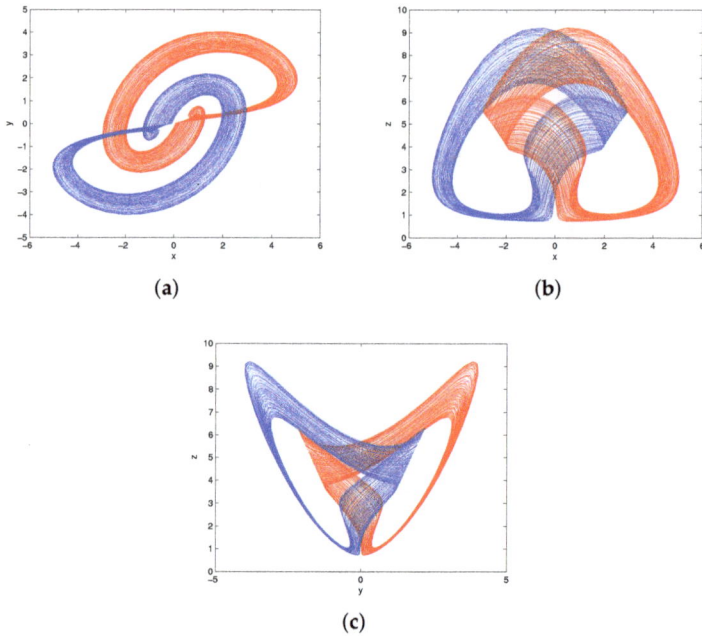

(a)

(b)

(c)

Figure 9. The phase portraits of the scaled system (8) for $(a, b, c, k) = (2, 8, 2.9, 4)$: (a) $x - y$; (b) $x - z$; (c) $y - z$.

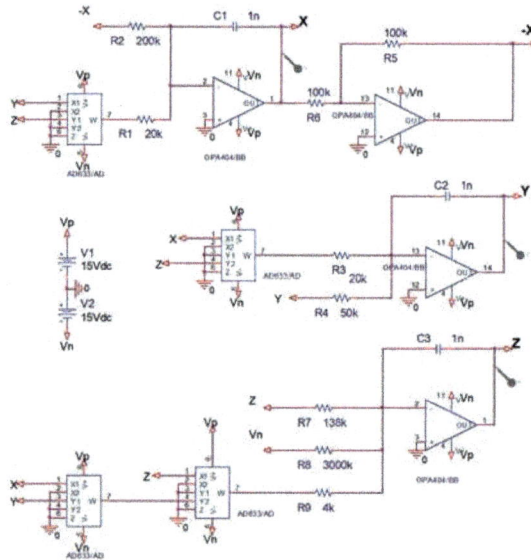

Figure 10. The circuit diagram of system (8).

Figure 11. The experimental circuit of system (8).

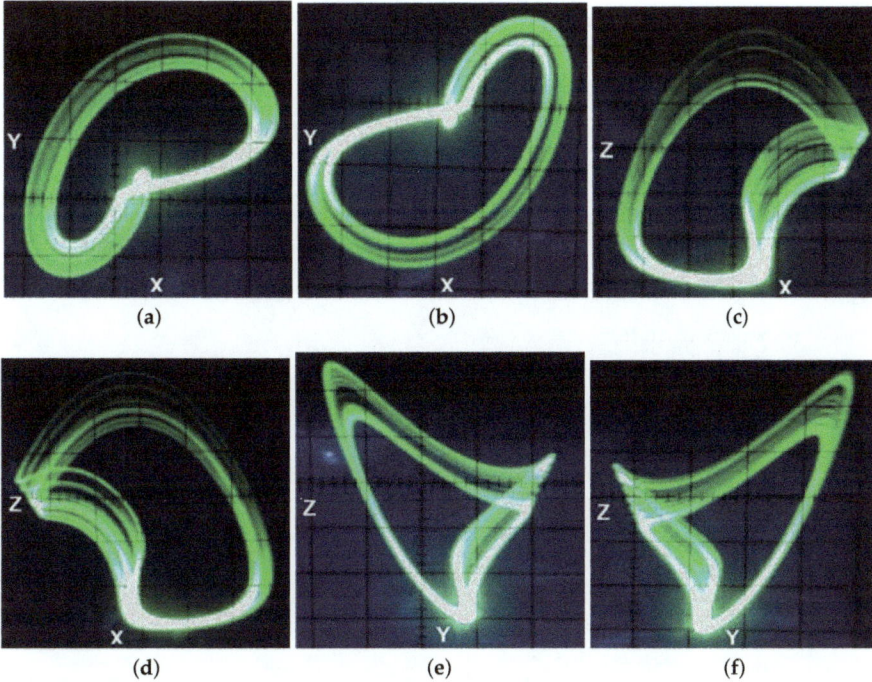

Figure 12. The phase portraits of two coexisting attractors of system (8) on the oscilloscope for $(a, b, c, k) = (2, 8, 2.9, 4)$: (**a,b**) $x - y$; (**c,d**) $x - z$; (**e,f**) $y - z$.

6. S-Box Design and Its Performance Analysis

This section aims to raise a new chaotic S-Box algorithm by applying system (1). In the algorithm design, first random number generation is performed and then S-Box is produced. The S-Box generation algorithm pseudo code is shown in Algorithm 1. For establishing the S-Box production algorithm, we first input the parameters $(a, b, c, k) = (2, 8, 2.9, 4)$ and initial value $(x_0, y_0, z_0) = (-1, -1, 1)$ of the system and then the float number outputs are produced. In order to generate more random outputs in the analysis of the chaotic system, we select an appropriate step interval Δh and used it as a sample value. More random sequences are obtained by setting the appropriate step interval $\Delta h = 0.000001$. System (1) is solved by using the RK4 algorithm with the initial conditions and the specified sampling value, and time series are obtained. In our designed chaotic S-Box algorithm, the outputs of y, z phases of system (1) are used. Float number values (32 bits) obtained from these phases are converted to a binary system. By taking 8 bits from the low significance parts (LSB) of the 32-bit number sequences generated from both phases, these values are XORed. The obtained new 8-bit value is converted to a decimal number. This value is discarded if the decimal number was previously generated and included in the S-Box, if not produced before, it is added to the S-Box. In this way, this process continues until the distinct 256 values (between 0 and 255) are obtained on the S-Box. The generated S-Box is shown in Table 2.

Algorithm 1 The S-Box generation algorithm pseudo code.

1: **Start;**
2: Inputting parameters and initial value of the system;
3: Sampling with step interval Δh;
4: $i = 1$, S-Box = [];
5: **while** $(i < 257)$ **do**
6:      Solving system with RK4 algorithm and obtaining time series (y, z);
7:      Convert float to binary number;
8:      Take LSB-8 bit value from RNG $y \oplus z$ phase;
9:      Convert binary to decimal number (8 bit)
10:     **if** (Is there decimal value in S-Box = yes) **then**
11:          Go step 6.
12:     **else**(Is there decimal value in S-Box = no)
13:          Sbox[i] $\leftarrow$ decimal value;
14:          i++;
15:     **end**
16: **end**
17: S-Box $\leftarrow$ reshape(Sbox,16,16);
18: Ready to use 16×16 chaos-based S-Box;
19: **End.**

Table 2. The chaotic S-Box of system (1).

199	30	5	41	38	140	230	139	66	0	11	195	76	204	54	23
254	198	50	108	231	92	87	182	217	28	56	253	219	232	215	49
102	151	68	86	176	248	12	32	126	249	141	154	82	138	174	165
145	62	115	150	201	104	170	148	78	97	192	247	252	96	211	153
45	98	40	91	109	113	196	107	209	83	144	120	191	75	242	208
175	246	100	181	85	70	197	136	235	210	93	216	71	105	162	149
88	240	31	238	42	171	90	73	112	243	255	128	239	121	26	34
25	226	59	244	135	142	53	36	146	157	117	124	116	10	205	60
173	29	2	72	203	3	214	224	127	241	143	74	6	156	122	61
110	8	1	233	79	51	77	47	236	222	185	152	180	15	103	234
206	227	169	202	137	221	177	179	163	52	245	67	89	80	220	7
237	183	17	4	101	37	39	57	178	194	58	69	213	147	18	228
46	35	225	84	14	125	95	134	129	63	99	55	106	161	218	27
250	21	13	24	207	193	48	184	189	114	111	167	16	160	188	123
155	132	158	130	118	166	164	168	33	159	223	64	44	81	190	172
212	20	229	186	65	251	133	22	131	43	119	94	19	9	187	200

In order to determine that the produced S-Boxes are robust and strong against attack, some performance tests are applied. We mainly focus on these tests: nonlinearity, outputs' bit independence criterion (BIC), strict avalanche criterion (SAC), and differential approach probability (DP). In addition, the comparisons of the performance between this new S-Box and the existing chaotic S-Box proposed by Chen [37], Khan [38], Wang [39], Ozkaynak [40], Jakimoski [41], Hussain [42], Tang [43] are presented in Table 3.

Nonlinearity is regarded to be the most core part of all the performance tests. The nonlinearity values of the S-Box yielded by system (1) are obtained as 104, 106, 104, 104, 108, 104, 110 and 104. Accordingly, its average value, minimum value and maximum value are computed as 105, 104 and 110.

By comparing the nonlinearity of other S-Box shown in Table 3, we can claim that the new S-Box is better than others in some measure.

Table 3. The comparison of different chaotic S-Boxes (BIC: bit independence criterion; SAC: strict avalanche criterion; DP: differential approach probability).

S-Box	Nonlinearity			BIC-SAC	BIC	SAC			DP
	Min	Avg	Max		Nonlinearity	Min	Avg	Max	
Proposed S-Box	104	105	110	0.5028	102.75	0.3906	0.5014	0.5937	10
Chen [37]	100	103	106	0.5024	103.1	0.4218	0.5000	0.6093	14
Khan [38]	96	103	106	0.5010	100.3	0.3906	0.5039	0.6250	12
Wang [39]	102	104	106	0.5070	103.8	0.4850	0.5072	0.5150	12
Ozkaynak [40]	100	103.2	106	0.5009	103.7	0.4218	0.5048	0.5938	10
Jakimoski [41]	98	103.2	108	0.5031	104.2	0.3761	0.5058	0.5975	12
Hussain [42]	102	105.2	108	0.5053	104.2	0.4080	0.5050	0.5894	12
Tang [43]	99	103.4	106	0.4995	103.3	0.4140	0.4987	0.6015	10

SAC is put forward by Webster et al. [44]. Generally speaking, the establishment of SAC implies a possibility that half of each output bit will be changed with the change of a single bit. Table 3 tells the average, minimum and maximum SAC values of the new S-Box as 0.5014, 0.3906, 0.5937. Evidently, the average value of the new S-Box is close to the ideal value 0.5. BIC is also an important criterion found by Webster et al. [44]. It can partially measure the security of cryptosystems. The set of vectors generated by reversing one bit of the open text is tested to be independent of all the pairs of avalanche variables. While the relation between avalanches is measured, variable pairs are necessary to calculate the correlation value [45]. BIC-SAC and BIC-Nonlinearity values are calculated when the BIC value is calculated. When the values in Table 3 are examined, the BIC-SAC values are calculated as follows: average value 0.5028, minimum value 0.4394 and maximum value 0.5312. The average value is almost equal to the optimum value 0.5.

DP is another performance index for testing the S-Box, which is established by Biham et al. [46]. In this analysis, the XOR distribution balance between the input and output bits of the S-Box is determined. The very close probability of XOR distribution between input and output bits often indicates the ability to resist the differential attack of the S-Box. The low DP value suggests that the S-Box is more resistant to attack. The minimum and maximum DP values of the new S-Box are determined as 4.0 and 10. From Table 3, we know that the DP value of the new S-Box is the same as the S-Boxes presented by Tang and Ozkaynak.

After testing the performance of the new S-Box by using some important indices and comparing with other S-Boxes, we can determine that the new S-Box generated by system (1) has better performance than other S-Boxes. Thus, it will be more suitable for attack resistant and strong encryption.

7. Conclusions

A special chaotic system with multiple attractors was studied in this letter. The complex dynamic behaviors of the system were mainly presented by numerical simulations. Bifurcation diagrams and phase portraits indicated that the system exhibits a pair of point attractors, a pair of periodic attractors, and a pair of strange attractors with the variation of system parameters. In addition, an electronic circuit was designed for realizing the chaotic attractors of the system. Moreover, a new S-Box was generated by applying the chaotic system, and the performance evaluation and comparison of the S-Box were presented. It showed that the new S-Box has better performance than some existing S-Boxes. Actually, the study of chaotic system with multiple attractors is of recent interest. More important issues corresponding to this topic will be addressed in our future paper.

Acknowledgments: This work was supported by the National Natural Science Foundation of China (No. 61603137, 61603127),the Jiangxi Natural Science Foundation of China (No. 20171BAB212016) and Sakarya University Scientific Research Projects Unit (No. 2016-09-00-008, 2016-50-01-026).

Author Contributions: Qiang Lai designed the study and wrote the paper. Akif Akgul and Ünal Çavuşoğlu contributed to the experiment and algorithm design. Chunbiao Li and Guanghui Xu partially undertook the theoretical analysis and simulation work of the paper. All authors read and approved the manuscript.

Conflicts of Interest: The authors declare no conflict of interest.

References

1. Lorenz, E.N. Deterministic non-periodic flow. *J. Atmos. Sci.* **1963**, *20*, 130–141.
2. Rössler, O.E. An equation for continuous chaos. *Phys. Lett. A* **1976**, *57*, 397–398.
3. Chen, G.; Ueta, T. Yet another chaotic attractor. *Int. J. Bifurc. Chaos* **1999**, *9*, 1465–1466.
4. Lü, J.; Chen, G. A new chaotic attractor coined. *Int. J. Bifurc. Chaos* **2000**, *3*, 659–661.
5. Guan, Z.H.; Lai, Q.; Chi, M.; Cheng, X.M.; Liu, F. Analysis of a new three-dimensional system with multiple chaotic attractors. *Nonlinear Dyn.* **2014**, *75*, 331–343.
6. Jafari, S.; Sprott, J.C.; Molaie, M. A simple chaotic flow with a plane of equilibria. *Int. J. Bifurc. Chaos* **2016**, *26*, 1650098.
7. Lai, Q.; Guan, Z.H.; Wu, Y.; Liu, F.; Zhang, D. Generation of multi-wing chaotic attractors from a Lorenz-like system. *Int. J. Bifurc. Chaos* **2013**, *23*, 1350152.
8. Wang, X.; Chen, G. Constructing a chaotic system with any number of equilibria. *Nonlinear Dyn.* **2013**, *71*, 429–436.
9. Riecke, H.; Roxin, A.; Madruga, S.; Solla, S.A. Multiple attractors, long chaotic transients, and failure in small-world networks of excitable neurons. *Chaos* **2007**, *17*, 026110.
10. Schwartz, J.L.; Grimault, N.; Hupe, J.M.; Moore, B.C.; Pressnitzer, D. Multistability in perception: Binding sensory modalities, an overview. *Philos. Trans. R. Soc. B* **2012**, *367*, 896–905.
11. Yuan, G.; Wang, Z. Nonlinear mechanisms for multistability in microring lasers. *Phys. Rev. A* **2015**, *92*, 043833.
12. Li, C.B.; Sprott, J.C. Variable-boostable chaotic flows. *Optik* **2016**, *127*, 10389-10398.
13. Li, C.B.; Sprott, J.C. Multistability in the Lorenz system: A broken butterfly. *Int. J. Bifurc. Chaos* **2014**, *24*, 1450131.
14. Li, C.B.; Sprott, J.C.; Hu, W.; Xu, Y. Infinite multistability in a self-reproducing chaotic system. *Int. J. Bifurc. Chaos* **2017**, *27*, 1750160.
15. Li, C.B.; Sprott, J.C.; Xing, H. Hypogenetic chaotic jerk flows. *Phys. Lett. A* **2016**, *380*, 1172–1177.
16. Li, C.B.; Sprott, J.C.; Xing, H. Constructing chaotic systems with conditional symmetry. *Nonlinear Dyn.* **2017**, *87*, 1351–1358.
17. Kengne, J.; Tabekoueng, Z.N.; Tamba, V.K.; Negou, A.N. Periodicity, chaos, and multiple attractors in a memristor-based Shinriki's circuit. *Chaos* **2015**, *25*, 103126.
18. Kengne, J.; Tabekoueng, Z.N.; Fotsin, H.B. Dynamical analysis of a simple autonomous jerk system with multiple attractors. *Nonlinear Dyn.* **2016**, *83*, 751–765.
19. Bao, B.C.; Jiang, T.; Xu, Q.; Chen, M.; Wu, H.; Hu, H.Y. Coexisting infinitely many attractors in active band-pass filter-based memristive circuit. *Nonlinear Dyn.* **2016**, *86*, 1711–1723.
20. Xu, Q.; Lin, Y.; Bao, B.C.; Chen, M. Multiple attractors in a non-ideal active voltage-controlled memristor based Chua's circuit. *Chaos Solitons Fractals* **2016**, *83*, 186–200.
21. Lai, Q.; Chen, S. Research on a new 3D autonomous chaotic system with coexisting attractors. *Optik* **2016**, *127*, 3000–3004.
22. Lai, Q.; Chen, S. Coexisting attractors gnerated from a new 4D smooth chaotic system. *Int. J. Control Autom. Syst.* **2016**, *14*, 1124–1131.
23. Lai, Q.; Chen, S. Generating multiple chaotic atractors from Sprott B system. *Int. J. Bifurc. Chaos* **2016**, *26*, 1650177.
24. Wei, Z.C.; Yu, P.; Zhang, W.; Yao, M. Study of hidden attractors, multiple limit cycles from Hopf bifurcation and boundedness of motion in the generalized hyperchaotic Rabinovich system. *Nonlinear Dyn.* **2016**, *82*, 131–141.

25. Amigo, J.M.; Kocarev, L.; Szczepanski, J. Theory and practice of chaotic cryptography. *Phys. Lett. A* **2007**, *366*, 211–216.

26. Alvarez, G.; Li, S. Some basic cryptographic requirements for chaos-based cryptosystems. *Int. J. Bifurc. Chaos* **2006**, *16*, 2129–2151.

27. Wang, X.; Pham, V.T.; Jafari, S.; Volos, C.; Munoz-Pacheco, J.M.; Tlelo-Cuautle, E. A new chaotic system with stable equilibrium: From theoretical model to circuit implementation. *IEEE Access* **2017**, *5*, 8851–8858.

28. Tlelo-Cuautle, E.; Rangel-Magdaleno, J.J.; Pano-Azucena, A.D.; Obeso-Rodelo, P.J.; Nunez-Perez, J.C. FPGA realization of multi-scroll chaotic oscillators. *Commun. Nonlinear Sci. Numer. Simul.* **2015**, *27*, 66–80.

29. Tlelo-Cuautle, E.; Pano-Azucena, A.D.; Rangel-Magdaleno, J.J.; Carbajal-Gomez, V.H.; Rodriguez-Gomez, G. Generating a 50-scroll chaotic attractor at 66 MHz by using FPGAs. *Nonlinear Dyn.* **2016**, *85*, 2143–2157.

30. Trejo-Guerra, R.; Tlelo-Cuautle, E.; Carbajal-Gomez, V.H.; Rodriguez-Gomez, G. A survey on the integrated design of chaotic oscillators. *Appl. Math. Comput.* **2013**, *219*, 5113–5122.

31. Akgul, A.; Moroz, I.; Pehlivan, I.; Vaidyanathan, S. A new four-scroll chaotic attractor and its engineering applications. *Optik* **2016**, *127*, 5491–5499.

32. Cavusoglu, U.; Akgul, A.; Kacar, S.; Pehlivan, I.; Zengin, A. A novel chaos based encryption algorithm over TCP data packet for secure communication. *Secur. Commun. Netw.* **2016**, *9*, 1285–1296.

33. Akgul, A.; Hussain, S.; Pehlivan, I. A new three-dimensional chaotic system, its dynamical analysis and electronic circuit applications. *Optik* **2016**, *127*, 7062–7071.

34. Akgul, A.; Calgan, H.; Koyuncu, I.; Pehlivan, I.; Istanbullu, A. Chaos-based engineering applications with a 3D chaotic system without equilibrium points. *Nonlinear Dyn.* **2016**, *84*, 481–495.

35. Trejo-Guerra, R.; Tlelo-Cuautle, E.; Jimenez-Fuentes, J.M.; Sanchez-Lopez, C.; Munoz-Pacheco, J.M.; Espinosa-Flores-Verdad, G.; Rocha-Pereza, J.M. Integrated circuit generating 3- and 5-scroll attractors. *Commun. Nonlinear Sci. Numer. Simul.* **2012**, *17*, 4328–4335.

36. Tlelo-Cuautle, E.; Fraga, L.G.; Pham, V.T.; Volos, C.; Jafari, S.; Quintas-Valles, A.J. Dynamics, FPGA realization and application of a chaotic system with an infinite number of equilibrium points. *Nonlinear Dyn.* **2017**, *89*, 1129–1139.

37. Chen, G.; Chen, Y.; Liao, X. An extended method for obtaining S-boxes based on three-dimensional chaotic Baker maps. *Chaos Solitons Fractals* **2007**, *31*, 571–579.

38. Khan, M.; Shah, T.; Mahmood, H.; Gondal, M.A.; Hussain, I. A novel technique for the construction of strong S-boxes based on chaotic Lorenz systems. *Nonlinear Dyn.* **2012**, *70*, 2303–2311.

39. Wang, Y.; Wong, K.W.; Liao, X.; Xiang, T. A block cipher with dynamic S-boxes based on tent map. *Commun. Nonlinear Sci. Numer. Simul.* **2009**, *14*, 3089–3099.

40. Ozkaynak, F.; Ozer, A.B. A method for designing strong S-Boxes based on chaotic Lorenz system. *Phys. Lett. A* **2010**, *374*, 3733–3738.

41. Jakimoski, G.; Kocarev, L. Chaos and cryptography: Block encryption ciphers based on chaotic maps. *IEEE Trans. Circuits Syst. I Fundam. Theory Appl.* **2001**, *48*, 163–169.

42. Hussain, I.; Shah, T.; Gondal, M.A. A novel approach for designing substitution-boxes based on nonlinear chaotic algorithm. *Nonlinear Dyn.* **2012**, *70*, 1791–1794.

43. Tang, G.; Liao, X. A method for designing dynamical S-boxes based on discretized chaotic map. *Chaos Solitons Fractals* **2005**, *23*, 1901–1909.

44. Webster, A.F.; Tavares, S.E. On the design of S-Boxes. In *Advances in Cryptology-CRYPTO'85 Proceedings*; Williams H.C., Eds.; Springer: Berlin, Heidelberg, 1986; Volume 218, pp. 523–534.

45. Hussain, I.; Shah, T.; Mahmood, H.; Gondal, M.A. Construction of S_8 Liu J S-boxes and their applications. *Comput. Math. Appl.* **2012**, *64*, 2450–2458.

46. Biham, E.; Shamir, A. Differential cryptanalysis of DES-like cryptosystems. *J. Cryptol.* **1991**, *4*, 3–72.

entropy

MDPI

Article

A New Chaotic System with a Self-Excited Attractor: Entropy Measurement, Signal Encryption, and Parameter Estimation

Guanghui Xu [1,2], Yasser Shekofteh [3,*], Akif Akgül [4], Chunbiao Li [5,6] and Shirin Panahi [7]

[1] School of Electrical and Electronic Engineering, Hubei University of Technology, Wuhan 430068, China; xuguanghui29@163.com
[2] Hubei Collaborative Innovation Center for High-efficiency Utilization of Solar Energy, Hubei University of Technology, Wuhan 430068, China
[3] Faculty of Computer Science and Engineering, Shahid Beheshti University, Tehran 1983969411, Iran
[4] Department of Electrical and Electronic Engineering, Faculty of Technology, Sakarya University, Serdivan 54187, Turkey; aakgul@sakarya.edu.tr
[5] Jiangsu Key Laboratory of Meteorological Observation and Information Processing, Nanjing University of Information Science & Technology, Nanjing 210044, China; goontry@126.com
[6] School of Electronic & Information Engineering, Nanjing University of Information Science & Technology, Nanjing 210044, China
[7] Department of Biomedical Engineering, Amirkabir University of Technology, Tehran 1591634311, Iran; sh.panahi89@gmail.com
* Correspondence: y_shekofteh@sbu.ac.ir; Tel./Fax: +98-21-2243-1804

Received: 28 December 2017; Accepted: 21 January 2018; Published: 27 January 2018

Abstract: In this paper, we introduce a new chaotic system that is used for an engineering application of the signal encryption. It has some interesting features, and its successful implementation and manufacturing were performed via a real circuit as a random number generator. In addition, we provide a parameter estimation method to extract chaotic model parameters from the real data of the chaotic circuit. The parameter estimation method is based on the attractor distribution modeling in the state space, which is compatible with the chaotic system characteristics. Here, a Gaussian mixture model (GMM) is used as a main part of cost function computations in the parameter estimation method. To optimize the cost function, we also apply two recent efficient optimization methods: WOA (Whale Optimization Algorithm), and MVO (Multi-Verse Optimizer) algorithms. The results show the success of the parameter estimation procedure.

Keywords: chaotic systems; circuit design; parameter estimation; optimization methods; Gaussian mixture model

1. Introduction

A chaotic system has been considered with great potential in engineering applications, in which many chaotic systems with different properties have been studied. Specifically, some systems have the properties of amplitude control and offset boosting [1–4]. In this paper, we use a new three-dimensional (3D) chaotic system in random number generation and signal encryption, which are important engineering applications of chaotic systems [5–10]. To do this, an electronic design of the system is implemented as a real electronic circuit to generate random numbers. Finally, the one-dimensional (1D) and two-dimensional (2D) parameter estimation of the system is reported based on a non-traditional parametric model cost function and two new optimization methods.

The topic of self-excited and hidden attractors is a new attractive topic in dynamical systems [11–13]. Recent studies have classified dynamical attractors as self-excited or hidden [14–17]. A self-excited attractor has a basin of attraction which intersects with at least one unstable equilibrium. If that is not the case, the attractor is hidden [18–21]. According to the above definition, most of the classical chaotic attractors are self-excited [22,23]. It has been demonstrated that the attractors in dynamical systems with no equilibria [24–32], with stable equilibria [33,34], with lines [35,36] and curves of equilibria [37–41], and with plane [42] or surface of equilibria [43] are hidden attractors. Even some systems can belong to more than one category [3,44–47]. Hidden attractors cannot simply be located. There have been some efforts in literature to solve this problem [18,48,49].

As we know, modeling of the real world chaotic systems has received great attention in recent decades [50–55]. Choosing proper values for model parameters is essential in chaotic systems, since they are very sensitive, both to model parameters and initial conditions. A slight change in the parameters of the chaotic system may cause important bifurcation in its behavior, because of the butterfly effect of the chaotic system [56]. Therefore, the parameter estimation problem of chaotic system models is a complex problem [57–59].

There are some widely used methods for the parameter estimation of the chaotic systems which are based on optimization methods [60–62]. In these methods, the problem of the parameter estimation is generally formulated as a cost function based on an error function between a time series obtained from a real system and a time series obtained from a known model with unknown parameters of that system. The goal of the parameter estimation method will then be to find the best values of the unknown parameters of the model which minimize the cost function. In addition, the optimization approaches have been used algorithms for this problem to find the best values of the unknown parameters as quickly as possible. They are algorithms such as genetic [63], particle swarm optimization [64], and evolutionary programming [65]. However, approaches that utilize cost function based on the error function seem to bear major limitations because of the butterfly effect of the chaotic systems [57–59].

It was remarked that the state space would be a proper domain to analyze the chaotic systems rather than the time-domain. The time series generated by the chaotic systems have random-like behavior in the time-domain, but they are ordered in the state space. They can show specific topologies in the state space named strange attractors. In this paper, we use a non-conventional metric as a useful cost function for the parameter estimation method. Accordingly, we model the attractor distribution of a real chaotic system by a parametric model named the Gaussian mixture model (GMM). It can provide flexible and probabilistic modeling for data distributions. GMM is also a commonly used parametric model in the pattern recognition and machine learning domain [66]. For example, in the speech recognition field, a set of GMMs was introduced to model phone attractors in a reconstructed phase space (RPS) in which the RPS is a time-independent domain similar to the state space [67–69]. The phone classification results showed that the GMM could be a useful model to capture the structure and topology of the speech attractors in the RPS. In addition, models of Gaussian mixture were recently used as the parameter identification method for some chaotic systems [70–72].

Here, to optimize the cost function, two recent efficient optimization methods are applied, including the WOA (Whale Optimization Algorithm), and MVO (Multi-Verse Optimizer) algorithm. Also, for testing the parameter estimation method in the chaotic systems, a real circuit is utilized based on a new chaotic system in this paper. All the data (time series) are obtained from the circuit that is designed based on the new chaotic system.

The contributions of this paper are described as:

- A new 3D chaotic system with saddle equilibriums is proposed by a set of ordinary differential equations.
- Dynamical properties of the 3D chaotic system are then reported that exhibit its dynamics.
- The electronic circuit implementation of the 3D chaotic system is studied and used to present a random number generator (RNG), and its signal encryption is then introduced as an engineering application.
- 1D and 2D parameter estimation of the electronic circuit is done by a GMM based cost function.

- The cost function is optimized using two new efficient optimization methods called the WOA and the MVO algorithms.
- By comparing the experimental data with numerically generated time series, the best-fitting parameters are found because the circuit had (almost) the same dynamics as the 3D chaotic system.

The structure of the paper is organized as follows: in the next section we introduce and analyze the new chaotic system with saddle equilibriums. In Section 2, we investigate it carefully through bifurcation analysis, spectrum of Lyapunov exponents, and its entropy. Section 3 deals with the circuit implementation of this new system and a real circuit application based on mobile RNG design. In the next section, the cost function based on the GMM is introduced. Two meta-heuristic optimization algorithms (WOA and MVO) are presented in Section 5. Results of the cost function and the parameter estimation of the new chaotic system using the WOA and MVO methods are available in Section 6. Finally, Section 7 is the conclusion of the paper.

2. A New Chaotic System and Its Analysis

In this section we introduce a new 3D system which can show chaotic behavior. Consider a system described with the following ordinary differential equations:

$$
\begin{aligned}
\dot{x} &= gz \\
\dot{y} &= dx^2 + ey^2 - f \\
\dot{z} &= -ax - bx^2 + cy^2
\end{aligned}
\tag{1}
$$

This system is in the chaotic state when $a = 4.0$, $b = 1.0$, $c = 1.0$, $d = 1.0$, $e = 1.0$, $f = 4.0$ and $g = 1.0$. Different projections of the phase portrait for this system are plotted in Figure 1, which shows its strange attractor in 2D state spaces. System (1) is a new offset-boostable one [1–4] in which the variable z can be boosted with a direct constant in the first dimension.

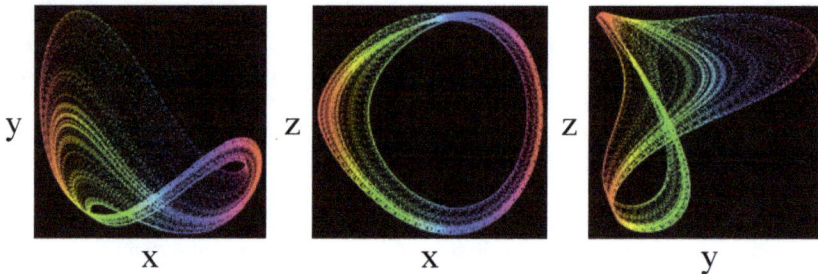

Figure 1. Different projections of the chaotic attractor of system (1) with the initial conditions $(-1.8, -1.5, -2.5)$.

This system has fixed-points in every (x^*, y^*, z^*) which satisfy the following equation,

$$
\begin{cases} \dot{x} = 0 \\ \dot{y} = 0 \\ \dot{z} = 0 \end{cases} \rightarrow
\begin{cases} 0 = z \\ 0 = x^2 + y^2 - 4 \\ 0 = -4x - x^2 + y^2 \end{cases}
\tag{2}
$$

According to Equation (2), the system (1) has two equilibria in $A = (0.7321, 3.4641, 0)$ and $B = (0.7321, -3.4641, 0)$. The Jacobian matrix of the system (1) is

$$
J = \begin{bmatrix} 0 & 0 & 1 \\ 2x & 2y & 0 \\ -4 - 2x & 2y & 0 \end{bmatrix}
\tag{3}
$$

and the corresponding eigenvalues for A and B are

$$A : \begin{cases} \lambda_1 = 3.9784 \\ \lambda_{2,3} = -0.12798 \pm i2.5428 \end{cases} \quad B : \begin{cases} \lambda_1 = -3.9784 \\ \lambda_{2,3} = 0.12798 \pm i2.5428 \end{cases} \tag{4}$$

Therefore, both equilibria are saddle-foci. Thus, the attractor is self-excited.

3. Bifurcation and Entropy Analysis

3.1. Bifurcation Analysis

In this part, we investigate the behaviors of the system (1) with respect to changing parameter g. In part (A) of Figure 2 the bifurcation diagram of the system is shown and in part (B) of this figure Lyapunov exponents can be observed. It is important to be careful about numerical calculation of Lyapunov exponents, since improper use of usual methods may cause some issues [14,15,73–77]. We have used the algorithm of [78] for computation of Lyapunov exponents.

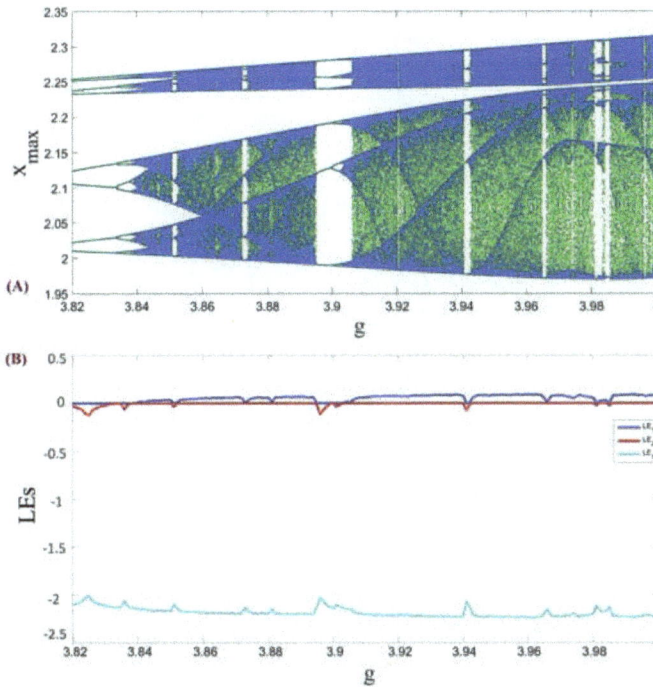

Figure 2. (A) Bifurcation diagram of the system (1) with respect to parameter g, and (B) Lyapunov exponents of the system (1) with respect to parameter g.

As can be seen in Figure 2A, changing parameter g causes a familiar period doubling route to chaos. In addition, positive values of the Lyapunov exponents in Figure 2B show that the underlying system is the chaotic system.

3.2. Entropy Analysis

There are many techniques to evaluate the system complexity from data. One of the most famous method which had been used since 1991 is Approximate Entropy (ApEn) [79]. ApEn can be applied to short and noisy data with outliers [80]. Therefore, many systems can be categorized by means of complexity [81]. Consider the N data sample $u(1)$, $u(2)$, $\ldots$, $u(N)$ with the vector sequence $x(1)$, $x(2)$, $\ldots$, $x(N - m + 1) \in R^m$ which can be defined as:

$$x(i) = [u(i), \, u(i+1), \, \ldots, \, u(i+m-1)] \tag{5}$$

where m is an integer and determines the dimension of $x(i)$ as the length of compared run of data. Then, for each i in the $1 \leq i \leq N - m + 1$, the following equation is defined:

$$c_i^m(r) = \frac{J}{N - m + 1} \tag{6}$$

$$d[x(i), x(j)] \leq r \, , \, 1 \leq j \leq N - m + 1 \tag{7}$$

$$d[x(i), x(j)] = \max_{k=1,2,\ldots,D} (|u(i+k-1) - u(j+k-1)|) \tag{8}$$

where J is the number of correct vectors in Equation (7), the number of vectors that the distance (infinity norm or maximum norm) between them and $x(i)$ is lower than r, and r is also a tolerance threshold that is defined by the product of a constant C to the standard deviation of data.

$$r = C \times std(u(t)) \quad 0.1 \leq C \leq 0.2 \tag{9}$$

Then, the *ApEn* can be written as:

$$\phi^m(r) = \frac{\sum_{i=1}^{N-m+1} log C_i^m(r)}{N - m + 1} \tag{10}$$

$$ApEn(m,r) = \lim_{N \to \infty} \left[\phi^m(r) - \phi^{m+1}(r) \right] \tag{11}$$

The estimation of Equation (11) for N data sample is as follows,

$$ApEn(m,r,N) = \phi^m(r) - \phi^{m+1}(r) \tag{12}$$

It can be derived that the *ApEn* values determine the similarity between chosen window and the sliding window of the data. Therefore, m determines the length of the window to be compared, and r is the tolerance threshold for accepting similar pattern between two windows. Figure 3 represents the *ApEn* diagram of the system (1) with respect to parameter g.

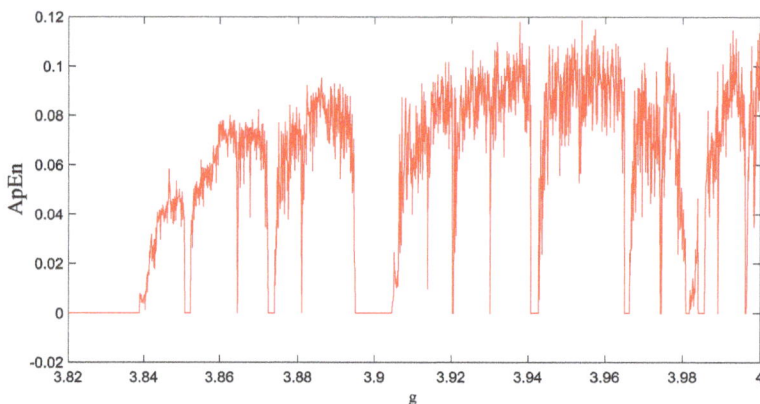

Figure 3. ApEn of the system (1) with respect to parameter g.

4. Real Circuit Design of the New Chaotic System as a Mobile RNG and Its Application for Signal Encryption

Random number generator (RNG) algorithms produce a sequence of numbers with properties of randomness and they are a research subject since a few decades. Chaotic systems are commonly used in the random numbers generation algorithms because they are complex and very sensitive. In this section, a mobile RNG design is implemented based on the introduced chaotic system (1) and then signal encryption application is realized with the RNG.

The micro-computer based mobile RNG can be used in many fields especially in encryption studies with low cost and high performance. It is aimed at encryption of multimedia data (audio, image, video, text etc.) with the realized mobile RNG to be flexible and user friendly.

As far as we know, random number generators require high cost hardware like computers and FPGA in order to successfully pass the universal tests [82–86]. In this paper, the design of a microcomputer-based mobile RNG and a signal encryption application with the designed RNG is realized without needing hardware such as FPGA, computers, etc. Therefore, "Raspberry Pi 3" is used here as hardware which supports 64-bit processing capability. Since the "Raspberry Pi 3" card has 64-bit processing capability, it can generate very sensitive decimal numbers; thus, randomness of these generated numbers is very high. BCM2837 SoC (system-on-chip) 64-bit ARMv8 quad core Cortex A53 processor running @1.2GHz produced by Broadcom is available on the card. The general view of "Raspberry Pi 3" is as given as in Figure 4.

Figure 4. The general outlook of "Raspberry Pi 3".

Our proposed circuit is used as an entropy source for RNG. Then, the NIST-800-22 tests are performed on random numbers to evaluate the performance of the designed RNG. In the next step, a signal encryption application is realized as an example application in "Raspberry Pi 3". Also, an electronic circuit implementation of the chaotic circuit is done in OrCAD-PSpice and on the oscilloscope.

4.1. Micro-Computer-Based Mobile RNG Design

As before mentioned, the "Raspberry Pi 3" board is used as a micro-computer for RNG design and encryption application. The chaotic system of (1) is also utilized in the RNG design. The RNG design steps are given in Algorithm 1 as a pseudo code.

Algorithm 1. Mobile RNG design algorithm pseudo code.	

1:	**Start**
2:	Entering parameters and initial condition of the chaotic system
3:	Determination of the value of Δh
4:	Sampling with determination Δh value
5:	**while** (least 1 M. Bit data) do
6:	Solving the chaotic system with RK4
7:	Convert float to binary number (32 bit)
8:	Select the bits (LSB-16 bit) from 32 bit binary number
9:	**end while**
10:	The implementation of NIST Tests for 1 M. Bit data
11:	**if** test results == pass **then**
12:	Successful results (Ready tested 1 M. Bit data)
13:	RNG applications (Cryptology, data hiding, watermarking, etc.)
14:	**else** (test results == false)
15:	return the previous steps and generate bits again
16:	**end if**
17:	**End**

After entering parameters and initial condition of the system (1), the outputs are discretized with the RK4 differential equation solving method. Then, float numbers are obtained and converted into 32 bits binary numbers. Later, the RNG design is executed with obtained binary numbers. The last 16 bits of the outputs (x, y and z variables) are used in the design. The NIST-800-22 statistical tests are also used to prove the success of the RNG design [87]. The NIST-800-22 tests consist of 16 different tests such as monobit, serial and discrete Fourier transform tests. The p-values of the test should be greater than 0.001 in order to be counted as successful in NISTS-800-22 tests.

Our experiments show that the random numbers generated from x, y and z outputs successfully passed all the tests with the last 16 bits. The NIST-800-22 tests results are given in Table 1. The ready tested random numbers that pass all of the NIST-800-22 tests can be used in applications that require high security such as cryptology, data hiding, watermarking, etc.

Table 1. RNG NIST-800-22 tests for x, y and z outputs.

Statistical Tests	p-Value-x (X_16bit)	p-Value-y (Y_16bit)	p-Value-z (Z_16bit)	Result
Frequency (Monobit) Test	0.5741	0.2209	0.9904	Successful
Block-Frequency Test	0.5692	0.2711	0.4011	Successful
Cumulative-Sums Test	0.6255	0.1218	0.4619	Successful
Runs Test	0.7012	0.1846	0.5313	Successful
Longest-Run Test	0.6207	0.1881	0.6901	Successful
Binary Matrix Rank Test	0.4378	0.9036	0.9755	Successful
Discrete Fourier Transform Test	0.0796	0.5819	0.6931	Successful
Non-Overlapping Templates Test	0.1685	0.0011	0.0803	Successful
Overlapping Templates Test	0.8824	0.1699	0.5441	Successful
Maurer's Universal Statistical Test	0.5665	0.3602	0.8932	Successful
Approximate Entropy Test	0.1364	0.7072	0.6264	Successful
Random-Excursions Test ($x = -4$)	0.9005	0.3467	0.6683	Successful
Random-Excursions Variant Test ($x = 9$)	0.5249	0.9845	0.5880	Successful
Serial Test-1	0.1784	0.6299	0.5716	Successful
Serial Test-2	0.5467	0.4709	0.7633	Successful
Linear-Complexity Test	0.7039	0.3601	0.2000	Successful

To obtain the random numbers, the pins x, y, and z GPIO (General purpose input/output) are utilized as shown in Figure 5. They are the 37th pin for x output, the 35th pin for y output, and the 38th pin for z output from "Raspberry Pi 3".

Figure 5. Pins of x, y and z for chaotic system outputs from "Raspberry Pi 3".

The generated x, y and z outputs (first 50 bits) are shown in Figure 6 as real-time oscilloscope outputs. The 35th, 37th, and 38th GPIO pins give x, y and z outputs in Figure 5, respectively. They are used for real-time oscilloscope outputs.

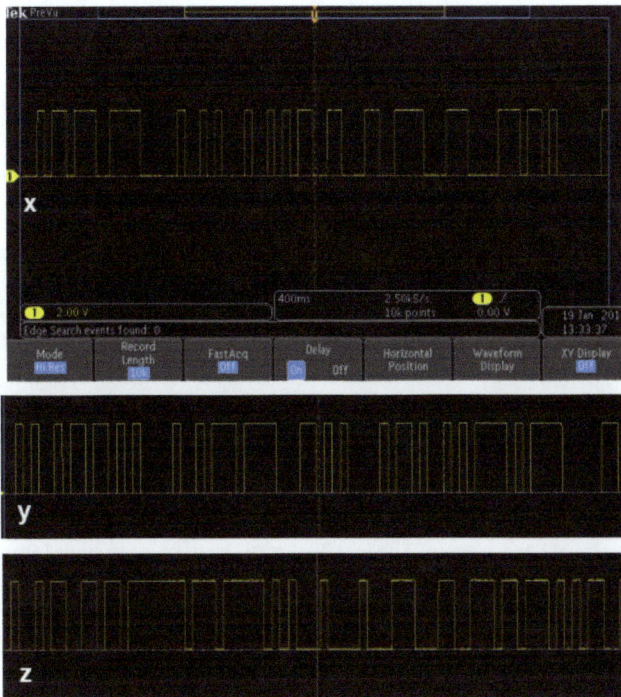

Figure 6. x, y and z outputs on the oscilloscope (first 50 bits).

4.2. Signal Encryption Application Using "Raspberry Pi 3"

In this section, a signal encryption application with RNG that was generated from the proposed chaotic system is realized in "Raspberry Pi 3". The steps of the encryption and decryption process are given in Algorithm 2. In the encryption application, a signal that consists of 512 bits is used and shown (first 50 bits) in Figure 7 as the real-time oscilloscope outputs.

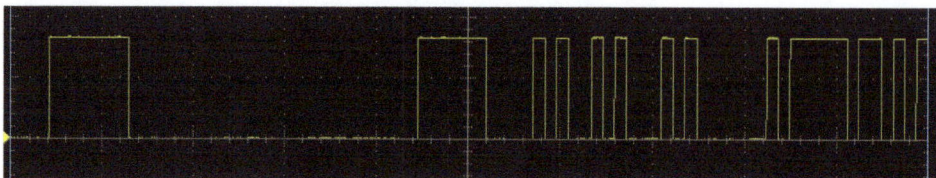

Figure 7. Original signal data (first 50 bits).

Algorithm 2. Chaos based encryption and decryption algorithm pseudo code.

1:	**Start**
2:	Getting ready to test random numbers for keys
3:	Getting signal data to be encrypted
4:	**for** i = 1 for all original data
5:	random number bit **xor** original data bit
6:	**end**
7:	Encrypted data
8:	**for** i = 1 for all encrypted data
9:	random number bit **xor** encrypted data bit
10:	**end**
11:	Decrypted data
12:	**End**

For the encryption process, the 'XOR' operator is used. Figure 8 shows the first 50 bits of the encrypted signal as real-time oscilloscope outputs. Since the encryption process is performed for each bit, the size of the encrypted data is also 512.

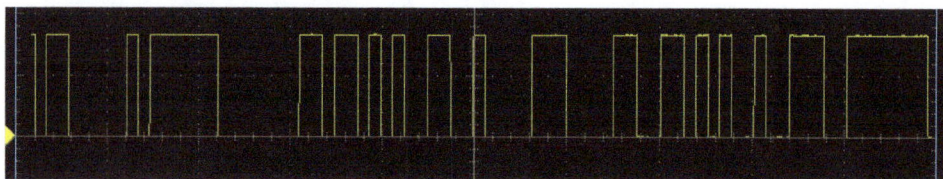

Figure 8. Encrypted signal data (first 50 bits).

The same keys generated from the chaotic system are needed for decryption. With these keys, the original data can be obtained, again. The first 50 bits of the decrypted signal are shown in Figure 9 as the real-time oscilloscope outputs. As can be seen, comparing Figures 7 and 9, for the first 50 bits, there is no deformation.

In the implemented method, a cryptoanalyser who wants to crack the encrypted data must know exactly all of the parameters and initial values of the chaotic system used in the encryption. Also, encrypted data will be not decrypted without "Raspberry Pi 3".

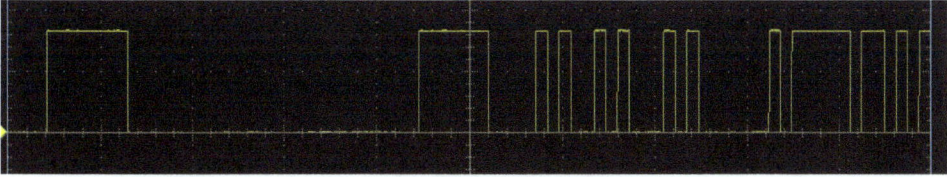

Figure 9. Decrypted Signal Data (first 50 bits).

4.3. Electronic Circuit Implementation of the Chaotic System in OrCAD-PSpice and on the Oscilloscope

In this part, we design an electronic circuit based on system (1) in OrCAD-PSpice (Figure 10) and on the board (Figure 11). The circuit includes simple electronic elements such as resistors, multipliers, capacitor, and opamps. Note that PSPICE simulation of chaotic circuits is quite trivial. In the literature, such systems are implemented with integrated circuit technology [88].

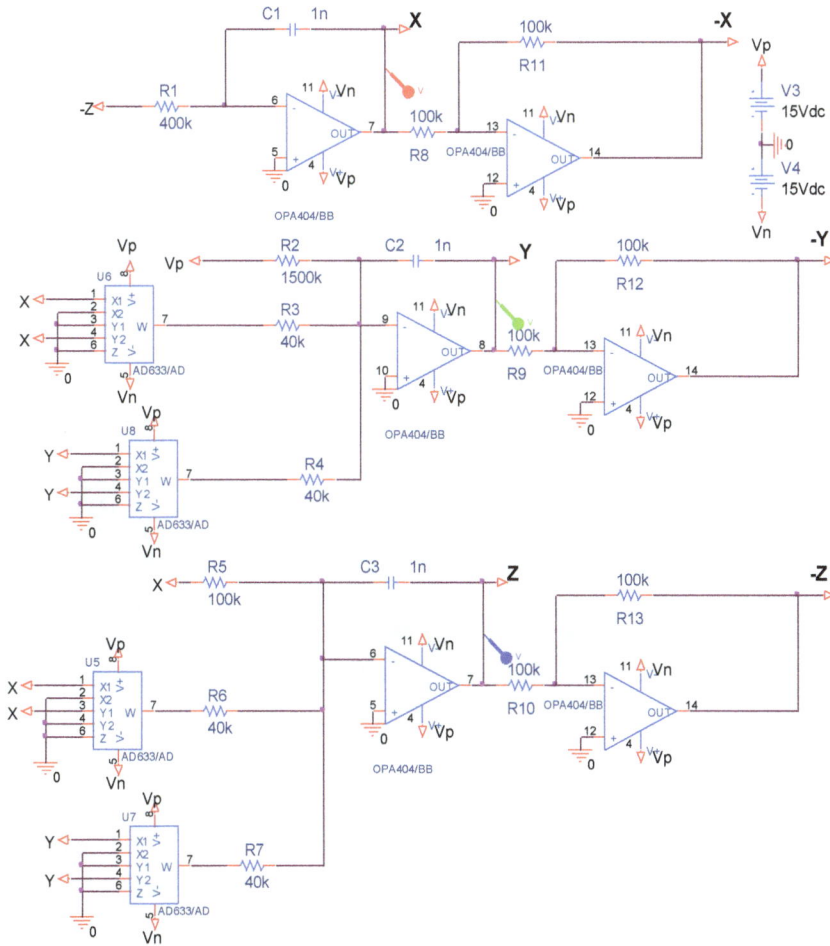

Figure 10. The electronic circuit schematic of system (1).

Figure 11. The experimental circuit of the chaotic circuit and the phase portraits of system (1) on the oscilloscope.

The OrCAD-PSpice simulation outputs, which are two-dimensional phase portraits of the chaotic system, are seen in Figures 12 and 13, respectively. As can be seen from the ORCAD-PSpice outputs in Figure 12 and oscilloscope outputs in Figure 13, the results are similar.

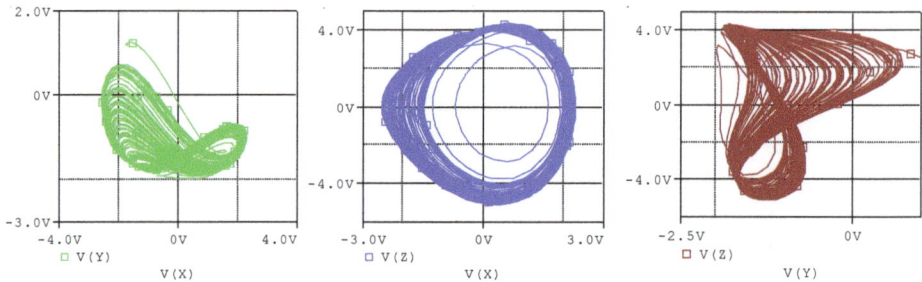

Figure 12. The phase portraits of the system (1) in ORCAD-Pspice.

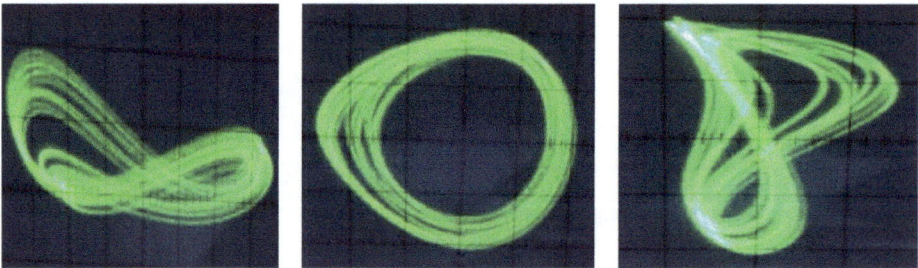

Figure 13. The phase portraits of system (1) on the oscilloscope.

5. Parameter Estimation of the Chaotic System

In this section, we introduce the parameter estimation method used for the chaotic circuit. This method utilizes a cost function which was adopted for the chaotic systems. The cost function of the parameter estimation method is based on a similarity metric using a parametric model of strange attractors in the state space. It was shown that this cost function could yield better results than the conventional error-based cost function over the time-domain [71]. The time-independent property of the state space is a sufficient reason to use this cost function because the state space can show complex behaviors of the strange attractor of chaotic systems [89].

As before mentioned, the utilized cost function is based on the attractor modeling; therefore, we need a model to represent the distribution of the attractor points in the state space. As a smooth parametric model, a Gaussian mixture model (GMM) can model the chaotic attractor geometry in the state space [67]. The GMM is a parametric probability density function represented by a weighted sum of Gaussian component densities [90]. It can model the distribution of the attractor points in the state space based on its powerful characteristics [91]. So far, metrics such as Kullback–Leibler divergence (also called relative entropy) were defined to measure distance between GMMs [90]. In addition, similarity-based metrics such as likelihood functions have been used to measure distance between a time series and a GMM. This idea was recently used as phone classification methods by parametric models of the distribution points of the speech signal in a high-dimension domain named RPS [67–69].

The GMMs have also been used for parameter estimation of some chaotic systems [70,71]. They were utilized similar to the task of the phone classification method. Suppose we have a chaotic system with a known model and its trajectory was recorded. We can then generate a GMM for the strange attractor of the chaotic system in the state space. Utilizing a distance-like metric over a likelihood function, we can compute dissimilarity between the learned GMM model of the real system attractor (with unknown parameters) and a distribution of a new attractor obtained by a system's model (with known parameters) in the state space to complete the parameter estimation method. Therefore, the score of the distance-based metric will be equal to the cost function of the parameter estimation method.

5.1. The GMM Computation as a Cost Function

A GMM with M mixtures is a weighted sum of M individual Gaussian densities. Each Gaussian density as a component of the GMM is represented by three main factors, mixture weight, mean vector, and covariance matrix. Therefore, they can be shown by a set of parameters, λ, as follows,

$$\lambda = \{w_m, \mu_m, \Sigma_m\}, \; m = 1, \ldots, M p(v|\lambda) = \sum_{m=1}^{M} w_m \frac{1}{(2\pi)^{D/2}} \frac{1}{|\Sigma_m|^{1/2}} \exp\left\{\frac{-1}{2}(v - \mu_m)^T \Sigma_m^{-1} (v - \mu_m)\right\} \quad (13)$$

and $p(v|\lambda)$ is the conditional probability of a D-dimensional single observation vector v given the GMM of λ. The $p(v|\lambda)$ can show a likelihood score. It expresses how probable the observed vector v is for the GMM of λ. In Equation (13) $|.|$ is the determinant operator, exp(.) denotes the exponential function, and M is the number of mixtures (Gaussian components). In addition, for m-th mixture, $w_m \in [0, 1]$ is an scalar and named m-th mixing coefficient or mixture weight, μ_m is the m-th D-dimensional mean vector, and Σ_m is the m-th $D \times D$ covariance matrix. The mean vector and covariance matrix of a Gaussian component can show the center and the shape of points distribution around those of the component. It should be noted that the mixing coefficients w_m are constrained to sum to 1, i.e., $\sum_{m=1}^{M} w_m = 1$.

As a problem which depends on the complexity of the data distribution, there is no analytical solution to determine the optimum number of GMM mixtures, M, needed for modeling of the attractors. Therefore, it is common to use a trial-and-error method to choose an adequate value of M. In our attractor modeling problem, to obtain a proper GMM model of the attractor in the state space, we evaluate some values of M. Generally, we need a higher value of M for attractor modeling if it has a very complex dynamic in the state space. One should note that while a higher number of mixtures can increase the performance of the cost function, it also increases the computational cost.

To find the similarity score between the attractor of a real system and the state space points of a specific model obtained from a chaotic system with known parameters (for example chaotic system (1)), the likelihood score can be calculated. Therefore, the parameter estimation of a known chaotic system with unknown parameters can be performed using the following two phases; a learning phase, here named "phase A", which includes fitting the GMM to the attractor of the real system, and an evaluation phase, named "phase B", to select the best values of parameters for the known chaotic model which causes the maximum similarity score or equally minimum distance score (cost function) over the learned GMM. Following are those phases in details:

5.2. Phase A

The first phase of the parameter estimation approach is the learning phase to find the GMM parameters, λ in Equation (13). The GMM learns the attractor's distribution of a real system, e.g., a chaotic circuit. Suppose $S = \{s_1, s_2, \ldots, s_N\}$ is an $N \times D$ matrix consisting of N-samples of the time series of the real data in the D-dimensional state space. Therefore, each sample is a D-dimensional observation vector. To find the GMM parameters, an iterative expectation-maximization (EM) algorithm is utilized as follows:

5.2.1. Initialization Step

Initialize the mean vector μ_m, covariance matrix Σ_m and mixing coefficients w_m in Equation (13) and evaluate the initial value of the logarithm of the likelihood score obtained from the input time series as follows,

$$log\, p(S|\lambda) = \sum_{n=1}^{N} log(p(s_n|\lambda)) \tag{14}$$

5.2.2. Expectation Step

Evaluate values of $r(s_i, m)$, named responsibility of i-th sample of S given the m-th Gaussian component, using the current values of the GMM parameters:

$$r(s_i, m) = \frac{w_m \frac{1}{(2\pi)^{D/2}} \frac{1}{|\Sigma_m|^{1/2}} exp\left\{ \frac{-1}{2}(s_i - \mu_m)^T \Sigma_m^{-1}(s_i - \mu_m) \right\}}{\sum_{j=1}^{M} w_j \frac{1}{(2\pi)^{D/2}} \frac{1}{|\Sigma_j|^{1/2}} exp\left\{ \frac{-1}{2}\left(s_i - \mu_j\right)^T \Sigma_j^{-1}\left(s_i - \mu_j\right) \right\}} \tag{15}$$

5.2.3. Maximization Step

Re-estimate the parameters of the GMM utilizing the estimated values of the responsibilities as follows:

$$N_m = \sum_{i=1}^{N} r(s_i, m) \tag{16}$$

$$\mu_m = \frac{1}{N_m} \sum_{i=1}^{N} r(s_i, m) s_i \tag{17}$$

$$\Sigma_m = \frac{1}{N_m} \sum_{i=1}^{N} r(s_i, m)(s_i - \mu_m)(s_i - \mu_m)^T \tag{18}$$

$$w_m = \frac{N_m}{N} \tag{19}$$

5.2.4. Likelihood Computation Step

Evaluate the logarithm of the likelihood score in Equation (14) and check for convergence criterion. If the convergence criterion is not satisfied, return to Section 5.2.2.

5.3. Phase B

The second phase is finding the best parameters of the known model of the chaotic system (with unknown parameters) using the learned GMM in the phase A. Here, the search space will be formed from a set of acceptable values of the model parameters. Now we suppose that the values of the parameters $a\&b$ of the system (1) are unknown. Then, for each pair of parameters (a, b), the chaotic system of (1) will be simulated, and a trajectory $T(a,b) = (t_1, t_2, t_3, \ldots, t_K | a, b)$ with K samples will be obtained where each t_K is D-dimensional measured data in the state space. Finally, using an average point-by-point log-likelihood score obtained from the learned GMM, λ, a similarity-based score is computed as follows,

$$log\Big(p(T^{(a,b)}|\lambda)\Big) = \frac{1}{K}\sum_{K=1}^{K} log(p(t_k|\lambda)) \tag{20}$$

where $T^{(a,b)}$ is a matrix whose rows are composed from the state space vectors of the system trajectory with the model's parameters (a,b), and K is the number of the state space point. The parameter estimation method of the model is accomplished by computing Equation (20) and selecting the parameters of the model that can obtain the best similarity-based score, which here means the maximum score. If we use the negative of the similarity-based score, then the parameter estimation becomes a cost function minimization. Therefore, the best parameter selection, $(a,b)^*$, would be conducted by the following criteria, $J(.)$, based on the negative of mean log-likelihood score,

$$(a,b)^* = argmin\{J((a,b))\} \& J((a,b)) = -p\Big(T^{(a,b)}\Big|\lambda\Big) \tag{21}$$

Equation (21) shows the utilized cost function and (a,b) is the set of the system parameters (1). Here, λ is the learned GMM of the real system attractor obtained from the phase A. The objective of the parameter estimation method is to determine the parameters of the system, (a,b) when the cost function is minimized to result in the minimum value of $J((a,b))$. The minimum value of the cost function guarantees the best solution with the proper parameters.

5.4. The GMM of Chaotic Circuit

Based on the observation vector v of the chaotic circuit, in this work, $D = 3$ is selected for the dimension of the state space according to system (1). Using the prepared real training data for the attractor by the chaotic circuit as a real system, the GMM will be specialized in order to model the geometry of that attractor. Figure 14 shows the attractor of the chaotic system in a three-dimensional state space with its GMM modeling using 256 Gaussian components, $M = 256$, where every three-dimensional ellipsoid corresponds to one of the Gaussian components.

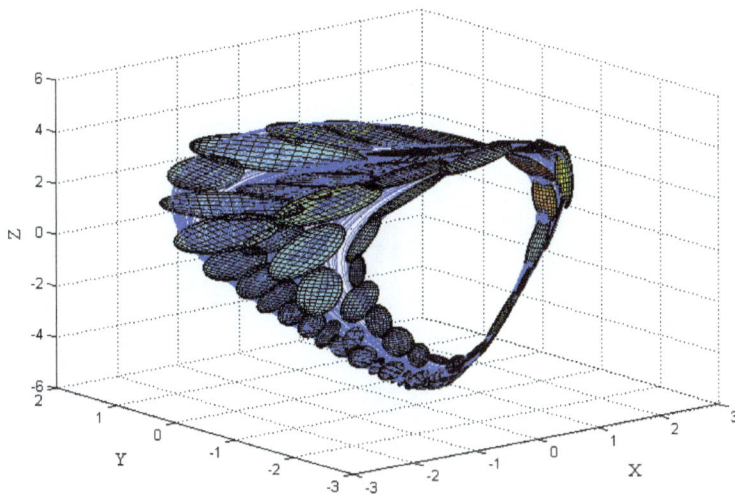

Figure 14. Plot of the attractor and its GMM modeling with $M = 256$ components for the chaotic system (1) with $a = 4$ & $b = 1$, in the 3-D state space.

6. Optimization Algorithm

There are four major categories to which different kinds of optimization methods belong: Enumerative methods, Calculus-Based methods, Heuristic methods, and Meta-heuristic methods [62]. Meta-heuristic optimization algorithms, which cover a wide range of problems, are becoming more and more popular in engineering applications [92,93]. Nature can be named as one of the most important sources of inspiration for new meta-heuristic algorithms. On this subject, black and white holes in cosmology and Humpback whales in the sea aid in constructing the MVO (Multi-Verse Optimizer) and WOA (Whale Optimization Algorithm) meta-heuristic algorithm [94,95]. We now introduce these methods:

6.1. The Whale Optimization Algorithm

The WOA algorithm is based on the hunting behavior of Humpback whales which can encircle the recognized location of prey. The WOA algorithm assumes that the current best candidate solution is the target prey or is close to the optimum. The next step is about the attacking strategy which is the bubble-net strategy. Putting it all together, the proposed WOA method includes three major steps in the simulation: the search for prey, encircling prey, and the bubble-net foraging behavior of humpback whales. For complete details see [94].

6.2. Multi-Verse Optimizer: A Nature-Inspired Algorithm for GlobalOptimization

Another novel nature-inspired algorithm is Multi-Verse Optimizer (MVO). Cosmology (white hole, black hole, and wormhole) is the main inspiration of this algorithm. As mentioned before, every search process in the optimization algorithm consists of two phase: exploration and exploitation. The MVO supports this by white and black holes in order to respond to the exploration phase and wormholes for the exploitation phase. Further details are described in [95].

6.3. Experimental Results

In this section, some simulations are done to investigate the acceptability of the parameter estimation method of the chaotic circuit. We have used a fourth-order Runge-Kutta method with a step size of 10 ms and a total of $30,000$ samples corresponding to a time of 300 s. Here, we assume that the original chaotic system of (1) should be estimated by minimization of the GMM-based cost function.

First, using some 1D parameter estimation methods, different number of the GMM's components, $M = (64, 96, 128, 192, 256)$, are used to show the sufficiency of the cost function. The experimental results of the cost function versus the values of the parameters a&b are depicted in Figures 15 and 16, respectively.

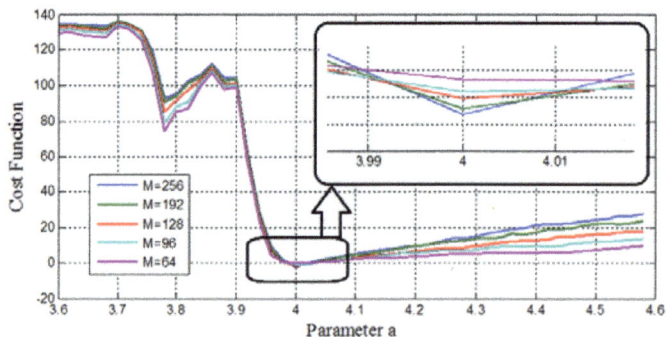

Figure 15. Cost function versus parameter *a*, with different number GMM components (M) for the 1D parameter estimation method.

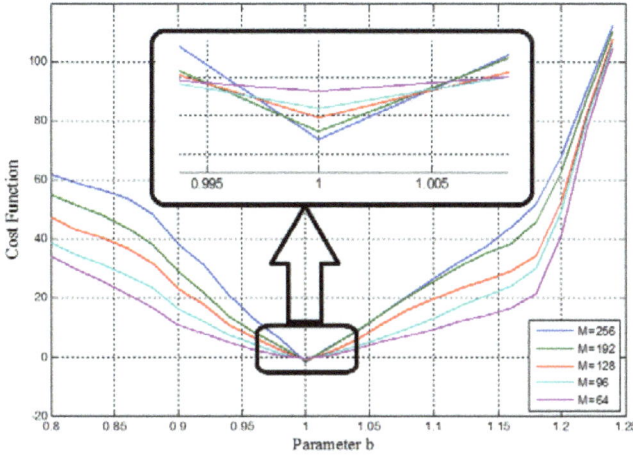

Figure 16. Cost function versus parameter b, with different number of GMM components (M) for the 1D parameter estimation method.

As can be seen, all of the cost functions show convex functions around the desired point. Therefore, they are acceptable for the parameter estimation methods. Specifically, they show the effect of changing the parameter of the model as a monotonically trend along with a global minimum at the exact expected value of the desired parameters ($a = 4.00$, $b = 1.00$). Therefore, the GMM—based cost function has the desired ideal properties for the parameter estimation problem. Moreover, Figures 15 and 16 show the effect of increasing the number of GMM components, M, used in the GMM modeling. In this case, $M = 256$ represents better performance to identify the parameters $a\&b$.

In Figures 17 and 18, a contour plot of the cost function and its "cost surface" are respectively shown for the chaotic system (1) with $M = 256$ along with variation in the parameters, $a\&b$. They show dissimilarity between the real system attractor and each model attractor for a 2D parameter estimation problem. The minimum value of the point on those plots gives the parameters for the best model.

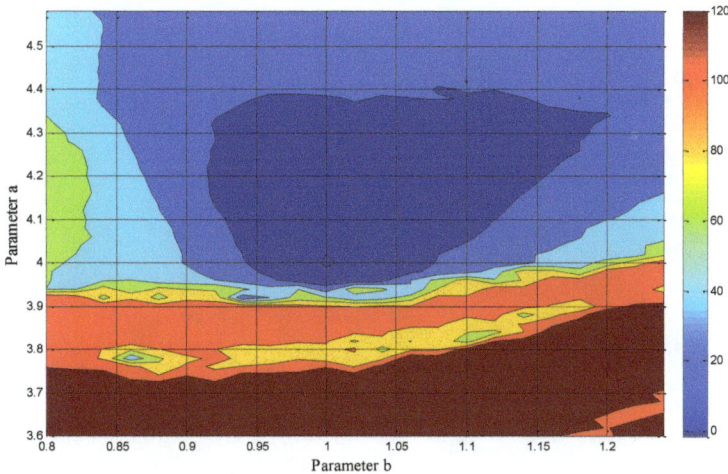

Figure 17. The contour plot of the GMM-based cost function for the introduced chaotic system ($M = 256$) along with variations in the parameters, $a\&b$.

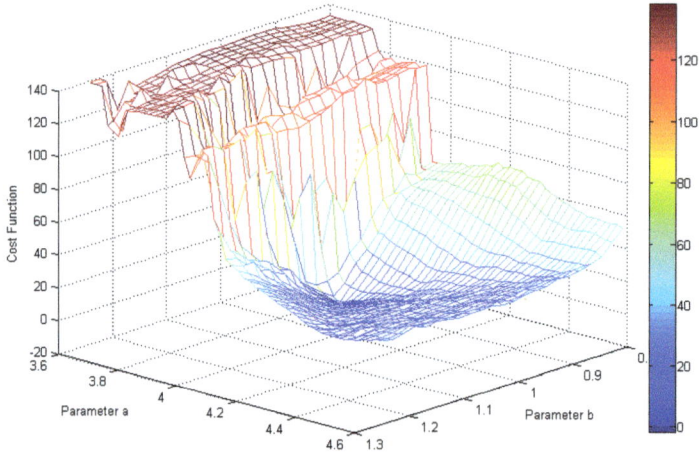

Figure 18. The "cost surface" of the GMM-based cost function for the introduced chaotic system ($M = 256$) along with variations in the parameters, *a&b*.

As can be seen in Figures 17 and 18, the global minimum of the cost function is in the right place ($a = 4.00$ and $b = 1.00$). Furthermore, the surface of the cost function is almost convex near the best parameters, which makes it an easy case for any optimization approach that moves downhill.

In order to examine the efficiency of the cost function in the parameter estimation, two mentioned meta-heuristic optimization methods are applied. All the basic parameters, such as maximum number of iterations (50) and number of search agent (25), are the same in both algorithms. For further details about the algorithms and their particular parameters, see [96]. Comparison between the performances of MVO and WOA optimization algorithm is shown in Figure 19.

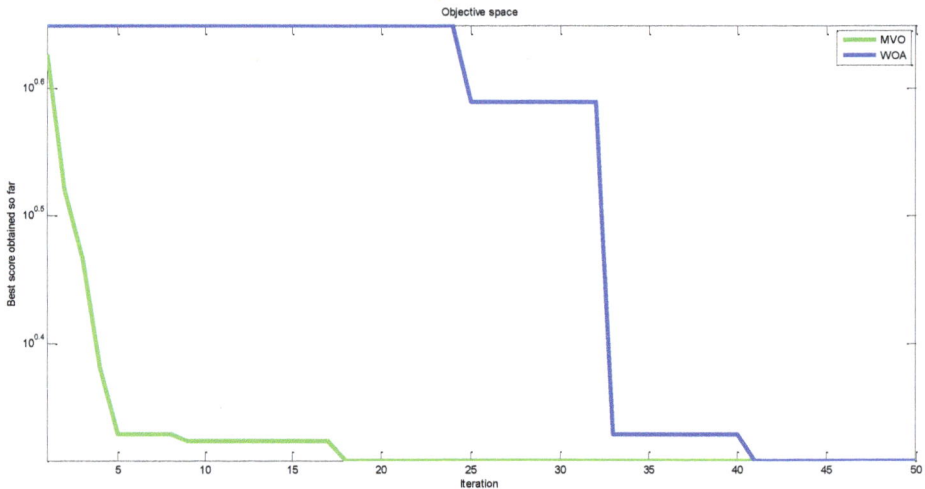

Figure 19. Comparison between the performances of the MVO and WOA optimization algorithm.

Based on the results of Figure 19, the MVO optimization method showed a superior performance in comparison with the WOA algorithms. In addition, Figure 20 represents the process of finding the best parameters using the WOA algorithm performed once for every 10 iterations. As can be seen, the individuals converge to the optimum area ($a = 4.00$ and $b = 1.00$).

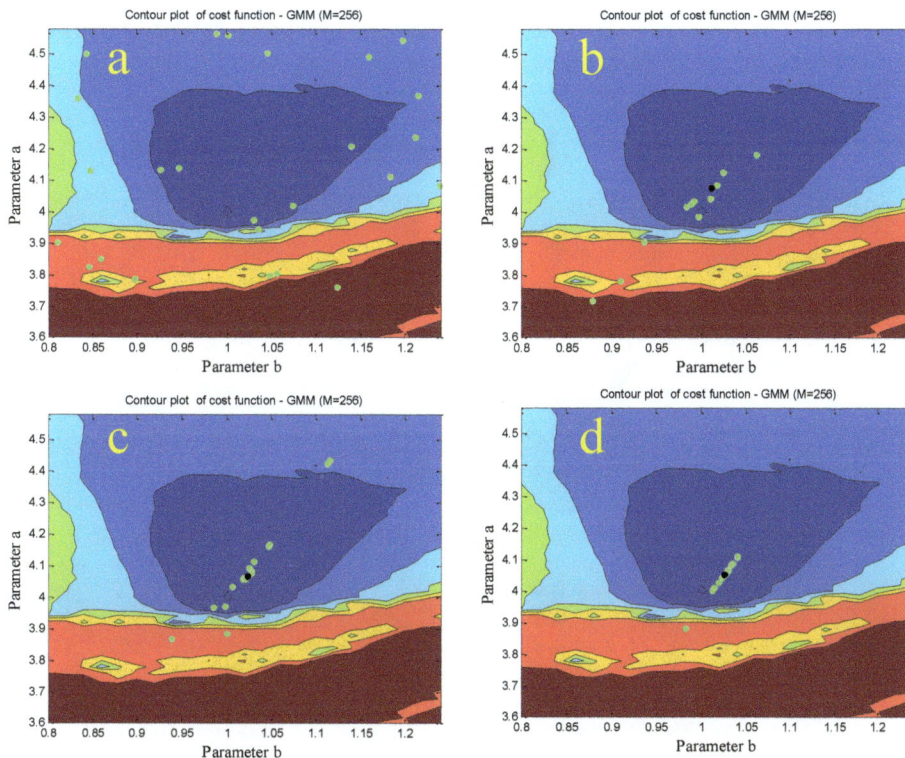

Figure 20. Process of finding the best parameters using the WOA algorithm. (**a–d**) represent the first, 10th, 20th, and 30th iteration, respectively.

7. Conclusions

In this paper, a new chaotic system has been investigated carefully through bifurcation, the largest Lyapunov exponent, ApEn, and stability analysis. Then, an engineering application of that system was proposed using a random number generator and its signal encryption application. After that, a GMM-based cost function was utilized in the parameter estimation of the chaotic circuit designed from the chaotic system. The cost function was based on the minimization of dissimilarity between the phase portrait obtained from the real system and that obtained from the model of the chaotic system. In order to minimize the cost function and to obtain the correct parameters, we used two new efficient optimization methods, the Whale Optimization Algorithm (WOA), and Multi-Verse Optimizer (MVO) algorithm. The MVO optimization method showed superior performance in comparison with the WOA algorithm.

Acknowledgments: This work was supported by the PhD launch fund of Hubei University of Technology (BSQD 2015024).

Author Contributions: Guanghui Xu and Yasser Shekofteh designed the study and wrote the paper. Akif Akgül and Shirin Panahi contributed to the experiment and algorithm design. Yasser Shekofteh and Chunbiao Li partially undertook the theoretical analysis and simulation work of the paper. All authors read and approved the manuscript.

Conflicts of Interest: The authors declare no conflict of interest.

References

1. Li, C.; Sprott, J.C. Variable-boostable chaotic flows. *Opt.-Int. J. Light Electron Opt.* **2016**, *127*, 10389–10398. [CrossRef]
2. Li, C.; Sprott, J.C.; Akgul, A.; Iu, H.H.; Zhao, Y. A new chaotic oscillator with free control. *Chaos Interdiscip. J. Nonlinear Sci.* **2017**, *27*, 083101. [CrossRef] [PubMed]
3. Jafari, M.A.; Mliki, E.; Akgul, A.; Pham, V.-T.; Kingni, S.T.; Wang, X.; Jafari, S. Chameleon: The most hidden chaotic flow. *Nonlinear Dyn.* **2017**, *88*, 1–15. [CrossRef]
4. Li, C.; Sprott, J.C.; Xing, H. Hypogenetic chaotic jerk flows. *Phys. Lett. A* **2016**, *380*, 1172–1177. [CrossRef]
5. Tlelo-Cuautle, E.; Carbajal-Gomez, V.; Obeso-Rodelo, P.; Rangel-Magdaleno, J.; Nuñez-Perez, J.C. FPGA realization of a chaotic communication system applied to image processing. *Nonlinear Dyn.* **2015**, *82*, 1879–1892. [CrossRef]
6. De la Fraga, L.G.; Torres-Pérez, E.; Tlelo-Cuautle, E.; Mancillas-López, C. Hardware implementation of pseudo-random number generators based on chaotic maps. *Nonlinear Dyn.* **2017**, *90*, 1661–1670. [CrossRef]
7. Pano-Azucena, A.D.; de Jesus Rangel-Magdaleno, J.; Tlelo-Cuautle, E.; de Jesus Quintas-Valles, A. Arduino-based chaotic secure communication system using multi-directional multi-scroll chaotic oscillators. *Nonlinear Dyn.* **2017**, *87*, 2203–2217. [CrossRef]
8. Valtierra, J.L.; Tlelo-Cuautle, E.; Rodríguez-Vázquez, Á. A switched-capacitor skew-tent map implementation for random number generation. *Int. J. Circuit Theory Appl.* **2017**, *45*, 305–315. [CrossRef]
9. García-Martínez, M.; Ontañón-García, L.; Campos-Cantón, E.; Čelikovský, S. Hyperchaotic encryption based on multi-scroll piecewise linear systems. *Appl. Math. Comput.* **2015**, *270*, 413–424. [CrossRef]
10. Tlelo-Cuautle, E.; Rangel-Magdaleno, J.; Pano-Azucena, A.; Obeso-Rodelo, P.; Nuñez-Perez, J.C. FPGA realization of multi-scroll chaotic oscillators. *Commun. Nonlinear Sci. Numer. Simul.* **2015**, *27*, 66–80. [CrossRef]
11. Danca, M.-F.; Kuznetsov, N. Hidden chaotic sets in a Hopfield neural system. *Chaos Solitons Fractals* **2017**, *103*, 144–150. [CrossRef]
12. Danca, M.-F.; Kuznetsov, N.; Chen, G. Unusual dynamics and hidden attractors of the Rabinovich–Fabrikant system. *Nonlinear Dyn.* **2017**, *88*, 791–805. [CrossRef]
13. Kuznetsov, N.; Leonov, G.; Yuldashev, M.; Yuldashev, R. Hidden attractors in dynamical models of phase-locked loop circuits: Limitations of simulation in MATLAB and SPICE. *Commun. Nonlinear Sci. Numer. Simul.* **2017**, *51*, 39–49.
14. Leonov, G.; Kuznetsov, N.; Mokaev, T. Hidden attractor and homoclinic orbit in Lorenz-like system describing convective fluid motion in rotating cavity. *Commun. Nonlinear Sci. Numer. Simul.* **2015**, *28*, 166–174. [CrossRef]
15. Leonov, G.A.; Kuznetsov, N.V.; Mokaev, T.N. Homoclinic orbits, and self-excited and hidden attractors in a Lorenz-like system describing convective fluid motion. *Eur. Phys. J. Spec. Top.* **2015**, *224*, 1421–1458. [CrossRef]
16. Sharma, P.; Shrimali, M.; Prasad, A.; Kuznetsov, N.; Leonov, G. Control of multistability in hidden attractors. *Eur. Phys. J. Spec. Top.* **2015**, *224*, 1485–1491. [CrossRef]
17. Sharma, P.R.; Shrimali, M.D.; Prasad, A.; Kuznetsov, N.; Leonov, G. Controlling Dynamics of Hidden Attractors. *Int. J. Bifurc. Chaos* **2015**, *25*, 1550061. [CrossRef]
18. Leonov, G.; Kuznetsov, N.; Vagaitsev, V. Localization of hidden Chua's attractors. *Phys. Lett. A* **2011**, *375*, 2230–2233. [CrossRef]
19. Leonov, G.; Kuznetsov, N.; Vagaitsev, V. Hidden attractor in smooth Chua systems. *Phys. D Nonlinear Phenom.* **2012**, *241*, 1482–1486. [CrossRef]
20. Leonov, G.A.; Kuznetsov, N.V. Hidden attractors in dynamical systems. From hidden oscillations in Hilbert–Kolmogorov, Aizerman, and Kalman problems to hidden chaotic attractor in Chua circuits. *Int. J. Bifurc. Chaos* **2013**, *23*, 1330002. [CrossRef]
21. Leonov, G.; Kuznetsov, N.; Kiseleva, M.; Solovyeva, E.; Zaretskiy, A. Hidden oscillations in mathematical model of drilling system actuated by induction motor with a wound rotor. *Nonlinear Dyn.* **2014**, *77*, 277–288. [CrossRef]

22. Dudkowski, D.; Jafari, S.; Kapitaniak, T.; Kuznetsov, N.V.; Leonov, G.A.; Prasad, A. Hidden attractors in dynamical systems. *Phys. Rep.* **2016**, *637*, 1–50. [CrossRef]

23. Tlelo-Cuautle, E.; de la Fraga, L.G.; Pham, V.-T.; Volos, C.; Jafari, S.; de Jesus Quintas-Valles, A. Dynamics, FPGA realization and application of a chaotic system with an infinite number of equilibrium points. *Nonlinear Dyn.* **2017**, *89*, 1–11. [CrossRef]

24. Pham, V.-T.; Volos, C.; Jafari, S.; Kapitaniak, T. Coexistence of hidden chaotic attractors in a novel no-equilibrium system. *Nonlinear Dyn.* **2017**, *87*, 2001–2010. [CrossRef]

25. Pham, V.-T.; Kingni, S.T.; Volos, C.; Jafari, S.; Kapitaniak, T. A simple three-dimensional fractional-order chaotic system without equilibrium: Dynamics, circuitry implementation, chaos control and synchronization. *AEU Int. J. Electron. Commun.* **2017**, *78*, 220–227. [CrossRef]

26. Pham, V.-T.; Jafari, S.; Volos, C.; Gotthans, T.; Wang, X.; Hoang, D.V. A chaotic system with rounded square equilibrium and with no-equilibrium. *Opt.-Int. J. Light Electron Opt.* **2017**, *130*, 365–371. [CrossRef]

27. Pham, V.-T.; Volos, C.; Gambuzza, L.V. A memristive hyperchaotic system without equilibrium. *Sci. World J.* **2014**, *2014*. [CrossRef] [PubMed]

28. Jafari, S.; Sprott, J.C.; Hashemi Golpayegani, S.M.R. Elementary quadratic chaotic flows with no equilibria. *Phys. Lett. A* **2013**, *377*, 699–702. [CrossRef]

29. Pham, V.-T.; Jafari, S.; Volos, C.; Wang, X.; Hashemi Golpayegani, S.M.R. Is that really hidden? The presence of complex fixed-points in chaotic flows with no equilibria. *Int. J. Bifurc. Chaos* **2014**, *24*, 1450146. [CrossRef]

30. Pham, V.-T.; Volos, C.; Jafari, S.; Wang, X. Generating a novel hyperchaotic system out of equilibrium. *Optoelectron. Adv. Mater.-Rapid Commun.* **2014**, *8*, 535–539.

31. Pham, V.-T.; Volos, C.; Jafari, S.; Wei, Z.; Wang, X. Constructing a novel no-equilibrium chaotic system. *Int. J. Bifurc. Chaos* **2014**, *24*, 1450073. [CrossRef]

32. Tahir, F.R.; Jafari, S.; Pham, V.-T.; Volos, C.; Wang, X. A Novel No-Equilibrium Chaotic System with Multiwing Butterfly Attractors. *Int. J. Bifurc. Chaos* **2015**, *25*, 1550056. [CrossRef]

33. Pham, V.-T.; Jafari, S.; Kapitaniak, T.; Volos, C.; Kingni, S.T. Generating a Chaotic System with One Stable Equilibrium. *Int. J. Bifurc. Chaos* **2017**, *27*, 1750053. [CrossRef]

34. Wang, X.; Pham, V.-T.; Jafari, S.; Volos, C.; Munoz-Pacheco, J.M.; Tlelo-Cuautle, E. A new chaotic system with stable equilibrium: From theoretical model to circuit implementation. *IEEE Access* **2017**, *5*, 8851–8858. [CrossRef]

35. Kingni, S.T.; Pham, V.-T.; Jafari, S.; Woafo, P. A chaotic system with an infinite number of equilibrium points located on a line and on a hyperbola and its fractional-order form. *Chaos Solitons Fractals* **2017**, *99*, 209–218. [CrossRef]

36. Jafari, S.; Sprott, J.C. Simple chaotic flows with a line equilibrium. *Chaos Solitons Fractals* **2013**, *57*, 79–84. [CrossRef]

37. Pham, V.-T.; Jafari, S.; Volos, C. A novel chaotic system with heart-shaped equilibrium and its circuital implementation. *Opt.-Int. J. Light Electron Opt.* **2017**, *131*, 343–349. [CrossRef]

38. Pham, V.T.; Volos, C.; Kapitaniak, T.; Jafari, S.; Wang, X. Dynamics and circuit of a chaotic system with a curve of equilibrium points. *Int. J. Electron.* **2017**, *105*, 1–13. [CrossRef]

39. Pham, V.-T.; Jafari, S.; Wang, X.; Ma, J. A Chaotic System with Different Shapes of Equilibria. *Int. J. Bifurc. Chaos* **2016**, *26*, 1650069. [CrossRef]

40. Pham, V.-T.; Jafari, S.; Volos, C.; Giakoumis, A.; Vaidyanathan, S.; Kapitaniak, T. A chaotic system with equilibria located on the rounded square loop and its circuit implementation. *IEEE Trans. Circuits Syst. II Express Br.* **2016**, *63*, 878–882. [CrossRef]

41. Pham, V.-T.; Jafari, S.; Volos, C.; Vaidyanathan, S.; Kapitaniak, T. A chaotic system with infinite equilibria located on a piecewise linear curve. *Opt.-Int. J. Light Electron Opt.* **2016**, *127*, 9111–9117. [CrossRef]

42. Jafari, S.; Sprott, J.C.; Molaie, M. A simple chaotic flow with a plane of equilibria. *Int. J. Bifurc. Chaos* **2016**, *26*, 1650098. [CrossRef]

43. Jafari, S.; Sprott, J.C.; Pham, V.-T.; Volos, C.; Li, C. Simple chaotic 3D flows with surfaces of equilibria. *Nonlinear Dyn.* **2016**, *86*, 1349–1358. [CrossRef]

44. Rajagopal, K.; Akgul, A.; Jafari, S.; Karthikeyan, A.; Koyuncu, I. Chaotic chameleon: Dynamic analyses, circuit implementation, FPGA design and fractional-order form with basic analyses. *Chaos Solitons Fractals* **2017**, *103*, 476–487. [CrossRef]

45. Rajagopal, K.; Jafari, S.; Laarem, G. Time-delayed chameleon: Analysis, synchronization and FPGA implementation. *Pramana* **2017**, *89*, 92. [CrossRef]

46. Pham, V.-T.; Wang, X.; Jafari, S.; Volos, C.; Kapitaniak, T. From Wang–Chen System with Only One Stable Equilibrium to a New Chaotic System without Equilibrium. *Int. J. Bifurc. Chaos* **2017**, *27*, 1750097. [CrossRef]

47. Pham, V.-T.; Volos, C.; Jafari, S.; Vaidyanathan, S.; Kapitaniak, T.; Wang, X. A chaotic system with different families of hidden attractors. *Int. J. Bifurc. Chaos* **2016**, *26*, 1650139. [CrossRef]

48. Nazarimehr, F.; Saedi, B.; Jafari, S.; Sprott, J.C. Are perpetual points sufficient for locating hidden attractors? *Int. J. Bifurc. Chaos* **2017**, *28*, 1750037. [CrossRef]

49. Dudkowski, D.; Prasad, A.; Kapitaniak, T. Perpetual Points: New Tool for Localization of Coexisting Attractors in Dynamical Systems. *Int. J. Bifurc. Chaos* **2017**, *27*, 1750063. [CrossRef]

50. Faure, P.; Korn, H. Is there chaos in the brain? I. Concepts of nonlinear dynamics and methods of investigation. *C. R. l'Acad. Sci.-Seri. III-Sci.* **2001**, *324*, 773–793. [CrossRef]

51. Korn, H.; Faure, P. Is there chaos in the brain? II. Experimental evidence and related models. *C. R. Biol.* **2003**, *326*, 787–840. [CrossRef] [PubMed]

52. Molaie, M.; Falahian, R.; Gharibzadeh, S.; Jafari, S.; Sprott, J.C. Artificial neural networks: Powerful tools for modeling chaotic behavior in the nervous system. *Front. Comput. Neurosci.* **2014**, *8*. [CrossRef] [PubMed]

53. Falahian, R.; Mehdizadeh Dastjerdi, M.; Molaie, M.; Jafari, S.; Gharibzadeh, S. Artificial neural network-based modeling of brain response to flicker light. *Nonlinear Dyn.* **2015**, *81*, 1951–1967. [CrossRef]

54. Jafari, S.; Sprott, J.C.; Hashemi Golpayegani, S.M.R. Layla and Majnun: A complex love story. *Nonlinear Dyn.* **2016**, *83*, 615–622. [CrossRef]

55. Aram, Z.; Jafari, S.; Ma, J.; Sprott, J.C.; Zendehrouh, S.; Pham, V.-T. Using chaotic artificial neural networks to model memory in the brain. *Commun. Nonlinear Sci. Numer. Simul.* **2017**, *44*, 449–459. [CrossRef]

56. Hilborn, R.C. *Chaos and Nonlinear Dynamics: An Introduction for Scientists and Engineers*; Oxford University Press: Oxford, UK, 2000.

57. Jafari, S.; Hashemi Golpayegani, S.M.R.; Daliri, A. Comment on 'Parameters identification of chaotic systems by quantum-behaved particle swarm optimization' [Int. J. Comput. Math. 86(12) (2009), pp. 2225–2235]. *Int. J. Comput. Math.* **2013**, *90*, 903–905. [CrossRef]

58. Jafari, S.; Hashemi Golpayegani, S.M.R.; Rasoulzadeh Darabad, M. Comment on "Parameter identification and synchronization of fractional-order chaotic systems" [Commun Nonlinear Sci Numer Simulat 2012; 17: 305–16]. *Commun. Nonlinear Sci. Numer. Simul.* **2013**, *18*, 811–814. [CrossRef]

59. Jafari, S.; Hashemi Golpayegani, S.M.R.; Jafari, A.H.; Gharibzadeh, S. Some remarks on chaotic systems. *Int. J. Gen. Syst.* **2012**, *41*, 329–330. [CrossRef]

60. He, Q.; Wang, L.; Liu, B. Parameter estimation for chaotic systems by particle swarm optimization. *Chaos Solitons Fractals* **2007**, *34*, 654–661. [CrossRef]

61. Tang, Y.; Guan, X. Parameter estimation for time-delay chaotic system by particle swarm optimization. *Chaos Solitons Fractals* **2009**, *40*, 1391–1398. [CrossRef]

62. Wang, L.; Xu, Y. An effective hybrid biogeography-based optimization algorithm for parameter estimation of chaotic systems. *Expert Syst. Appl.* **2011**, *38*, 15103–15109. [CrossRef]

63. Weile, D.S.; Michielssen, E. Genetic algorithm optimization applied to electromagnetics: A review. *IEEE Trans. Antennas Propag.* **1997**, *45*, 343–353. [CrossRef]

64. Kennedy, J. Particle swarm optimization. In *Encyclopedia of the Sciences of Learning*; Springer: Berlin, Germany, 2011; pp. 760–766.

65. Yao, X.; Liu, Y. Fast Evolutionary Programming. *Evolut. Program.* **1996**, *3*, 451–460.

66. Arı, Ç.; Aksoy, S.; Arıkan, O. Maximum likelihood estimation of Gaussian mixture models using stochastic search. *Pattern Recognit.* **2012**, *45*, 2804–2816. [CrossRef]

67. Povinelli, R.J.; Johnson, M.T.; Lindgren, A.C.; Roberts, F.M.; Ye, J. Statistical models of reconstructed phase spaces for signal classification. *IEEE Trans. Signal Process.* **2006**, *54*, 2178–2186. [CrossRef]

68. Shekofteh, Y.; Almasganj, F. Feature extraction based on speech attractors in the reconstructed phase space for automatic speech recognition systems. *ETRI J.* **2013**, *35*, 100–108. [CrossRef]

69. Shekofteh, Y.; Almasganj, F.; Daliri, A. MLP-based isolated phoneme classification using likelihood features extracted from reconstructed phase space. *Eng. Appl. Artif. Intell.* **2015**, *44*, 1–9. [CrossRef]

70. Lao, S.-K.; Shekofteh, Y.; Jafari, S.; Sprott, J.C. Cost function based on gaussian mixture model for parameter estimation of a chaotic circuit with a hidden attractor. *Int. J. Bifurc. Chaos* **2014**, *24*, 1450010. [CrossRef]

71. Shekofteh, Y.; Jafari, S.; Sprott, J.C.; Golpayegani, S.M.R.H.; Almasganj, F. A gaussian mixture model based cost function for parameter estimation of chaotic biological systems. *Commun. Nonlinear Sci. Numer. Simul.* **2015**, *20*, 469–481. [CrossRef]

72. Jafari, S.; Sprott, J.C.; Pham, V.-T.; Hashemi Golpayegani, S.M.R.; Jafari, A.H. A New Cost Function for Parameter Estimation of Chaotic Systems Using Return Maps as Fingerprints. *Int. J. Bifurc. Chaos* **2014**, *24*, 1450134. [CrossRef]

73. Kuznetsov, N.; Mokaev, T.; Vasilyev, P. Numerical justification of Leonov conjecture on Lyapunov dimension of Rossler attractor. *Commun. Nonlinear Sci. Numer. Simul.* **2014**, *19*, 1027–1034. [CrossRef]

74. Leonov, G.; Kuznetsov, N.; Mokaev, T. Homoclinic orbit and hidden attractor in the Lorenz-like system describing the fluid convection motion in the rotating cavity. *arXiv* **2014**, arXiv:1412.7667.

75. Kuznetsov, N.; Leonov, G.; Mokaev, T. The Lyapunov dimension and its computation for self-excited and hidden attractors in the Glukhovsky-Dolzhansky fluid convection model. *arXiv* **2015**, arXiv:1509.09161v2.

76. Leonov, G.; Kuznetsov, N.; Mokaev, T. The Lyapunov dimension formula of self-excited and hidden attractors in the Glukhovsky-Dolzhansky system. *arXiv* **2015**, arXiv:1509.09161v1.

77. Kuznetsov, N. The Lyapunov dimension and its estimation via the Leonov method. *Phys. Lett. A* **2016**, *380*, 2142–2149. [CrossRef]

78. Wolf, A.; Swift, J.B.; Swinney, H.L.; Vastano, J.A. Determining Lyapunov exponents from a time series. *Phys. D Nonlinear Phenom.* **1985**, *16*, 285–317. [CrossRef]

79. Pincus, S.M. Approximate entropy as a measure of system complexity. *Proc. Natl. Acad. Sci. USA* **1991**, *88*, 2297–2301. [CrossRef] [PubMed]

80. Pincus, S. Approximate entropy (ApEn) as a complexity measure. *Chaos Interdiscip. J. Nonlinear Sci.* **1995**, *5*, 110–117. [CrossRef] [PubMed]

81. Chon, K.H.; Scully, C.G.; Lu, S. Approximate entropy for all signals. *IEEE Eng. Med. Biol. Mag.* **2009**, *28*. [CrossRef]

82. Koyuncu, İ.; Özcerit, A.T. The design and realization of a new high speed FPGA-based chaotic true random number generator. *Comput. Electr. Eng.* **2016**, *58*, 203–214. [CrossRef]

83. Akgul, A.; Moroz, I.; Pehlivan, I.; Vaidyanathan, S. A new four-scroll chaotic attractor and its engineering applications. *Opt.-Int. J. Light Electron Opt.* **2016**, *127*, 5491–5499. [CrossRef]

84. Çavuşoğlu, Ü.; Akgül, A.; Kaçar, S.; Pehlivan, İ.; Zengin, A. A novel chaos-based encryption algorithm over TCP data packet for secure communication. *Secur. Commun. Netw.* **2016**, *9*, 1285–1296. [CrossRef]

85. Avaroğlu, E.; Koyuncu, İ.; Özer, A.B.; Türk, M. Hybrid pseudo-random number generator for cryptographic systems. *Nonlinear Dyn.* **2015**, *82*, 239–248. [CrossRef]

86. Akgul, A.; Calgan, H.; Koyuncu, I.; Pehlivan, I.; Istanbullu, A. Chaos-based engineering applications with a 3D chaotic system without equilibrium points. *Nonlinear Dyn.* **2016**, *84*, 481–495. [CrossRef]

87. Rukhin, A.; Soto, J.; Nechvatal, J.; Barker, E.; Leigh, S.; Levenson, M.; Banks, D.; Heckert, A.; Dray, J.; Vo, S. *Statistical Test Suite for Random and Pseudorandom Number Generators for Cryptographic Applications*; NIST Special Publication; Booz-Allen and Hamilton Inc.: McLean, VA, USA, 2010.

88. Trejo-Guerra, R.; Tlelo-Cuautle, E.; Jimenez-Fuentes, J.; Sánchez-López, C.; Muñoz-Pacheco, J.; Espinosa-Flores-Verdad, G.; Rocha-Pérez, J. Integrated circuit generating 3-and 5-scroll attractors. *Commun. Nonlinear Sci. Numer. Simul.* **2012**, *17*, 4328–4335. [CrossRef]

89. Kantz, H.; Schreiber, T. *Nonlinear Time Series Analysis*; Cambridge University Press: Cambridge, UK, 2004; Volume 7.

90. Bishop, C.M. Pattern recognition. *Mach. Learn.* **2006**, *128*, 1–58.

91. Nakagawa, S.; Wang, L.; Ohtsuka, S. Speaker identification and verification by combining MFCC and phase information. *IEEE Trans. Audio Speech Lang. Process.* **2012**, *20*, 1085–1095. [CrossRef]

92. Yang, X.-S. *Engineering Optimization: An Introduction with Metaheuristic Applications*; John Wiley & Sons: Hoboken, NJ, USA, 2010.

93. De la Fraga, L.G.; Tlelo-Cuautle, E. Optimizing the maximum Lyapunov exponent and phase space portraits in multi-scroll chaotic oscillators. *Nonlinear Dyn.* **2014**, *76*, 1503–1515. [CrossRef]

94. Mirjalili, S.; Lewis, A. The whale optimization algorithm. *Adv. Eng. Softw.* **2016**, *95*, 51–67. [CrossRef]

95. Mirjalili, S.; Mirjalili, S.M.; Hatamlou, A. Multi-verse optimizer: A nature-inspired algorithm for global optimization. *Neural Comput. Appl.* **2016**, *27*, 495–513. [CrossRef]

96. Tang, W.; Wu, Q. Biologically inspired optimization: A review. *Trans. Inst. Meas. Control* **2009**, *31*, 495–515. [CrossRef]

![entropy logo] *entropy*

MDPI

Article

A Novel Algorithm to Improve Digital Chaotic Sequence Complexity through CCEMD and PE

Chunlei Fan [1], Zhigang Xie [2] and Qun Ding [1,*]

[1]   Electrical Engineering College, Heilongjiang University, Harbin 150080, China; 1172053@s.hlju.edu.cn
[2]   Department of Electronic and Information Engineering, Polytechnic University, Hong Kong 999077, China; encktsel@polyu.edu.hk
*   Correspondence: 1984008@hlju.edu.cn; Tel.: +86-0451-8660-8504

Received: 18 March 2018; Accepted: 12 April 2018; Published: 18 April 2018

Abstract: In this paper, a three-dimensional chaotic system with a hidden attractor is introduced. The complex dynamic behaviors of the system are analyzed with a Poincaré cross section, and the equilibria and initial value sensitivity are analyzed by the method of numerical simulation. Further, we designed a new algorithm based on complementary ensemble empirical mode decomposition (CEEMD) and permutation entropy (PE) that can effectively enhance digital chaotic sequence complexity. In addition, an image encryption experiment was performed with post-processing of the chaotic binary sequences by the new algorithm. The experimental results show good performance of the chaotic binary sequence.

Keywords: chaotic system; empirical mode decomposition; permutation entropy; image encryption

1. Introduction

With the rapid development of computer technology and network communication technology, information has become an important asset in today's society. Therefore, the confidentiality of personal information has become more and more essential. For example, internet data transmission and confidential phone and bank cards require adequate security and confidentiality measures. Therefore, the study of secret communication and cryptography has become an urgent issue. At present, the chaotic signal has benefits such as intrinsic stochasticity, initial value sensitivity, and synchronizing characteristics. Therefore, some traditional chaotic systems with a self-excited attractor are widely used in secret communication and have significant advantages [1–5]. Further, in recent years a hidden chaos attractor has been found, which makes the development of a high-dimensional nonlinear system an attractive challenge [6–9]. At present, most scholars primarily study the dynamic characteristics of hidden attractors. In this paper, we aimed to study chaos with a hidden attractor from the perspective of secure communication and cryptography. Chaos with a hidden attractor is used as a digital chaotic sequence generator with the purpose of encrypting private data. However, in the process of quantization, calculation precision is a crucial factor that degenerates the dynamic characteristics of a chaotic system so that the complexity of a digital chaotic sequence does not satisfy the requirements of information security and cryptography [10,11]. Aiming to solve this problem, Du [12] put forward an algorithm to improve the performance of chaotic binary sequences based on Karhunen–Loève (K–L) transformation. Zhou [13] proposed to scramble the chaotic binary sequence by m sequence in order to improve the complexity of the digital chaotic sequence. Cernak [14] came up with a method to improve the randomness and periodic length of the chaotic binary sequence by perturbing parameters of the chaotic system. Based on the above analysis, these algorithms improve the performance and complexity of digital chaotic sequences by reconstructing the binary sequence method. In this paper, we attempted to generate high complexity in the chaotic sequence based on digital signal processing technology.

Empirical mode decomposition (EMD) in digital signal processing has been extensively applied in nonlinear signal processing [15–18]. EMD was first proposed by Huang et al. [19–21]. It is an effective tool for analyzing nonlinear and non-stationary signals. The EMD method is closely related to the corresponding Hilbert transform method. Through the decomposition of nonlinear and non-stationary signals, a series of intrinsic mode functions (IMFs) are obtained, which makes each IMF a stable signal for narrowband [22]. The IMFs play a crucial role in the analysis of non-stationary or nonlinear signals. However, there are some problems with the EMD method, of which the main one is mode mixing. Complementary ensemble empirical mode decomposition (CEEMD) can effectively restrain the mode mixing of EMD at a certain level [23–25]. Based on the above considerations, we proposed a new algorithm which combines CEEMD with permutation entropy (PE) [26] to effectively improve the complexity of the digital chaotic sequence.

The rest of this paper is organized as follows: Section 2 describes a hidden chaos attractor with no equilibria. The dynamic characteristics of a complex chaotic system are studied by means of numerical simulation and theoretical analysis. Section 3 proposes a new algorithm to improve the complexity of the digital chaotic sequence. Section 4 considers image encryption with post-processing of the chaotic binary sequences by the algorithm outlined in Section 3. The security of the encrypted image is analyzed through key sensitivity, information entropy, and histogram analysis. Section 5 summarizes the discussions of this paper.

2. The Characteristic Analysis of a Chaotic System

In this section, a system can be expressed as the following set of differential equations:

$$\begin{cases} x = -y \\ y = cx + z \\ z = ay^2 + xz - b \end{cases} \tag{1}$$

where a, b, c are real parameters. When $a = 2$, $b = 0.35$, $c = 1$ and the initial value is $(-1.6, 0.82, 1.9)$, the system displays a single-scroll chaotic system [27]. Different projections of the chaotic attractor for this system are shown in Figure 1.

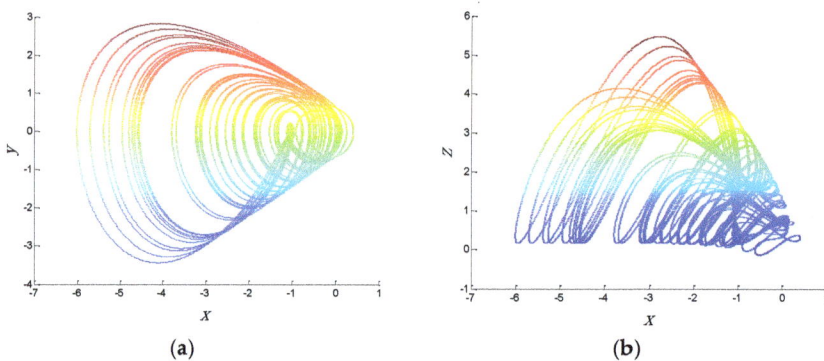

(a)

(b)

Figure 1. *Cont.*

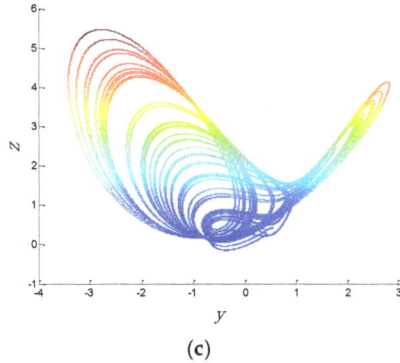

(c)

Figure 1. The different projections of chaotic attractor with: (**a**) *x-y*; (**b**) *x-z*; (**c**) *y-z*.

Equilibria of the chaotic system can be obtained by solving $x = y = z = 0$. The equation is shown as follows:

$$\begin{cases} x = 0 \\ y = 0 \\ z = 0 \end{cases} \rightarrow \begin{cases} 0 = -y \\ 0 = cx + z \\ 0 = ay^2 + xz - b \end{cases}, \tag{2}$$

However, it is easy to see in the chaotic system that when $a = 2$, $b = 0.35$, $c = 1$, Equation (2) has no solution. Therefore, the chaotic system has no equilibria in this case. For the classification of chaotic attractors, if the basin of chaotic attraction intersects with any open neighborhood of an equilibrium, this attractor is called a self-excited attractor. However, if the basin of chaotic attraction is not connected with any equilibrium, this attractor is called a hidden attractor [28–30]. Therefore, the above chaotic system displays a hidden attractor in this case because it is a system with no equilibria. In addition, the Poincaré map of the system can be obtained in the $P = \{y = 0 | (x, z) \in R^2\}$ plane. For the above three-dimensional chaotic system $(x, y, z) \in R^3$, all $(x, 0, z)$ points were calculated by a MATLAB (R2012a, MathWorks, Natick, MA, USA) numerical simulation to obtain the Poincaré map. The Poincaré cross section projected in *x-z* is shown in Figure 2. The Poincaré cross section indicates that the system is a chaotic system through some dense points. Further, for the above chaotic system, the maximal Lyapunov exponent was calculated by a MATLAB numerical simulation. The maximal Lyapunov exponent can indicate the degree of the average divergence of the chaotic trajectory. If the exponent is more than zero, it denotes that the system has the sensitivity of the initial value. According to the result of the MATLAB calculation, this exponent is 0.081. For instance, the time series of *x* generated from two very close initial values $(-1.6, 0.82, 1.9)$ and $(-1.601, 0.82, 1.9)$ are shown in Figure 3, with the purpose of verifying the initial value sensitivity for the chaotic system. Figure 3 is plotted by the MATLAB numerical simulation. According to the Differential Equation (1), the "*t*" presents the number of iterations. As can be seen from Figure 3, the chaotic system is sensitive dependence on initial value.

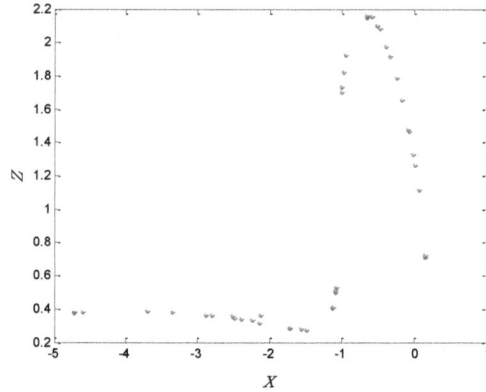

Figure 2. Poincaré map in the *x-z* plane.

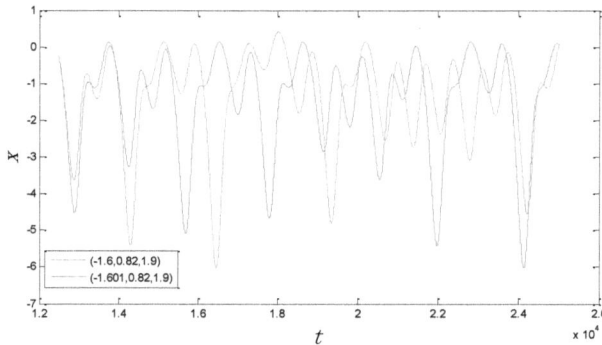

Figure 3. Initial value sensitivity for the time series x with the initial values $(-1.6, 0.82, 1.9)$ and $(-1.601, 0.82, 1.9)$.

3. A New Algorithm to Improve the Complexity of Digital Chaotic Sequences

In this section, we designed a novel algorithm based on CEEMD that can effectively enhance the complexity of digital chaotic sequences. CEEMD can adaptively decompose a non-stationary or non-linear signal into different IMFs. The oscillating frequency of each IMF decreases according to the decomposition order of each IMF. We present a new algorithm to enhance the complexity of chaotic discrete sequences by combining CEEMD with permutation entropy (PE). At the same time, the digital chaotic sequences are converted into chaotic binary sequences through a quantitative method with the purpose of encrypting images or private data. The essential novelty of this algorithm is to eliminate all low complexity IMF components in a chaotic time series, with the purpose of improving the randomness and complexity of the sequence.

3.1. The Basic Principles of EMD

Empirical mode decomposition (EMD) is an adaptive method to decompose non-stationary and non-linear signals into a set of IMFs (intrinsic mode functions) and a residual component. Each IMF should satisfy the following two conditions: (1) For the whole data set, the number of zero crossing and extrema must either be equal or differ at most by one. (2) For any data point, the mean value of the upper and lower envelope determined by the local maxima and minima is zero [31]. The implementation process of the EMD method is shown as follows:

1. All the local maxima and minima of the signal $s(t)$ are calculated to construct the upper envelopes $e_+(t)$ and lower envelopes $e_-(t)$ by the cubic spline interpolation. Further, $m_{11}(t)$ represents the mean of the upper and lower envelopes and is shown as follows:

$$m_{11}(t) = \frac{e_+(t) + e_-(t)}{2} \tag{3}$$

$$s(t) - m_{11}(t) = h_{11}(t) \tag{4}$$

where $h_{11}(t)$ denotes a temporary signal. If $h_{11}(t)$ satisfies the above two crucial factors, it is a first-order IMF component. Otherwise, $h_{11}(t)$ will serve as an initial signal and the above procedures are repeated until the $h_{1k}(t)$ is an IMF and sets the $h_{1k}(t)$ as $c_1(t)$.

$$c_1(t) = h_{1k}(t) \tag{5}$$

2. Next, the first-order IMF has a high frequency, which can be extracted from $s(t)$ by

$$s(t) - c_1(t) = R_1(t) \tag{6}$$

$R_1(t)$ is processed as the new signal and the above procedures are repeated so that the other IMFs can be generated $R_i(t)$, $i = 2, \cdots, n$.

3. When the residual $R_n(t)$ becomes a monotonic function or constant, EMD decomposition is terminated. The $s(t)$ can finally be shown as follows:

$$s(t) = \sum_{i=1}^{n} c_i(t) + R_n(t) \tag{7}$$

Thus, a non-linear signal $s(t)$ can be decomposed into n IMFs and a residual $R_n(t)$. However, there are some problems with the EMD method, and one of these is mode mixing. Generally speaking, each IMF component represents a specific physical quantity. If an IMF component contains a large number of different frequencies of signals then this phenomenon is called mode mixing, which seriously affects the performance of EMD decomposition. Aiming to resolve this issue, the complementary ensemble empirical mode decomposition (CEEMD) method can effectively restrain mode mixing of EMD at a certain level. The CEEMD method was used by adding two opposite white noise signals to an original signal $s(t)$, and to the adopted EMD, with the purpose of restraining mode mixing.

3.2. The Implementation of the New Algorithm

First, suppose $x(t)$ is a time series of chaotic systems. The white noise signal $w_i(t)$ and $-w_i(t)$ with a zero mean value are added to the signal $x(t)$, and the following equation is defined:

$$\begin{cases} x_i^+(t) = x(t) + a_i w_i(t) \\ x_i^-(t) = x(t) - a_i w_i(t) \end{cases} \quad 1 \le i \le N_p, \tag{8}$$

where $w_i(t)$ shows the added white noise signal, and a_i and N_p denote the amplitude and number of the noise signals, respectively. In addition, the variance of the white noise is 1. $\{I_{1i}^+(t)\}$ and $\{I_{1i}^-(t)\}$ $(1 \le i \le N_p)$ represent the first order component sequence, which can be generated by decomposing $x_i^+(t)$ and $x_i^-(t)$ with the EMD method. The mean value of all components is defined as follows:

$$I_1(t) = \frac{1}{2N} \sum_{i=1}^{N_p} [I_{1i}^+(t) + I_{1i}^-(t)]. \tag{9}$$

$I_1(t)$ is sampled to generate a discrete time sequence $I_1(n)$. Then, it is checked whether $I_1(n)$ is a low complexity discrete sequence based on the PE value. The PE is widely applied in the measurement of discrete sequence complexity because of its high robustness and rapid and simple algorithm characteristics. PE can be described as follows:

1. For a discrete time sequence $X_N = \{\ X_1, \ X_2, \ \cdots \ \ X_N\ \}$, where m and τ represent the embedding dimension and a delay factor, respectively, the sequence X_N can be reconstructed as

$$X(n), X(n+\tau), \cdots, X(n+(m-1)\tau) \quad 1 \le n \le N-m+1 \ , \tag{10}$$

2. Each sequence of Equation (10) is placed depending on an ascending order.

$$X(n+(k_1-1)\tau) \le X(n+(k_2-1)\tau) \le \cdots \le X(n+(k_m-1)\tau), \tag{11}$$

3. Further, $\pi_n = \{\ k_1, \ k_2, \ \cdots, \ \ k_m\ \}$ displays the original position index of each element, which is one of the possible order types of all $m!$ permutations. Suppose P_g is a symbol permutation and $\sum_{g=1}^{w} P_g = 1$, where $g = 1, 2, \cdots, w$, $w \le m!$. Then, PE H_p is defined as

$$H_p = -\sum_{g=1}^{w} P_g \ln P_g. \tag{12}$$

When $H_p = 1/m!$, then H_p obtains the maximum value $\ln(m!)$. Further, the normalized PE h_p is defined as $h_p = H_p / \ln(m!)$.

Based on a large amount of MATLAB simulation data, when the PE value of the $I_1(n)$ is less than $\theta \in [0.5, 0.6]$, the amplitude of $I_1(n)$ changes slowly and takes on a lower frequency. After this, the above method is used to find all the low complexity signals in the IMFs. All low complexity IMF signals are separated from the target signal $x(t)$ to generate the signal $r(t)$. Then, the $r(t)$ can be written as

$$r(t) = x(t) - \sum_{j=1}^{p} I_j(t). \tag{13}$$

where p is the sum total of low complexity signals in the IMFs.

3.3. Experimental Results

The time series $(x(t), y(t), z(t))$ are generated from the chaotic system as experimental data. The generated $x(t)$, $y(t)$ and $z(t)$ time series signals are shown in Figure 4.

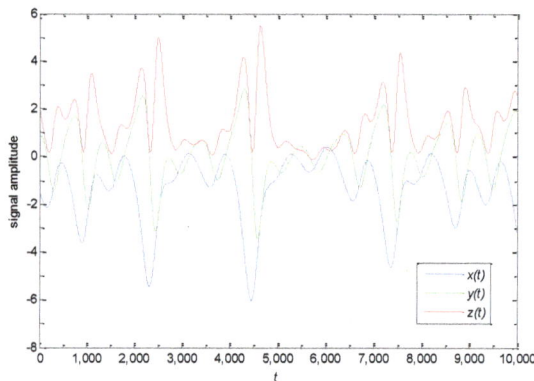

Figure 4. Chaotic time series with $x(t)$ (blue color), $y(t)$ (green color), and $z(t)$ (red color).

Next, these chaotic time series are processed by the above method. All the low complexity signals in the IMFs are shown in Figure 5a–c, where RS (Logogram of Residual $R_n(t)$) is a residual signal. As can be seen from the figure, the amplitude of these IMF signals changes slowly with time and the frequency of the signals reduces. These IMF components are sampled to generate discrete time sequences with the purpose of calculating the PE value. For the $x(t)$, $y(t)$ and $z(t)$ time series, the calculation results of the PE value of each IMF component are shown in Table 1. This table shows that the PE values of these IMFs are less than $\theta \in [0.5, 0.6]$. Therefore, based on the essential novelty of the above method, these IMFs will be removed from the original chaotic time series.

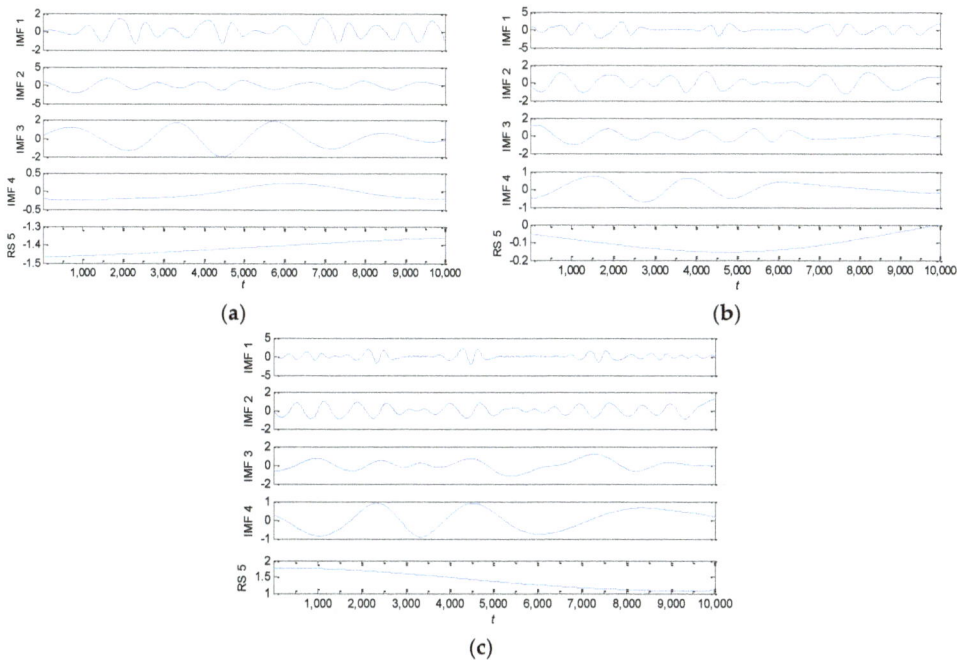

Figure 5. All the low complexity signals in the intrinsic mode functions (IMFs) with: (**a**) $x(t)$; (**b**) $y(t)$; (**c**) $z(t)$.

Table 1. The permutation entropy (PE) value of each intrinsic mode function (IMF) with $x(t)$, $y(t)$, and $z(t)$.

IMF Component	$x(t)$	$y(t)$	$z(t)$
IMF1	0.1181	0.1959	0.1658
IMF2	0.1116	0.1153	0.1198
IMF3	0.1096	0.1113	0.1102
IMF4	0.1069	0.1076	0.1072
RS5	0.0542	0.0997	0.1066

The time series $r_x(t)$, $r_y(t)$ and $r_z(t)$ will be generated by removing the low complexity IMF components from the original signals in $x(t)$, $y(t)$ and $z(t)$. The time series $r_x(t)$, $r_y(t)$ and $r_z(t)$ are shown in Figure 6. Moreover, these time series are also sampled to generate discrete time sequences with the purpose of calculating the PE values, and Figure 7 denotes the comparison of the PE values to the original signals $x(t)$, $y(t)$, $z(t)$ and the post-processing signals $r_x(t)$, $r_y(t)$, $r_z(t)$. It can be seen from Figure 7 that the entropy value of the latter is significantly greater than that of the former and shows a good level of complexity. These high-complexity discrete time sequences can be quantized to generate a good performance in the chaotic binary sequences. These binary sequences will serve as useful key stream sequences of the stream cipher to encrypt private data.

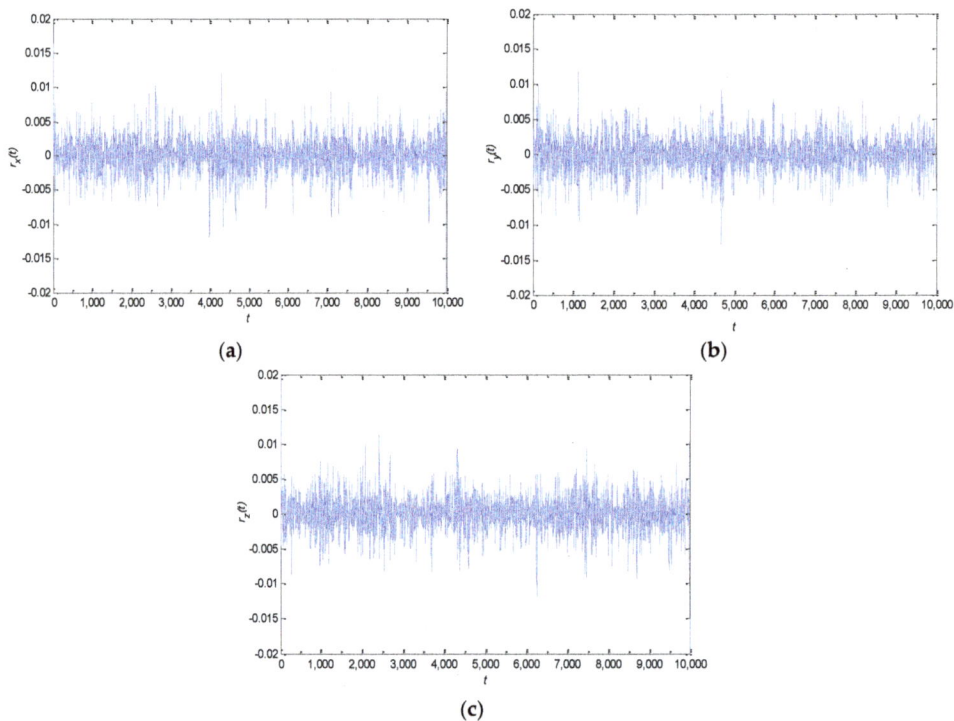

(a)

(b)

(c)

Figure 6. Time series after algorithm processing with: (**a**) $x(t)$; (**b**) $y(t)$; (**c**) $z(t)$.

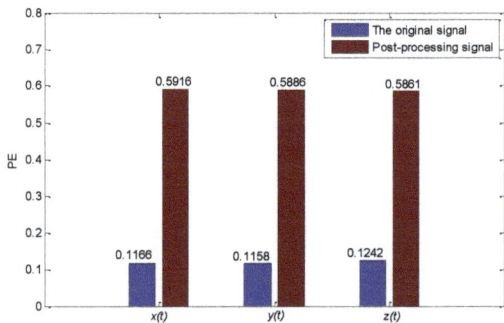

Figure 7. Permutation entropy (PE) value comparisons between the original signal and post-processing signal.

3.4. The Generation and Performance Test of the Chaotic Binary Sequence

The three outputs $r_x(t)$, $r_y(t)$, and $r_z(t)$ are quantized by the interval quantization method, and its mathematical equation is shown below.

$$Q_{0-1}(t) = \begin{cases} 1, & x(t) \in \bigcup_{k=0}^{2^m-1} D_{2k}^m \\ 0, & x(t) \in \bigcup_{k=0}^{2^m-1} D_{2k+1}^m \end{cases} \quad ; \quad k = 0, 1, 2, \cdots , \quad (14)$$

where $Q_{0-1}(t)$ and m are a quantized chaotic binary sequence and arbitrary integer, and $D_0^m, D_1^m, D_2^m \cdots$ are 2^m consecutive equal intervals on the range of the real value of $x(t)$. If the real value falls on the odd range the result of quantization is 0, otherwise it is 1. $r_x(t)$, $r_y(t)$, and $r_z(t)$ are quantized as $Q_x(t)$, $Q_y(t)$, and $Q_z(t)$ through the interval quantization method. Then, the NIST-800-22 test suite is performed to evaluate the performance of the random binary sequences $Q_x(t)$, $Q_y(t)$, and $Q_z(t)$. The NIST-800-22 is composed of 16 different tests, including approximate entropy, linear complexity, and the discrete Fourier transform tests [32,33]. If the *p*-value of the test is greater than 0.01, the test is successful. The NIST-800-22 test results are shown in Table 2. As can be seen from the table, the chaotic random sequences $Q_x(t)$, $Q_y(t)$, and $Q_z(t)$ passed all the tests. These chaotic sequences can be used in high security fields such as network security and multimedia encryption.

Table 2. NIST-800-22 tests.

Test Item	$Q_x(t)$ *p*-Value	$Q_y(t)$ *p*-Value	$Q_z(t)$ *p*-Value	Result
Approximate Entropy	0.28711	0.01063	0.41042	Success
Block Frequency	0.02501	0.43924	0.64085	Success
Cumulative Sums	0.14372	0.56658	0.64761	Success
FFT	0.52063	0.37221	0.11875	Success
Frequency	0.28014	0.48392	0.87461	Success
Linear Complexity	0.22374	0.46932	0.78321	Success
Longest Run	0.70665	0.51078	0.26541	Success
Non-Overlapping Template	0.32974	0.75331	0.11253	Success
Overlapping Template	0.24088	0.70399	0.32227	Success
Random Excursions	0.43747	0.51791	0.82733	Success
Random Excursions Variant	0.64578	0.11253	0.66691	Success
Binary Matrix Rank	0.15319	0.58700	0.44130	Success
Runs	0.88206	0.84530	0.71884	Success
Serial Test-1	0.10056	0.17826	0.81473	Success
Serial Test-2	0.15538	0.15538	0.69926	Success
Maurer's Universal	0.75331	0.14268	0.56553	Success

4. Image Encryption with a Chaotic Binary Sequence

This subsection describes the experiments used to demonstrate the performance of the chaotic binary sequence by encrypting images. The Lena and Baboon images, with a size of 256 × 256, are encrypted by the above chaotic random sequences—$Q_x(t)$, $Q_y(t)$, and $Q_z(t)$. Then, $Q_x(t)$, $Q_y(t)$, and $Q_z(t)$ serve as the key stream sequences of the stream cipher with the purpose of encrypting the R, G, and B components of the color images.

4.1. Key Sensitivity

The sensitivity of chaos to the initial value can support the effective avoidance of tentative attacks. Using the Lena and Baboon images with a size of 256 × 256 as examples, Figure 8a,d shows the plain-images, while the cipher-images are given in Figure 9b,e. However, a 10^{-5} change of the initial value will lead to incorrect decryption results, as shown in Figure 9c,f. The experimental results show that the chaotic binary sequence shows high key sensitivity.

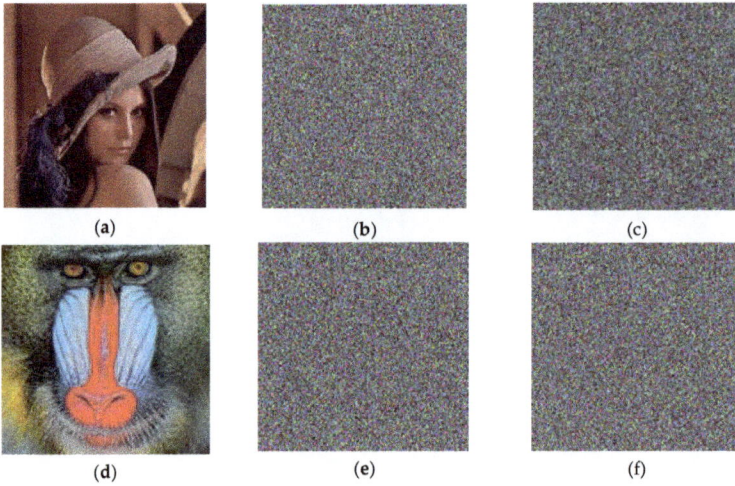

Figure 8. Key sensitivity test with: (**a**) plain-image for Lena; (**b**) cipher-image for Lena; (**c**) incorrect decryption using a 10^{-5} change of the initial value for Lena; (**d**) plain-image for Baboon; (**e**) cipher-image for Baboon; (**f**) incorrect decryption using a 10^{-5} change of the initial value for Baboon.

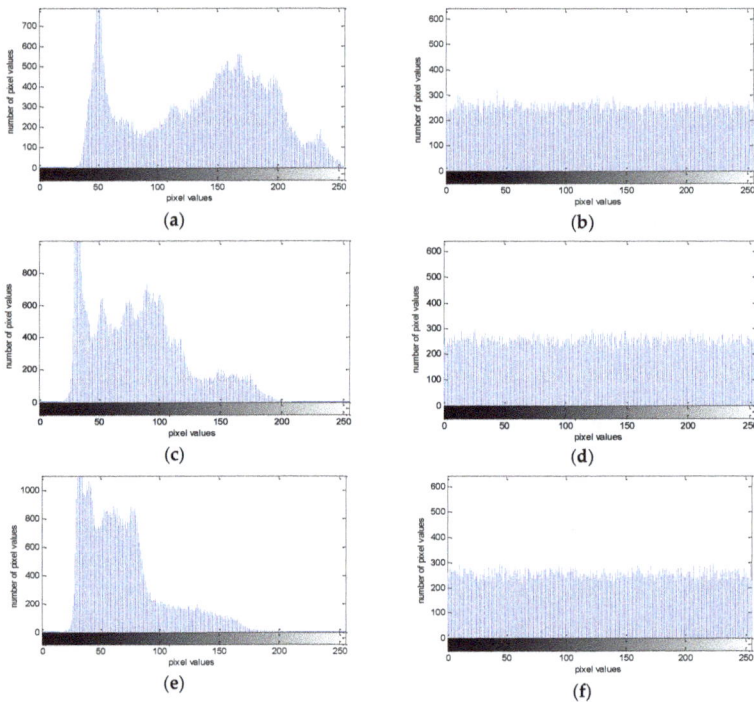

Figure 9. Histogram test with: (**a**) R component of the plain-image; (**b**) R component of the cipher-image; (**c**) B component of the plain-image; (**d**) B component of the cipher-image; (**e**) G component of the plain-image; (**f**) G component of the cipher-image.

4.2. Histogram Analysis

The image histogram can be approximated as the density function of the gray value, which is an important indicator in the analysis of an image's statistical properties [34]. The histogram test is shown in Figure 9, and the horizontal and vertical coordinates of the histogram represent the pixel values and number of pixel values, respectively. Figure 9 show that the gray histogram of the encrypted image is relatively uniform, which indicates that the security performance of this key sequence is relatively high, and the image is not easily able to be tampered with and decrypted during transmission.

4.3. Correlation Analysis of Adjacent Pixels

Generally speaking, the smaller the adjacent pixel correlation of the cipher-image, the more obvious the effect of resisting statistical attack [35]. The mathematical equation can be shown as follows:

$$\rho_{xy} = \frac{\text{cov}(x, y)}{\sqrt{D(x)D(y)}}. \tag{15}$$

where $\text{cov}(x, y) = \frac{1}{N} \sum_{i=1}^{N} (x_i - E(x))(y_i - E(y))$, $D(x) = \frac{1}{N} \sum_{i=1}^{N} (x_i - E(x))^2$, $E(x) = \frac{1}{N} \sum_{i=1}^{N} x_i$, x_i and y_i represent the different gray values of two adjacent pixels and N denotes the number of randomly selected adjacent pixels.

The above equation was used and some pairs of adjacent pixels in different directions were randomly chosen, and the test results are listed in Table 3. It can be seen from the experimental data that the correlation of adjacent pixels of a cipher-image tends to be zero.

Table 3. Correlation analysis of adjacent pixels for the Lena and Baboon images.

Direction	Plain-Image for Lena	Cipher-Image for Lena	Plain-Image for Baboon	Cipher-Image for Baboon
Horizontal	0.9712	0.0392	0.9287	0.0133
Vertical	0.9655	0.0091	0.9004	0.0522
Diagonal	0.9401	0.0215	0.8711	0.0093

4.4. Information Entropy Analysis

Information entropy can reflect the randomness of the information in images, namely the uncertainty of the distribution of pixel values in a cipher-image. Its mathematical equation is shown below [36].

$$H(\phi) = \sum_{i=0}^{2^L - 1} p(\phi_i) \log_2 \frac{1}{p(\phi_i)}. \tag{16}$$

where L is the number of bits required to store each pixel value, and $p(\phi_i)$ presents the probability of the symbol ϕ_i. When the probability of each symbol ϕ_i is equal, the information entropy ($H(\phi) = 8$) is at its largest. When the information entropy is closer to 8, the gray value tends to be distributed randomly. Table 4 provides a comparison of this data with other experiments. This comparison shows that the information entropy of our method is closer to 8. Therefore, it can effectively resist information entropy attacks.

Table 4. Information entropy analysis for the Lena and Baboon images.

Methods	R Component	G Component	B Component
The paper for Lena	7.9972	7.9971	7.9972
The paper for Baboon	7.9970	7.9968	7.9971
Reference [37]	7.9914	7.9914	7.9915
Reference [38]	7.9851	7.9852	7.9832

5. Discussion

Some traditional chaotic systems with a self-excited attractor have been widely used in secret communication. However, for chaotic systems with hidden attractors, most of the current research has focused on studying the dynamic characteristics of the system rather than its application in the field of information security. Therefore, in this paper, we aimed to study chaos with a hidden attractor from the perspective of secure communication and data encryption. First, we introduced the dynamic characteristics of a chaotic system with hidden attractors by means of a numerical simulation and theoretical analysis, including equilibria, a Poincaré cross section, and initial value sensitivity. After that, a new algorithm was designed to enhance the complexity of digital chaotic sequences with the purpose of satisfying the requirements of data encryption. The essential novelty of the algorithm is to eliminate all low complexity IMF components of a chaotic time series by using digital signal processing technology. PE value comparisons between the original signal and post-processing signal show the performance of the algorithm is good. In addition, the NIST-800-22 test was performed to demonstrate the randomness and complexity of the chaotic binary sequence. The chaotic binary sequence can serve as a good key stream sequence of a stream cipher to encrypt private data. Furthermore, an image encryption experiment was undertaken to show the security of the above method. However, some weaknesses in this technique remain, and we believe that the new algorithm should be optimized in operation efficiency.

Acknowledgments: This work was supported by the Natural Science Foundation of China (No. 61471158) and the "modern sensing technology" innovation team project of Heilongjiang province (No. 2012TD007).

Author Contributions: Chunlei Fan conceived and wrote the paper. Zhigang Xie gave some theoretical guidance. Chunlei Fan and Qun Ding contributed to the MATLAB numerical simulation. All authors have read and approved the final manuscript.

Conflicts of Interest: The authors declare no conflict of interest.

References

1. Xu, H.; Tong, X.J.; Meng, X.W. An efficient chaos pseudo-random number generator applied to video encryption. *Optik* **2016**, *127*, 9305–9319. [CrossRef]
2. Wang, Q.X.; Yu, S.M.; Li, C.Q.; Lu, J.H.; Fang, X.L.; Guyeux, C.; Bahi, J.M. Theoretical Design and FPGA-Based Implementation of Higher-Dimensional Digital Chaotic Systems. *IEEE Trans. Circuits Syst. I* **2016**, *63*, 401–412. [CrossRef]
3. Valli, D.; Ganesan, K. Chaos based video encryption using maps and Ikeda time delay system. *Eur. Phys. J. Plus* **2017**, *132*, 542. [CrossRef]
4. Vaidyanathan, S.; Akgul, A.; Kacar, S.; Cavusoglu, U. A new 4-D chaotic hyperjerk system, its synchronization, circuit design and applications in RNG, image encryption and chaos-based steganography. *Eur. Phys. J. Plus* **2018**, *133*, 46. [CrossRef]
5. Pan, J.; Ding, Q.; Du, B.X. A New Improved Scheme of Chaotic Masking Secure Communication Based on Lorenz System. *Int. J. Bifurc. Chaos* **2012**, *22*, 1250125. [CrossRef]
6. Ren, S.L.; Panahi, S.; Rajagopal, K.; Akgul, A.; Pham, V.T.; Jafari, S. A New Chaotic Flow with Hidden Attractor: The First Hyperjerk System with No Equilibrium. *Z. Naturforsch. A* **2018**, *73*, 239–249. [CrossRef]
7. Kamal, N.K.; Varshney, V.; Shrimali, M.D.; Prasad, A.; Kuznetsov, N.V.; Leonov, G.A. Shadowing in hidden attractors. *Nonlinear Dyn.* **2018**, *91*, 2429–2434. [CrossRef]

8. Jafari, S.; Pham, V.T.; Golpayegani, S.M.R.H.; Moghtadaei, M.; Kingni, S.T. The Relationship Between Chaotic Maps and Some Chaotic Systems with Hidden Attractors. *Int. J. Bifurc. Chaos* **2016**, *26*, 1650211. [CrossRef]

9. Dudkowski, D.; Jafari, S.; Kapitaniak, T.; Kuznetsov, N.V.; Leonov, G.A.; Prasad, A. Hidden attractors in dynamical systems. *Phys. Rep.* **2016**, *637*, 1–50. [CrossRef]

10. Zhang, H.; Xiang, S.Y.; Zhang, Y.H.; Guo, X.X. Complexity-enhanced polarization-resolved chaos in a ring network of mutually coupled vertical-cavity surface-emitting lasers with multiple delays. *Appl. Opt.* **2017**, *56*, 6728–6734. [CrossRef] [PubMed]

11. Rontani, D.; Mercier, E.; Wolfersberger, D.; Sciamanna, M. Enhanced complexity of optical chaos in a laser diode with phase-conjugate feedback. *Opt. Lett.* **2016**, *41*, 4637–4640. [CrossRef] [PubMed]

12. Du, B.X.; Geng, X.L.; Chen, F.Y.; Pan, J.; Ding, Q. Generation and Realization of Digital Chaotic Key Sequence Based on Double K-L Transform. *Chin. J. Electron.* **2013**, *22*, 131–134.

13. Zhou, H.; Ling, X.T. Realizing Finite Precision Chaotic Systems via Perturbation of m-Sequences. *Acta Electron. Sin.* **1997**, *25*, 95–97.

14. Cernak, J. Digital generators of chaos. *Phys. Lett. A* **1996**, *214*, 151–160. [CrossRef]

15. Liu, D.; Zeng, H.T.; Xiao, Z.H.; Peng, L.H.; Malik, O.P. Fault diagnosis of rotor using EMD thresholding-based de-noising combined with probabilistic neural network. *J. Vibroeng.* **2017**, *19*, 5920–5931.

16. Li, J.L.; Lindemann, J.; Egelhaaf, M. Local motion adaptation enhances the representation of spatial structure at EMD arrays. *PLoS Comput. Biol.* **2017**, *13*, e1005919. [CrossRef] [PubMed]

17. Su, J.S.; Wang, Y.Q.; Yang, X.Y.; Wang, X.F. Enhancement of Weak Lidar Signal Based on Variable Frequency Resolution EMD. *IEEE Photonic Technol. Lett.* **2016**, *28*, 2882–2885. [CrossRef]

18. Singh, D.S.; Zhao, Q. Pseudo-fault signal assisted EMD for fault detection and isolation in rotating machines. *Mech. Syst. Signal Process.* **2016**, *81*, 202–218. [CrossRef]

19. Huang, N.E.; Wu, Z.H. A review on Hilbert-Huang transform: Method and its applications to geophysical studies. *Rev. Geophys.* **2008**, *46*, 1–23. [CrossRef]

20. Mandic, D.P.; Rehman, N.U.; Wu, Z.H.; Huang, N.E. Empirical Mode Decomposition-Based Time-Frequency Analysis of Multivariate Signals. *IEEE Signal Process. Mag.* **2013**, *30*, 74–86. [CrossRef]

21. Tsai, P.H.; Lin, C.; Tsao, J.; Lin, P.F.; Wang, P.C.; Huang, N.E.; Lo, M.T. Empirical mode decomposition based detrended sample entropy in electroencephalography for Alzheimer's disease. *J. Neurosci. Methods* **2012**, *210*, 230–237. [CrossRef] [PubMed]

22. Zhang, X.; Liu, Z.W.; Miao, Q.; Wang, L. An optimized time varying filtering based empirical mode decomposition method with grey wolf optimizer for machinery fault diagnosis. *J. Sound Vib.* **2018**, *418*, 55–78. [CrossRef]

23. Xu, Y.; Zhang, M.; Zhu, Q.; He, Y. An improved multi-kernel RVM integrated with CEEMD for high-quality intervals prediction construction and its intelligent modeling application. *Chemom. Intell. Lab. Syst.* **2017**, *171*, 151–160. [CrossRef]

24. Vrochidou, E.; Alvanitopoulos, P.; Andreadis, I.; Elenas, A. Artificial accelerograms composition based on the CEEMD. *Trans. Inst. Meas. Control* **2016**, *40*, 239–250. [CrossRef]

25. Jia, J.; Goparaju, B.; Song, J.L.; Zhang, R.; Westover, M.B. Automated identification of epileptic seizures in EEG signals based on phase space representation and statistical features in the CEEMD domain. *Biomed. Signal Process. Control* **2017**, *38*, 148–157. [CrossRef]

26. Bandt, C.; Pompe, B. Permutation Entropy: A Natural Complexity Measure for Time Series. *Phys. Rev. Lett.* **2002**, *88*, 174102. [CrossRef] [PubMed]

27. Yuan, F.; Wang, G.Y.; Wang, X.W. Extreme multistability in a memristor-based multi-scroll hyper-chaotic system. *Chaos* **2016**, *26*, 507–519. [CrossRef] [PubMed]

28. Leonov, G.A.; Kuznetsov, N.V.; Mokaev, T.N. Homoclinic orbits, and self-excited and hidden attractors in a Lorenz-like system describing convective fluid motion. *Eur. Phys. J. Spec. Top.* **2015**, *224*, 1421–1458. [CrossRef]

29. Kuznetsov, N.V.; Leonov, G.A.; Mokaev, T.N.; Prasad, A.; Shrimali, M.D. Finite-time Lyapunov dimension and hidden attractor of the Rabinovich system. *Nonlinear Dyn.* **2017**, *92*, 267–285. [CrossRef]

30. Leonov, G.A.; Kuznetsov, N.V. Hidden Attractors in Dynamical Systems from Hidden Oscillations in Hilbert–Kolmogorov, Aizerman, and Kalman Problems to Hidden Chaotic Attractor in Chua Circuits. *Int. J. Bifurc. Chaos* **2014**, *23*, 1330002. [CrossRef]

31. Li, Y.; Xu, M.; Wei, Y.; Huang, W. An improvement EMD method based on the optimized rational Hermite interpolation approach and its application to gear fault diagnosis. *Measurement* **2015**, *63*, 330–345. [CrossRef]
32. Liu, L.F.; Miao, S.X. The complexity of binary sequences using logistic chaotic maps. *Complexity* **2016**, *21*, 121–129. [CrossRef]
33. Nian-Sheng, L. Pseudo-randomness and complexity of binary sequences generated by the chaotic system. *Commun. Nonlinear Sci.* **2011**, *16*, 761–768. [CrossRef]
34. Murillo-Escobar, M.A.; Cruz-Hernandez, C.; Abundiz-Perez, F.; Lopez-Gutierrez, R.M.; Del Campo, O.R.A. A RGB image encryption algorithm based on total plain image characteristics and chaos. *Signal Process.* **2015**, *109*, 119–131. [CrossRef]
35. Wang, Y.; Lei, P.; Yang, H.Q.; Cao, H.Y. Security analysis on a color image encryption based on DNA encoding and chaos map. *Comput. Electr. Eng.* **2015**, *46*, 433–446. [CrossRef]
36. Ye, G.; Pan, C.; Huang, X.; Zhao, Z.; He, J. A Chaotic Image Encryption Algorithm Based on Information Entropy. *Int. J. Bifurc. Chaos* **2018**, *28*, 1850010. [CrossRef]
37. Liu, H.J.; Kadir, A.; Sun, X.B. Chaos-based fast colour image encryption scheme with true random number keys from environmental noise. *IET Image Process.* **2017**, *11*, 324–332. [CrossRef]
38. Liu, H.J.; Wang, X.Y. Color image encryption based on one-time keys and robust chaotic maps. *Comput. Math. Appl.* **2010**, *59*, 3320–3327. [CrossRef]

entropy

MDPI

Article

A New Two-Dimensional Map with Hidden Attractors

Chuanfu Wang [ID] **and Qun Ding** *

Electronic Engineering College, Heilongjiang University, Harbin 150080, China; 1172054@s.hlju.edu.cn
* Correspondence: 1984008@hlju.edu.cn

Received: 31 January 2018; Accepted: 24 April 2018; Published: 27 April 2018

Abstract: The investigations of hidden attractors are mainly in continuous-time dynamic systems, and there are a few investigations of hidden attractors in discrete-time dynamic systems. The classical chaotic attractors of the Logistic map, Tent map, Henon map, Arnold's cat map, and other widely-known chaotic attractors are those excited from unstable fixed points. In this paper, the hidden dynamics of a new two-dimensional map inspired by Arnold's cat map is investigated, and the existence of fixed points and their stabilities are studied in detail.

Keywords: hidden attractors; fixed point; stability

1. Introduction

The investigations of the chaotic system were greatly encouraged by the discovery of the Lorenz system [1]. The Lorenz system is one of the most wildly-studied continuous-time dynamic systems, and other classical continuous-time dynamic systems include the Rössler system, Chua system, Chen system, Lü system, and Sprott system [2–6]. Most attractors of those classical continuous-time dynamic systems are excited from unstable equilibria. However, hidden attractors imply the basin of attraction does not contain neighborhoods of equilibria [7]. For finding hidden attractors, a lot of systems improved when classical continuous-time dynamic systems were proposed [8–19]. These investigations of hidden attractors can be classified by the number and stability of equilibria, such as no equilibrium, finite stable equilibria, and infinite stable equilibria. In 2010, a special analytical-numerical algorithm of finding hidden attractors in Chua system was proposed [20]. The special algorithm can find the accuracy initial values that lead to hidden attractors and has promoted the development of finding hidden attractors in continuous-time dynamic systems [21–24]. However, most investigations of hidden attractors are mainly in continuous-time dynamic systems, such as the Chen system, Sprott system, Chua system, and Lü system [15–24]. There are only a few of investigations of hidden attractors in discrete-time systems.

The classical chaotic maps include the Logistic map, Tent map, Henon map, and Arnold's cat map [25–28]. In line with continuous-time dynamic systems, these hidden attractors in the classical chaotic maps are also excited from unstable fixed points. In 2016, Jafari et al. studied a new one-dimensional chaotic map with no fixed point inspired by Logistic map, and its bifurcation and period doubling were introduced [29]. At the same year, Jiang et al. performed a search to find hidden attractors in a new two-dimensional chaotic map inspired by the Henon map and analyzed several different cases on fixed points, such as no fixed point, single fixed point, and two fixed points [30].

In this paper, a new two-dimensional chaotic map inspired by Arnold's cat map is proposed. For the limitation in the form of the Arnold's cat map, the number of fixed points in the new two-dimensional chaotic map is not larger than two, and the fixed points are closely related to the Lyapunov exponents. Due to the restriction of the form of the Arnold's cat map, the new two-dimensional chaotic attractor can only appear in the case of no fixed point. Thus, our concern focuses on the case of no fixed point. The paper is arranged as follows: Section 2 describes the Arnold's cat map and shows its chaotic attractor. Section 3 analyzes the stability of the equilibria in the new two-dimensional map. Section 4 demonstrates the digitalization and hardware implementation of the new two-dimensional map. The complexity of output time series is tested by approximate entropy in Section 5, and Section 6 summarizes the conclusions of this paper.

2. Arnold's Cat Map

Arnold's cat map, also known as cat chaotic map, is a chaotic map of repeated folding and stretching in a limited area and is wildly used in multimedia chaotic encryption [28]. Arnold's cat map is a two-dimensional chaotic map and defined as

$$\begin{bmatrix} x(n+1) \\ y(n+1) \end{bmatrix} = \begin{bmatrix} 1 & 1 \\ 1 & 2 \end{bmatrix} \begin{bmatrix} x(n) \\ y(n) \end{bmatrix} (\mathrm{mod}1) \tag{1}$$

Equation (1) can be transformed into Equation (2) for calculating the fixed points.

$$\begin{cases} x(n) = x(n) + y(n)(\mathrm{mod}1) \\ y(n) = x(n) + 2y(n)(\mathrm{mod}1) \end{cases} \tag{2}$$

Equation (2) always has a fixed point, because it is composed of homogeneous linear equations. The fixed point of the Arnold's cat map is $\begin{cases} x^* = 0 \\ y^* = 0 \end{cases}$. The Jacobian matrix of Arnold's cat map is $\begin{bmatrix} 1 & 1 \\ 1 & 2 \end{bmatrix}$, and the two eigenvalues are calculated as $Eig_1 = \frac{3+\sqrt{5}}{2} \approx 2.618$ and $Eig_2 = \frac{3-\sqrt{5}}{2} \approx 0.382$. Therefore, the fixed point (0,0) is an unstable fixed point. It is a saddle point, because there is a positive Lyapunov exponent and a negative Lyapunov exponent. For a positive Lyapunov exponent, Arnold's cat map is a two-dimensional chaotic map. In Arnold's cat map, two eigenvalues of the Jacobian matrix are associated, respectively, to an expanding and a contracting eigenspace, which are also the stable and unstable manifolds [31]. The phase diagram of Arnold's cat map is shown in Figure 1.

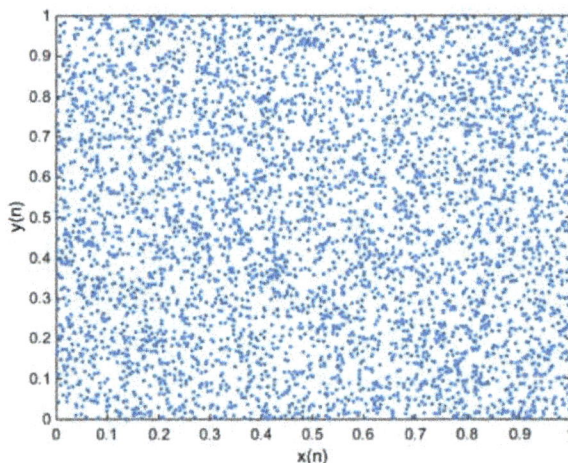

Figure 1. When $x(0) = 0.7$ and $y(0) = 0.6$, this represents the phase diagram of Arnold's cat map.

The generalized Arnold's cat map is defined as

$$
\begin{bmatrix} x(n+1) \\ y(n+1) \end{bmatrix} = \begin{bmatrix} 1 & a \\ b & ab+1 \end{bmatrix} \begin{bmatrix} x(n) \\ y(n) \end{bmatrix} (\mathrm{mod}1). \tag{3}
$$

The fixed point of the generalized Arnold's cat map also is $\begin{cases} x^* = 0 \\ y^* = 0 \end{cases}$. The Jacobian matrix of

the generalized Arnold's cat map is , and two eigenvalues are $\lambda_1 = 1 + \frac{ab + \sqrt{(ab+2)^2 - 4}}{2} > 1$ and
$\lambda_2 = 1 + \frac{ab - \sqrt{(ab+2)^2 - 4}}{2} < 1$. Therefore, the fixed point (0,0) also is an addle point. The generalized
Arnold's cat map also is chaotic map, because it has one positive Lyapunov exponent. Despite Arnold's
cat map or the generalized Arnold's cat map, they always have unstable saddle pint (0,0).

3. A New Two-Dimensional Chaotic Map without Fixed Points

In this paper, a new two-dimensional chaotic map inspired by Arnold's cat map is proposed. It is
defined as

$$
\begin{bmatrix} x(n+1) \\ y(n+1) \end{bmatrix} = \begin{bmatrix} a & b \\ c & d \end{bmatrix} \begin{bmatrix} x(n) \\ y(n) \end{bmatrix} + \begin{bmatrix} e \\ f \end{bmatrix} (\mathrm{mod}1). \tag{4}
$$

in which $a = kc + 1, b = k(d - 1), e \neq kf \neq 0, e \in (0,1), f \in (0,1)$, and $k \neq 0$. For calculating
the fixed point, Equation (4) can be transformed into two-dimensional equations.

$$
\begin{cases} x(n) = (kc+1)x(n) + k(d-1)y(n) + e(\mathrm{mod}1) \\ y(n) = cx(n) + dy(n) + f(\mathrm{mod}1) \end{cases} \tag{5}
$$

Equation (5) can be transformed into nonhomogeneous linear equations by collecting the
like terms.

$$
\begin{cases} kcx(n) + k(d-1)y(n) = -e(\mathrm{mod}1) \\ cx(n) + (d-1)y(n) = -f(\mathrm{mod}1) \end{cases} \tag{6}
$$

There is no solution to the nonhomogeneous linear equations. Thus, the map (4) has no fixed
point. However, the coefficients should be further limited for obtaining hidden chaotic attractors. The
Jacobian matrix of the map (4) is

$$
J_1 = \begin{bmatrix} kc+1 & kd-k \\ c & d \end{bmatrix}. \tag{7}
$$

The characteristic equation of the matrix J_1 is calculated as

$$
\det(\lambda I - J_1) = \lambda^2 - tr(J_1)\lambda + \det(J_1) = 0. \tag{8}
$$

in which $\det(J_1) = d + ck$ is the determinant of matrix J_1 and $tr(J_1) = kc + 1 + d$ is the trace
of matrix J_1. The characteristic equation of the matrix J_1 is a quadratic function. The roots of the
Equation (8) are

$$
\lambda_{1,2} = \frac{kc + 1 + d \pm (kc + d - 1)}{2} \tag{9}
$$

in which $\lambda_1 = 1$ and $\lambda_2 = kc + d$. The two corresponding Lyapunov exponents are $LE_1 = \ln|\lambda_1| = 0$
and $LE_2 = \ln|kc + d|$. The non-chaos fixed point attractors have negative Lyapunov exponents.
The non-chaos periodic or limit cycle attractors have non-positive Lyapunov exponents. The chaotic
attractors have positive Lyapunov exponents. Therefore, the chaotic system exists at least a positive
Lyapunov exponent. For obtaining hidden chaotic attractors, the second eigenvalue $\lambda_2 = kc + d$

should be larger than 1. For example, the coefficients are set as $c = 1, d = 2,$ and $k = 2,$ and $|kc + d| = 4 > 1$. Combining with the map (4), the new two-dimensional map with no fixed point is defined as

$$\begin{bmatrix} x(n+1) \\ y(n+1) \end{bmatrix} = \begin{bmatrix} 3 & 2 \\ 1 & 2 \end{bmatrix} \begin{bmatrix} x(n) \\ y(n) \end{bmatrix} + \begin{bmatrix} 0.1 \\ 0.2 \end{bmatrix} (\text{mod}1). \tag{10}$$

in which $e = 0.1$ and $f = 0.2$. The phase diagram of attractors is shown in Figure 2, and the plot of the output time series is shown in Figure 3.

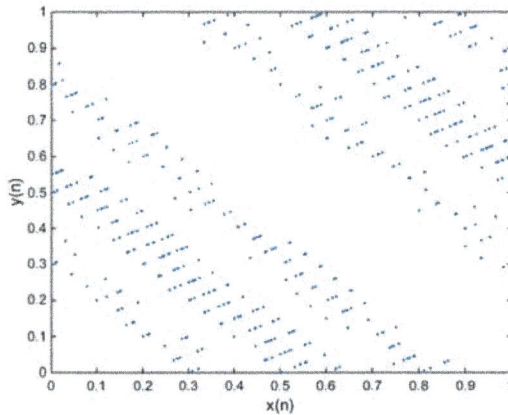

Figure 2. When $x(0) = 0.7$ and $y(0) = 0.6$, the phase diagram of the new 2-D map with $a = 3$, $b = 2, c = 1, d = 2, e = 0.1,$ and $f = 0.2$.

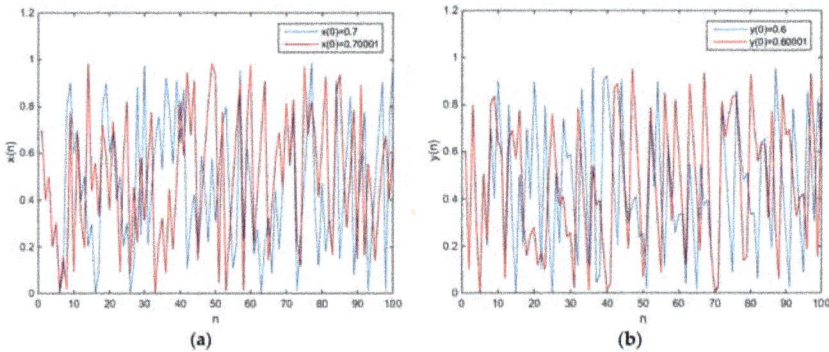

(a)

(b)

Figure 3. The plot of the output time series (a) $x(n)$, (b) $y(n)$.

The new two-dimensional map (10) is a chaotic map, because it has a positive Lyapunov exponent.

From Equation (9), the chaotic behavior in the map (4) is dependent only on the coefficients $c, d,$ and k, and the second eigenvalue of the Jacobian matrix J_1 is the simple combination of $c, d,$ and k. The Lyapunov exponent can be changed with different parameters $c, d,$ and k. When $c = 1.1, d = 2,$ $k = 2,$ and $t = 4.2 > 1$. Combining with the map (4), a new two-dimensional map without fixed points is defined as

$$\begin{bmatrix} x(n+1) \\ y(n+1) \end{bmatrix} = \begin{bmatrix} 3.2 & 2 \\ 1.1 & 2 \end{bmatrix} \begin{bmatrix} x(n) \\ y(n) \end{bmatrix} + \begin{bmatrix} 0.1 \\ 0.2 \end{bmatrix} (\text{mod}1). \qquad (11)$$

in which $e = 0.1$ and $f = 0.2$. The phase diagram of the chaotic attractors is shown in Figure 4, and the plot of the output time series is shown in Figure 5.

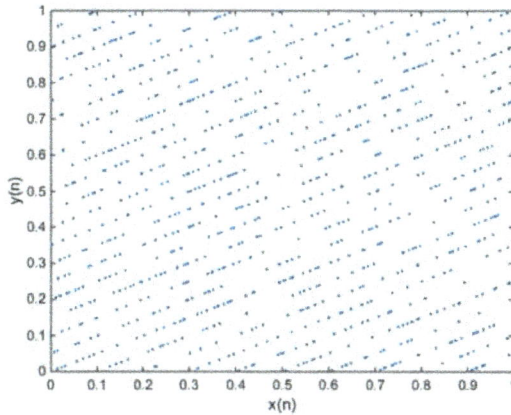

Figure 4. When $x(0) = 0.7$ and $y(0) = 0.6$, the phase diagram of the new 2-D map with $a = 3.2$, $b = 2, c = 1.1, d = 2, e = 0.1$, and $f = 0.2$.

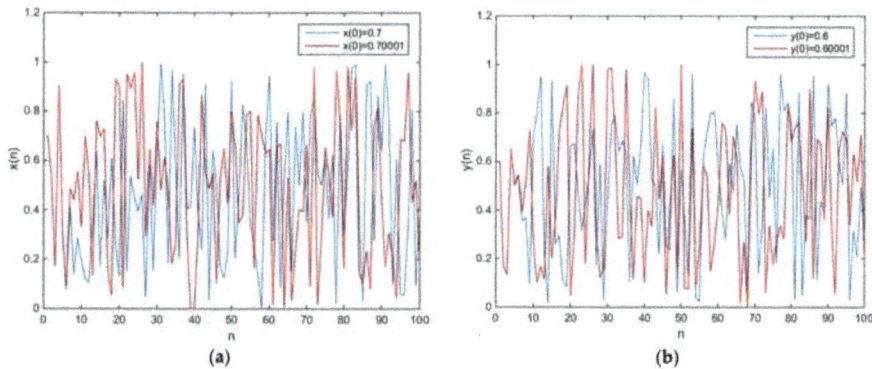

Figure 5. The plot of the output time series (a) $x(n)$, (b) $y(n)$.

As can be observed from the plots of the time series, it is obvious that the map (11) has initial value sensitivity, randomness, and so on. For no fixed point in the map (11), the map (11) is a chaotic map and has a hidden chaotic attractor. As can be seen from Figure 1, the chaotic attractor of the map (11) is not similar to that of Arnold's cat map. When $c = -0.25, d = 2, k = 2$, and $t = 1.5 > 1$. Combining with the map (4), a new two-dimensional map is defined as

$$\begin{bmatrix} x(n+1) \\ y(n+1) \end{bmatrix} = \begin{bmatrix} 0.5 & 2 \\ -0.25 & 2 \end{bmatrix} \begin{bmatrix} x(n) \\ y(n) \end{bmatrix} + \begin{bmatrix} 0.1 \\ 0.2 \end{bmatrix} (\text{mod}1). \qquad (12)$$

in which $e = 0.1$ and $f = 0.2$. The phase diagram of the chaotic attractors is shown in Figure 6, and the plot of the output time series is shown in Figure 7.

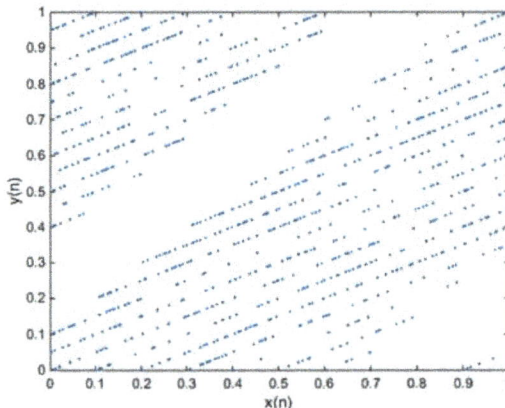

Figure 6. When $x(0) = 0.7$ and $y(0) = 0.6$, the phase diagram of the new 2-D map with $a = 0.5$, $b = 2, c = -0.25, d = 2, e = 0.1$, and $f = 0.2$.

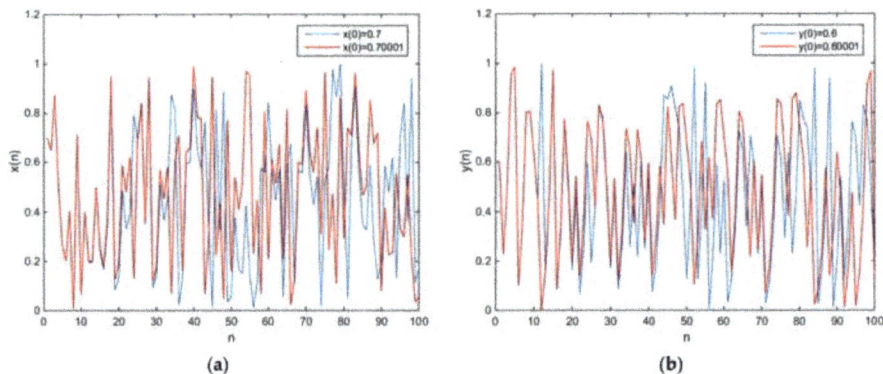

Figure 7. The plot of the output time series (**a**) $x(n)$, (**b**) $y(n)$.

Compared with Figure 1, the phase diagram of the map (12) is dissimilar to the Figures 2 and 4. When initial value is changed, $x(n)$ has the same output time series in the first 20 iterations, and $y(n)$ has the same output time series in the first 10 iterations.

4. Digitalization and Hardware Implementation

For the digitalization of information, chaotic systems need to be digitized before they are used [32,33]. The new two-dimensional chaotic map can be digitalized in two ways: one is the floating-point representation, the other is fixed-point representation. According to IEEE 754-2008 [34], floating-point is divided into single precision and double precision. The form of floating-point is shown in Table 1.

Table 1. Floating-point representation.

Floating-Point	Sign	Exponent	Fraction
Single precision	1 bit	8 bits	23 bits
Double precision	1 bit	11 bits	52 bits

The hardware consumptions of the fixed-point representation are not very high, because it includes sign bit and fraction bits. However, there is no standard form in fixed-point representation. Compared with floating-point computing, fixed-point computing is faster, and hardware implementation is smaller [35]. For a better performance in hardware implementation, the new two-dimensional map is represented by fixed-point in this paper. Since $x(n)$ and $y(n)$ are decimal numbers, they are more easily represented by fixed-point. Therefore, $x(n)$ and $y(n)$ are changed from decimals to integers. The digitized N bits of decimal β can be represented as

$$\widetilde{\beta} = \left\lfloor 2^N \beta \right\rfloor 2^{-N}. \tag{13}$$

in which $\widetilde{\beta}$ is an N bits approximation of β. The digitized N bits of map $\varphi(\beta) = \lambda\beta \bmod 1$ can be written as $\widetilde{\varphi}(\widetilde{\beta}) = \left\lfloor 2^N \lambda\widetilde{\beta} \bmod 2^N \right\rfloor 2^{-N}$, and the digitized N bits of map (4) can be written as

$$\begin{cases} \widetilde{x}(n+1) = \left\lfloor 2^N a\widetilde{x}(n) + 2^N b\widetilde{y}(n) + 2^N e \bmod 2^N \right\rfloor 2^{-N} \\ \widetilde{y}(n+1) = \left\lfloor 2^N c\widetilde{x}(n) + 2^N d\widetilde{y}(n) + 2^N f \bmod 2^N \right\rfloor 2^{-N} \end{cases}. \tag{14}$$

Multiplying both sides by 2^N, map (14) can be represented as

$$\begin{cases} 2^N \widetilde{x}(n+1) = \left\lfloor 2^N a\widetilde{x}(n) + 2^N b\widetilde{y}(n) + 2^N e \bmod 2^N \right\rfloor \\ 2^N \widetilde{y}(n+1) = \left\lfloor 2^N c\widetilde{x}(n) + 2^N d\widetilde{y}(n) + 2^N f \bmod 2^N \right\rfloor \end{cases}. \tag{15}$$

Denote $2^N \widetilde{x}(n), 2^N \widetilde{y}(n), 2^N e$, and $2^N f$ by $x(n), y(n), e$, and f, respectively. The digitized N bits of map (15) can be written as

$$\begin{bmatrix} x(n+1) \\ y(n+1) \end{bmatrix} = \left\lfloor \begin{bmatrix} a & b \\ c & d \end{bmatrix} \begin{bmatrix} x(n) \\ y(n) \end{bmatrix} + \begin{bmatrix} e \\ f \end{bmatrix} \right\rfloor (\bmod 2^N), \tag{16}$$

in which $x(n)$ and $y(n)$ is in the interval $[0, 2^N - 1]$, N presents the length of finite precision, $x(n)$ and $y(n)$ is represented by N bits. When $N = 32$, the hardware implementation by FPGA is shown in Figure 8.

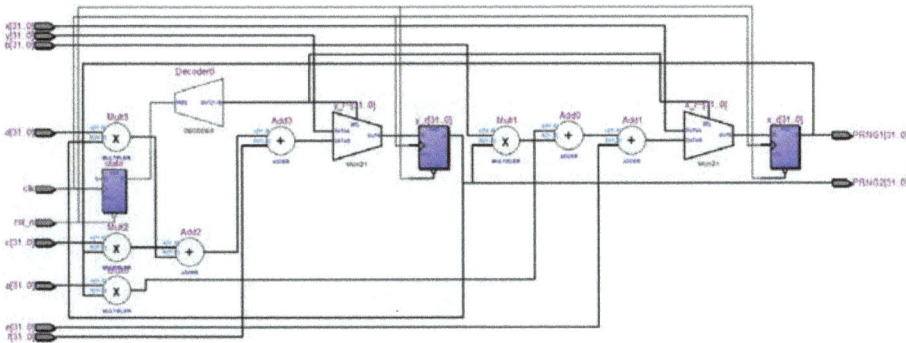

Figure 8. The block diagram of the hardware implementation by FPGA.

5. The Analysis of Complexity

The approximate entropy algorithm is proposed from the angle of measuring the complexity of time series [36–39]. The main idea of the approximate entropy algorithm is using a non-negative value to quantify the complexity and irregularity of the time series, and the value increases with the increase of sequence complexity. The calculation process of the approximate entropy is shown as follows:

1   Suppose the initial data is the sequence $x(1), x(2), ... x(N)$, and then divide them into m-dimensional vectors

$$X(i) = [x(i), x(i + 1), ..., x(i + m - 1)], \tag{17}$$

in which $i = 1, 2, 3...N - m + 1$.

2   The distance between $x(i)$ and $x(j)$ is defined as

$$d(i, j) = \max_{k=1-m-1}[||x(i + k) - x(j + k)||]. \tag{18}$$

3   Setting a threshold value $r(r > 0)$, for each i, we can obtain the statistics of $d(i, j)$.

$$C_i^m(r) = \frac{1}{N - m + 1}Sum\{d(i, j) < r\} \tag{19}$$

4   The mean of logarithm of $C_i^m(r)$ is written as $\phi^m(r)$ and can be calculated by

$$\phi^m(r) = \frac{1}{N - m + 1}\sum_{i=1}^{N-m+1} \ln C_i^m(r) \tag{20}$$

5   Changing dimension and repeating step 1 to step 4, we can obtain the approximate entropy

$$ApEn(m, r) = \lim_{N \to \infty}[\phi^m(r) - \phi^{m+1}(r)] \tag{21}$$

However, in practical terms, the length of the data sequence is bounded. Therefore, the approximate entropy algorithm is changed into

$$ApEn(m, r, N) = \phi^m(r) - \phi^{m+1}(r)] \tag{22}$$

Pincus found that there exists a minimal dependency between $ApEn$ and N when $m = 2$ and $r \in [0.1SD(x), 0.2SD(x)]$ [36]. $SD(x)$ is the standard deviation of x. The complexity of the output time series of Arnold's cat map and new two-dimensional chaotic maps are tested by approximate entropy algorithm, and the consequence show that the output time series of Equation (11) has a higher complexity. The specific results are shown in Table 2.

Table 2. Approximate entropy test.

Chaotic Map	Time Series	m	$r = 0.15SD$	N	ApEn
Equation (1)	$x(n)$	2	0.0435	2000	0.9787
	$y(n)$	2	0.0436	2000	0.9963
Equation (10)	$x(n)$	2	0.0426	2000	0.7591
	$y(n)$	2	0.0437	2000	0.8841
Equation (11)	$x(n)$	2	0.0436	2000	1.4171
	$y(n)$	2	0.0434	2000	1.2831
Equation (12)	$x(n)$	2	0.0438	2000	0.5433
	$y(n)$	2	0.0428	2000	0.8429

6. Conclusions

In this paper, the hidden attractors for the two-dimensional chaotic map are studied, and the existence of fixed points and their stability are considered. Due to the restriction of the form of the Arnold's cat map, the new two-dimensional chaotic attractor can only appear in the case of no fixed point. The selection of coefficients directly affects whether the two-dimensional map has chaotic behavior, because the eigenvalue of the Jacobian matrix of the new two-dimensional chaotic map is a simple combination of the coefficients. Three concrete examples are given to illustrate the relationship between the coefficients and the chaotic behavior. The different coefficients can not only determine the chaotic behavior of the new two-dimensional chaotic map but also affect the shape of the chaotic attractor.

Author Contributions: C.W. conceived and wrote the paper. Q.D. gave some theoretical guidance. All authors have read and approved the final manuscript.

Acknowledgments: This work is supported financially by Natural Science Foundation of China (No. 61471158) and the Innovative Team of the Heilongjiang Province (No. 2012TD007).

Conflicts of Interest: The authors declare no conflict of interest.

References

1. Lorenz, E.N. Deterministic non-periodic flow. *J. Atmos. Sci.* **1963**, *20*, 130–141. [CrossRef]
2. Rössler, O.E. An equation for continuous chaos. *Phys. Lett. A* **1976**, *57*, 397–398. [CrossRef]
3. Chua, L.O.; Lin, G.N. Canonical realization of Chua's circuit family. *IEEE Trans. Circuits Syst.* **1990**, *37*, 885–902. [CrossRef]
4. Chen, G.; Ueta, T. Yet another chaotic attractor. *Int. J. Bifurc. Chaos* **1999**, *9*, 1465–1466. [CrossRef]
5. Lü, J.; Chen, G. A new chaotic attractor coined. *Int. J. Bifurc. Chaos* **2000**, *3*, 659–661. [CrossRef]
6. Jafari, S.; Sprott, J.C.; Molaie, M. A simple chaotic flow with a plane of equilibria. *Int. J. Bifurc. Chaos* **2016**, *26*, 1650098. [CrossRef]
7. Dawid, D.; Sajad, J.; Tomasz, K.; Nikolay, V.K.; Gennady, A.L.; Awadhesh, P. Hidden attractors in dynamical systems. *Phys. Rep.* **2016**, *637*, 1–50.
8. Lai, Q.; Guan, Z.H.; Wu, Y.; Liu, F.; Zhang, D. Generation of multi-wing chaotic attractors from a Lorenz-like system. *Int. J. Bifurc. Chaos* **2013**, *23*, 1350152. [CrossRef]
9. Wang, X.; Chen, G. Constructing a chaotic system with any number of equilibria. *Nonlinear Dyn.* **2013**, *71*, 429–436. [CrossRef]
10. Jafari, S.; Sprott, J.C. Simple chaotic flows with a line equilibrium. *Chaos Solitons Fractals* **2013**, *57*, 79–84. [CrossRef]
11. Pham, V.-T.; Jafari, S.; Volos, C.; Wang, X.; Golpayegani, S.M.R.H. Is that really hidden? The presence of complex fixed-points in chaotic flows with no equilibria. *Int. J. Bifurc. Chaos* **2014**, *24*, 1450146. [CrossRef]
12. Pham, V.-T.; Volos, C.K.; Jafari, S.; Wei, Z.; Wang, X. Constructing a novel no–equilibrium chaotic system. *Int. J. Bifurc. Chaos* **2014**, *24*, 1450073. [CrossRef]
13. Pham, V.-T.; Vaidyanathan, S.; Volos, C.K.; Jafari, S. Hidden attractors in a chaotic system with an exponential nonlinear term. *Eur. Phys. J. Spec. Top.* **2015**, *224*, 1507–1517. [CrossRef]
14. Tahir, F.R.; Pham, V.-T.; Volos, C.K.; Wang, X. A novel no–equilibrium chaotic system with multiwing butterfly attractors. *Int. J. Bifurc. Chaos* **2015**, *25*, 1550056. [CrossRef]
15. Molaie, M.; Jafari, S.; Sprott, J.C.; Golpayegani, S.M.R.H. Simple chaotic flows with one stable equilibrium. *Int. J. Bifurc. Chaos* **2013**, *23*, 1350188. [CrossRef]
16. Lao, S.-K.; Shekofteh, Y.; Jafari, S.; Sprott, J.C. Cost function based on Gaussian mixture model for parameter estimation of a chaotic circuit with a hidden attractor. *Int. J. Bifurc. Chaos* **2014**, *24*, 1450010. [CrossRef]
17. Shahzad, M.; Pham, V.-T.; Ahmad, M.A.; Jafari, S.; Hadaeghi, F. Synchronization and circuit design of a chaotic system with coexisting hidden attractors. *Eur. Phys. J. Spec. Top.* **2015**, *224*, 1637–1652. [CrossRef]
18. Kingni, S.T.; Jafari, S.; Simo, H.; Woafo, P. Three–dimensional chaotic autonomous system with only one stable equilibrium: Analysis, circuit design, parameter estimation, control, synchronization and its fractional–order form. *Eur. Phys. J.* **2014**, *129*, 76. [CrossRef]
19. Wei, Z. Dynamical behaviors of a chaotic system with no equilibria. *Phys. Lett. A* **2011**, *376*, 102–108. [CrossRef]
20. Leonov, G.A.; Kuznetsov, N.V.; Vagaitsev, V.I. Localization of hidden Chua's attractors. *Phys. Lett. A* **2011**, *23*, 2230–2233. [CrossRef]

21. Bragin, V.O.; Vagaitsev, V.I.; Kuznetsov, N.V.; Leonov, G.A. Algorithms for finding hidden oscillations in nonlinear systems: The Aizerman and Kalman conjectures and Chua's circuits. *J. Comput. Syst. Sci. Int.* **2011**, *50*, 511–543. [CrossRef]

22. Leonov, G.A.; Vagaitsev, V.I.; Kuznetsov, N.V. Algorithm for localizing Chua attractors based on the harmonic linearization method. *Dokl. Math.* **2010**, *82*, 693–696. [CrossRef]

23. Kuznetsov, N.V.; Leonov, G.A.; Vagaitsev, V.I. Analytical-numerical method for attractor localization of generalized Chua's system. *IFAC Proc. Vol.* **2010**, *4*, 29–33. [CrossRef]

24. Sprott, J.C.; Jafari, S.; Pham, V.-T.; Hosseini, Z.S. A chaotic system with a single unstable node. *Phys. Lett. A* **2015**, *379*, 2030–2036. [CrossRef]

25. May, R.M. Simple mathematical models with very complicated dynamics. *Nature* **1976**, *261*, 459–467. [CrossRef] [PubMed]

26. Pierre, C.; Jean-Pierre, E. *Iterated Map on the Interval as Dynamical Systems*; Springer: Berlin, Germany, 1980.

27. Hénon, M. A two-dimensional mapping with a strange attractor. *Commun. Math. Phys.* **1976**, *50*, 69–77. [CrossRef]

28. Chen, G.; Mao, Y.; Chui, C.K. A symmetric image encryption scheme based on 3D chaotic cat maps. *Chaos Soliton Fractals* **2004**, *21*, 749–761. [CrossRef]

29. Jafari, S.; Pham, T.; Moghtadaei, M.; Kingni, S.T. The relationship between chaotic maps and some chaotic systems with hidden attractors. *Int. J. Bifurc. Chaos* **2016**, *26*, 1650211. [CrossRef]

30. Jiang, H.B.; Liu, Y.; Wei, Z.; Zhang, L.P. Hidden chaotic attractors in a class of two-dimensional maps. *Nonlinear Dyn.* **2016**, *85*, 2719–2727. [CrossRef]

31. Franks, J.M. Invariant sets of hyperbolic toral automorphisms. *Am. J. Math.* **1977**, *99*, 1089–1095. [CrossRef]

32. José, L.V.; Esteban, T.C.; Ángel, R.V. A switched-capacitor skew-tent map implementation for random number generation. *Int. J. Circ. Theor. Appl.* **2017**, *45*, 305–315.

33. Luis, G.F.; Esteban, T.P.; Esteban, T.C.; Cuauhtemoc, M.L. Hardware implementation of pseudo-random number generators based on chaotic maps. *Nonlinear Dyn.* **2017**, *90*, 1661–1670.

34. Ferrenberg, A.M.; Landau, D.P.; Wong, Y.J. Monte Carlo simulations: Hidden errors from "good" random number generators. *Phys. Rev. Lett.* **1992**, *69*, 3382–3384. [CrossRef] [PubMed]

35. Kyle, J.P.; Richard, J.P. Analyzing logistic map pseudorandom number generators for periodicity induced by finite precision floating-point representation. *Chaos Solut. Fractals* **2012**, *43*, 238–245.

36. Pincus, S.M. Approximate entropy as a measure of system complexity. *Proc. Natl. Acad. Sci. USA* **1991**, *88*, 2297–2301. [CrossRef] [PubMed]

37. Pincus, S. Approximate entropy (ApEn) as a complexity measure. *Chaos Interdiscipl. J. Nonlinear Sci.* **1995**, *5*, 110–117. [CrossRef] [PubMed]

38. Xu, G.H.; Shekofteh, Y.; Akgül, A.; Li, C.B.; Panahi, S. A New Chaotic System with a Self-Excited Attractor: Entropy Measurement, Signal Encryption, and Parameter Estimation. *Entropy* **2018**, *20*, 86. [CrossRef]

39. Koyuncu, İ.; Özcerit, A.T. The design and realization of a new high speed FPGA-based chaotic true random number generator. *Comput. Electr. Eng.* **2016**, *58*, 203–214. [CrossRef]

entropy

MDPI

Article

Stochastic Entropy Solutions for Stochastic Nonlinear Transport Equations

Rongrong Tian and Yanbin Tang *

School of Mathematics and Statistics, Hubei Key Laboratory of Engineering Modeling and Scientific Computing, Huazhong University of Science and Technology, Wuhan 430074, China; tianrr2015@hust.edu.cn
* Correspondence: tangyb@hust.edu.cn

Received: 26 April 2018; Accepted: 21 May 2018; Published: 23 May 2018

check for
updates

Abstract: This paper considers the existence and uniqueness of stochastic entropy solution for a nonlinear transport equation with a stochastic perturbation. The uniqueness is based on the doubling variable method. For the existence, we develop a new scheme of parabolic approximation motivated by the method of vanishing viscosity given by Feng and Nualart (J. Funct. Anal. 2008, 255, 313–373). Furthermore, we prove the continuous dependence of stochastic strong entropy solutions on the coefficient b and the nonlinear function f.

Keywords: nonlinear transport equation; stochastic (strong) entropy solution; uniqueness; existence

MSC: 60H15; 35R60

1. Introduction

In this paper, we consider the existence and uniqueness of the solutions to the nonlinear transport equation with a stochastic forcing:

$$\begin{cases} d\rho(t,x) + b(x) \cdot \nabla_x f(\rho(t,x))dt = A(\rho(t,x))dW_t, \ t > 0, \ x \in \mathbb{R}^d, \\ \rho(t,x)|_{t=0} = \rho_0(x), \ x \in \mathbb{R}^d, \end{cases} \tag{1}$$

where W_t is a one-dimensional Wiener process on a stochastic basis $(\Omega, \mathcal{F}, \mathbb{P}, \{\mathcal{F}_t\}_{t \geqslant 0})$ and $A : \mathbb{R} \to \mathbb{R}$ is a real valued function. $f : \mathbb{R} \to \mathbb{R}$ and $b : \mathbb{R}^d \to \mathbb{R}^d$ are Borel functions, and the initial data ρ_0 is non-random.

When $\text{div}_x b = 0$, then $b(x) \cdot \nabla_x f(\rho(t,x)) = \text{div}_x(b(x)f(\rho(t,x)))$, the equation in (1) models the phenomenon of complex fluid mixing in porous media flows and other problems in mathematics and physics [1–5]. A particular application of this model involves two-phase fluid flow, which has been used to study the flow of water through oil in a porous medium [6,7]. For the porous media flows, the spatial variations of porous formations occur on all length scales, but only the variations at the largest length scales are reliably reconstructed from data available. The heterogeneities occurring on the smaller length scales have to be incorporated stochastically. Consequently, the flows through such formations are stochastic [8].

There has been an interest in studying the effect of stochastic force on the corresponding deterministic equations, especially on the existence and uniqueness. Most of papers focus on the following Cauchy problem:

$$\begin{cases} d\rho(t,x) + \text{div}_x F(\rho(t,x))dt = A(t,x,\rho(t,x))dW_t, \ t > 0, \ x \in \mathbb{R}^d, \\ \rho(t,x)|_{t=0} = \rho_0(x), \ x \in \mathbb{R}^d, \end{cases} \tag{2}$$

where W_t is a one-dimensional standard Brownian motion or a cylindrical Brownian motion, or a space-time Gaussian white noise.

The various well-posedness results have been established for the Cauchy problem (2). When $d = 1$, the L^∞ solution has been established in [9,10] for $A = A(\rho)$ and $A = A(t, x)$, respectively, under hypotheses that $\rho_0 \in L^\infty$ and A has compact support. For general A, even for initial data $\rho_0 \in L^\infty$, the solution is not in L^∞ since the maximum principle is not available. Therefore, L^p $(1 \leqslant p < \infty)$ is a natural space on which the solutions are posed.

When $A = A(x, \rho)$, the framework of L^p-solutions $(2 \leqslant p < \infty)$ was first established by Feng and Nualart [11], but the existence was true only for $d = 1$. These solutions were generalized to weak-in-time by Bauzet, Vallet and Wittbold [12], Biswas and Majee [13], and Karlsen and Storrøsten [14]. For any dimension $d \geqslant 1$, the well-posedness of kinetic solutions was obtained by Debussche and Vovelle [15], and then the result was extended by Hofmanová [16]. Recently, due to the fact that uniform spatial BV-bound is preserved for problem (2) if A satisfies a Lipschitz condition, Chen, Ding and Karlsen [17] supplied a result on well-posedness of $\cap_{p \geqslant 1} L^p$ solutions in $\mathbb{R}^d$ for $d \geqslant 1$. Furthermore, there are many papers devoted to the study of the Cauchy problem (2), such as the study on bounded domains [18–20], invariant measures [21,22], Lévy noises [23–26] and long time behaviors [27]. For more details in this direction for random fluxes, we refer the readers to [28–31].

When F depends explicitly on x, so far as we know, there are few research works on the Cauchy problem (2). Even though for the problem (1), there are still few works since that the presence of b will bring us some new difficulties on the proof of existence and uniqueness of solutions. Moreover, from the viewpoint of conservations laws and numerical simulations, L^∞ is a natural space on which solutions are posed, how to get the boundedness of solutions is another difficulty. We would like to point out that there are two big difficulties arisen here. One is how to get the compactness of solutions for the viscosity equation, another is how to prove the boundedness of solutions. To overcome the first difficulty, we develop a new scheme of parabolic approximation, which sheds some new light on the method of vanishing viscosity. For the second difficulty, we use the Ito's formula and the cut-off technique. We know that there are probably three classical methods to deal with the compactness of solutions for the viscosity equation so far when F is independent of x. The first is based upon Young's relaxed measure [11,14], which is suitable to space-time Gaussian white noise. The second is to estimate the spatial BV-bound and temporal L^1-continuity [17], which is suitable to get the convergence of solutions for almost everywhere (t, x) and almost surely ω. The third is to use the kinetic formulation [15,16], which is suitable to cylindrical Brownian motion.

In this paper, we adapt the method given by [11,14], but there is a significant difference. The more important thing is that we obtain the continuity of solutions in the temporal variable. The arguments for problem (1) can be generalized to an equation in which the stochastic term is represented by

$$\int_{z \in Z} A(x, \rho(t, x), z) W(t, dz),$$

where Z is a metric space, and W is a space-time Gaussian white noise martingale random measure with respect to the filtration $\{\mathcal{F}_t\}_{t \geqslant 0}$, if one assumes in addition that A is Lipschitz continuous in x. Up to longer and more tedious calculations, the arguments for space-time Gaussian white noise is similar to problem (1). There is no new component except some minor changes, which is also similar to the proof given in [11]. To make the present proof more refined, we discuss the simple case and prove the existence and uniqueness of solutions to (1) in this paper. Encouraged and inspired by the definition given in [11], we first give a notion of stochastic entropy solution.

Definition 1. *Let* $|b|$, $\mathrm{div}\, b \in L^1_{loc}(\mathbb{R}^d)$, $f \in \mathcal{C}^2(\mathbb{R})$, $A \in \mathcal{C}(\mathbb{R})$, $\rho_0 \in L^1 \cap L^\infty(\mathbb{R}^d)$. *An* $\{\mathcal{F}_t\}_{t \geqslant 0}$-*adapted and* $L^2(\mathbb{R}^d)$-*valued stochastic process* $\rho = \rho(t, x, \omega)$ *is said to be a stochastic entropy solution of* (1), *if*

(i)   for every $T > 0$ and every $p \in [1, \infty)$,

$$\rho \in \mathcal{C}([0, T]; L^p(\Omega; L^p_{loc}(\mathbb{R}^d))) \tag{3}$$

and

$$\sup_{0 \leqslant t \leqslant T} \|\rho(t)\|_{L^1(\Omega \times \mathbb{R}^d)} + \sup_{0 \leqslant t \leqslant T} \|\rho(t)\|_{L^\infty(\Omega \times \mathbb{R}^d)} < \infty; \tag{4}$$

(ii)   for every entropy pair (η, q), $(\eta \in \mathcal{C}^\infty(\mathbb{R}), \eta'' \geqslant 0, q(v) = \int^v \eta'(s)f'(s)ds, f'(s) = df(s)/ds)$, every nonnegative function $\varphi \in \mathcal{C}^2_0(\mathbb{R}^d)$ and every $0 \leqslant s < t < \infty$,

$$\int_{\mathbb{R}^d} \varphi(x)\eta(\rho(t,x))dx - \int_{\mathbb{R}^d} \varphi(x)\eta(\rho(s,x))dx$$
$$\leqslant \int_s^t \int_{\mathbb{R}^d} \operatorname{div}_x(b(x)\varphi(x))q(\rho(r,x))dxdr + \frac{1}{2}\int_s^t \int_{\mathbb{R}^d} \eta''(\rho(r,x))A^2(\rho(r,x))\varphi(x)dxdr$$
$$+ \int_s^t dW_r \int_{\mathbb{R}^d} \eta'(\rho(r,x))A(\rho(r,x))\varphi(x)dx, \quad \mathbb{P} - a.s., \tag{5}$$

where the stochastic integral in the last term in (5) is interpreted in Itô's sense.

Furthermore, stochastic entropy solution ρ is called a stochastic strong entropy solution if the below conditions hold:

(iii)   for each $\{\mathcal{F}_t\}_{t \geqslant 0}$-adapted $L^2(\mathbb{R}^d)$-valued stochastic process $\tilde{\rho}(t,x,\omega)$, satisfying (3) and (4), we define $\tilde{\eta}$ through each entropy function η by

$$\tilde{\eta}(r,v,y) := \int_{\mathbb{R}^d} \eta'(\tilde{\rho}(r,x) - v)A(\tilde{\rho}(r,x))\psi(x,y)dx, \tag{6}$$

where $r \geqslant 0$, $v \in \mathbb{R}$, $y \in \mathbb{R}^d$ and $\psi \in \mathcal{C}^2_0(\mathbb{R}^{2d})$, there is a deterministic function $D(s,t)$, such that

$$\mathbb{E}\int_{\mathbb{R}^d}\left[\int_s^t \tilde{\eta}(r,v,y)dW_r\right]_{v=\rho(t,y)}dy \leqslant \mathbb{E}\int_s^t \int_{\mathbb{R}^d} \partial_v \tilde{\eta}(r,v=\rho(r,y),y)A(\rho(r,y))dydr + D(s,t); \tag{7}$$

(iv)   for each $T > 0$, there exist partitions $0 = t_0 < t_1 < \cdots < t_n = T$ such that

$$\lim_{\max(t_i - t_{i-1}) \to 0} \sum_{i=1}^n D(t_{i-1}, t_i) = 0. \tag{8}$$

We now state our main results. The first one is focused on the uniqueness.

Theorem 1. *Let $f \in \mathcal{C}^2(\mathbb{R})$, $\rho_0 \in L^1 \cap L^\infty(\mathbb{R}^d)$ and*

$$b \in BV_{loc}(\mathbb{R}^d; \mathbb{R}^d), \ \operatorname{div}b, \ \frac{|b(\cdot)|}{1 + |\cdot|} \in L^\infty(\mathbb{R}^d), \ A \in \mathcal{C}^{\frac{1}{2}}(\mathbb{R}). \tag{9}$$

Suppose that ρ_1 and ρ_2 are stochastic entropy solutions of (1), and one of them is a stochastic strong entropy solution. Then, for every $t > 0$,

$$\mathbb{E}\|\rho_1(t) - \rho_2(t)\|_{L^1(\mathbb{R}^d)} = 0. \tag{10}$$

Remark 1. *Compared with the uniqueness results given in [11,17], Theorem 1 is new since the 1/2-Hölder continuity of A is enough to ensure the uniqueness. Moreover, compared with the uniqueness result for stochastic differential equations in [32], the hypotheses of 1/2-Hölder continuity on A is optimal.*

If b, f and A are more regular, we also have the following existence results.

Theorem 2. *Let $f \in C^2(\mathbb{R})$ such that f' is bounded and $f(0) = 0$. Assume that $b \in W^{1,\infty}(\mathbb{R}^d; \mathbb{R}^d)$ and*

$$\rho_0 \in L^1 \cap L^\infty(\mathbb{R}^d), \ A \in Lip(\mathbb{R}), \ A(0) = 0, \ and \ \exists \ N > 0, \ A(u) = 0, \ \forall \ |u| \geqslant N. \tag{11}$$

Then,

(i) *(1) has a stochastic strong entropy solution.*
(ii) *Moreover, in addition $\rho_0 \in BV(\mathbb{R}^d)$, for every $T > 0$, we have $\rho \in L^\infty([0,T]; L^1(\Omega; BV(\mathbb{R}^d)))$ and there is a constant C depending only on $\|b\|_{W^{1,\infty}(\mathbb{R}^d)}$ and $\|f'\|_{L^\infty(\mathbb{R})}$ such that*

$$\sup_{0 \leqslant t \leqslant T} \mathbb{E}\|\rho(t)\|_{BV(\mathbb{R}^d)} \leqslant C(\|b\|_{W^{1,\infty}(\mathbb{R}^d)}, \|f'\|_{L^\infty(\mathbb{R})})\|\rho_0\|_{BV(\mathbb{R}^d)}. \tag{12}$$

Remark 2. (i) *If $\mathrm{div}\, b = 0$, then $f(0) = 0$ is not needed.*
(ii) *For a general function A, even for initial data $\rho_0 \in L^\infty$, the solution is not in L^∞. To maintain the boundedness of solutions, additional assumptions on A should be added. Inspired by [9,10], we can suppose that A has compact support.*

We now discuss the continuous dependence of the solutions on b, f and A. Some results for the continuity on A have established for the case of constant vector field b [17]. Here, we only give the continuous dependence of the solutions on b and f.

Theorem 3. *Let $\tilde{\rho}_0 \in L^1 \cap L^\infty(\mathbb{R}^d)$, $\rho_0 \in L^1 \cap L^\infty \cap BV(\mathbb{R}^d)$, $b, \tilde{b} \in W^{1,\infty}(\mathbb{R}^d; \mathbb{R}^d)$. $f, \tilde{f} \in C^2(\mathbb{R})$ such that $f', \tilde{f}'$ are bounded and $f(0) = \tilde{f}(0) = 0$. A meets the assumption (11). Let ρ be the unique stochastic strong entropy solution of (1) and $\tilde{\rho}$ be the unique stochastic strong entropy solution of*

$$\begin{cases} d\tilde{\rho}(t,x) + \tilde{b}(x) \cdot \nabla_x \tilde{f}(\tilde{\rho}(t,x))dt = A(\tilde{\rho}(t,x))dW_t, \ t > 0, \ x \in \mathbb{R}^d, \\ \tilde{\rho}(t,x)|_{t=0} = \tilde{\rho}_0(x), \ x \in \mathbb{R}^d. \end{cases}$$

For every $T > 0$, there exists a constant $C > 0$, which depends only on $\|b\|_{W^{1,\infty}(\mathbb{R}^d)}$, $\|f'\|_{L^\infty(\mathbb{R})}$, $\|\tilde{f}'\|_{L^\infty(\mathbb{R})}$, $\|\mathrm{div}\,\tilde{b}\|_{L^\infty(\mathbb{R}^d)}$, $\|\tilde{b}\|_{L^\infty(\mathbb{R}^d)}$ and T, such that

$$\sup_{0 \leqslant t \leqslant T} \mathbb{E} \int_{\mathbb{R}^d} |\rho(t,x) - \tilde{\rho}(t,x)|dx \leqslant \int_{\mathbb{R}^d} |\rho_0(x) - \tilde{\rho}_0(x)|dx$$
$$+C[\|b - \tilde{b}\|_{L^\infty(\mathbb{R}^d)} + \|f' - \tilde{f}'\|_{L^\infty(\mathbb{R})}]\|\rho_0\|_{BV(\mathbb{R}^d)}. \tag{13}$$

Remark 3. *Without the noise, (1) has been discussed by Chen and Karlsen. Some results on the existence and uniqueness of solutions as well as continuous dependence on b and f have been obtained in [33]. Here, we get an analogue of [33] (Theorem 3.2) but simplify some assumptions on the velocity fields b and $\tilde{b}$.*

The present paper is organized as follows. In Section 2, we give the proof of Theorem 1. Section 3 is devoted to the proof for Theorem 2. In Section 4, we prove the continuous dependence of solutions on b and f.

We end up this section by introducing some notations. $\mathbb{N}$ is natural numbers set. $m \in \mathbb{N}$ and $C_0^m(\mathbb{R}^d)$ stands for the vector space consisting of all functions ϕ, which, together with all their partial derivatives $\partial^\alpha \phi$ of order $|\alpha| \leqslant m$, are continuous and have compact supports in $\mathbb{R}^d$. Given a measurable function ς, $\varsigma^+ = \max\{\varsigma, 0\} = \varsigma \vee 0$ and $\varsigma^- = -\min\{\varsigma, 0\} = -[\varsigma \wedge 0]$. The symbols ∇, div, Δ, if not differently specified, are referred to derivatives in x. For every $R > 0$, $B_R := \{x \in \mathbb{R}^d : |x| < R\}$. It almost surely can be abbreviated to *a.s.*. The letter C will mean a positive constant, whose values may change in different places.

2. Proof of Theorem 1

Let ρ_1 be a stochastic entropy solution of (1) with the initial data ρ_0^1 and ρ_2 be a stochastic strong entropy solution of (1) with the initial data ρ_0^2, respectively. We set $\rho_{12}(t,x) := \rho_1(t,x) - \rho_2(t,x)$ for every $t > 0$ and $x \in \mathbb{R}^d$.

Let ϱ be a 1-dimensional standard mollifier,

$$\varrho(r) = C_0 \exp(\frac{1}{r^2-1})1_{|r|<1}(r), \int_{\mathbb{R}} \varrho(r)dr = 1. \tag{14}$$

For $\theta > 0$, we set $\varrho_\theta(r) = \varrho(r/\theta)/\theta$ and define

$$\eta_\theta(r) = \int_{-\infty}^r \int_{-\infty}^{s-\theta} \varrho_\theta(\tau)d\tau ds. \tag{15}$$

For any $\delta > 0$ and any $0 \leqslant \varphi \in \mathcal{C}_0^2(\mathbb{R}^d)$, we set

$$\psi_\delta(x,y) = (\delta^{-d} \prod_{k=1}^d \varrho(\frac{x_k - y_k}{\delta}))\varphi(\frac{x+y}{2}) \in \mathcal{C}_0^2(\mathbb{R}^{2d}). \tag{16}$$

If one chooses the entropy function by η_θ, the test function by $\psi_\delta(x,y)$, and $\delta = \theta^{2/3}$, in view of the assumption $b \in BV_{loc}(\mathbb{R}^d; \mathbb{R}^d)$, then all calculations from [11] (Lemma 3.1) to [11] (Lemma 3.3) are adapted to the present case. Furthermore, noting the fact that if $g \in L^1(\mathbb{R}^d)$, for every $\varepsilon > 0$, we define $g_\varepsilon(x) = 1_{|g| \leqslant \varepsilon}(x)|g(x)|/\varepsilon$, then for almost everywhere $x \in \mathbb{R}^d$,

$$g_\varepsilon(x) \to 0 \quad \text{as } \varepsilon \downarrow 0. \tag{17}$$

Hence, Ref. [11] (Lemma 3.4) holds true as well if $A \in \mathcal{C}^{\frac{1}{2}}(\mathbb{R})$.

Therefore, for every $t > 0$, we conclude that

$$\mathbb{E} \int_{\mathbb{R}^d} \varphi(x)[\rho_{12}]_+(t,x)dx - \int_{\mathbb{R}^d} \varphi(x)[\rho_{12}]_+(0,x)dx$$
$$\leqslant \mathbb{E} \int_0^t \int_{\mathbb{R}^d} [\mathrm{div}b(x)\varphi(x) + b(x) \cdot \nabla\varphi(x)]1_{[0,\infty)}(\rho_{12}(r,x))[f(\rho_1(r,x)) - f(\rho_2(r,x))]dxdr$$
$$\leqslant \|f'\|_{L^\infty([-a,a])}\mathbb{E} \int_0^t \int_{\mathbb{R}^d} |\mathrm{div}b(x)\varphi(x) + b(x) \cdot \nabla\varphi(x)|[\rho_{12}]_+(r,x)dxdr, \tag{18}$$

where $a = [\sup_{0 \leqslant t \leqslant T} \|\rho_1(t)\|_{L^\infty(\Omega \times \mathbb{R}^d)}] \vee [\sup_{0 \leqslant t \leqslant T} \|\rho_2(t)\|_{L^\infty(\Omega \times \mathbb{R}^d)}]$.

Let the test function φ in (16) satisfy that $supp(\varphi) \subset B_2$, $\varphi = 1$ on $|x| \leqslant 1$. Let $R > 0$ be a real number, and set $\varphi_R = \varphi(\cdot/R)$. With the help of (9): $\mathrm{div}b$, $|b(\cdot)|/(1 + |\cdot|) \in L^\infty(\mathbb{R}^d)$, if one takes $\varphi(\cdot/R)$ instead of φ in (18) and lets R tend to infinity, then

$$\mathbb{E} \int_{\mathbb{R}^d} [\rho_{12}]_+(t,x)dx - \int_{\mathbb{R}^d} [\rho_{12}]_+(0,x)dx$$
$$\leqslant \|f'\|_{L^\infty([-a,a])}\|\mathrm{div}b\|_{L^\infty(\mathbb{R}^d)}\mathbb{E} \int_0^t \int_{\mathbb{R}^d} [\rho_{12}]_+(r,x)dxdr. \tag{19}$$

Thus, by the Grönwall inequality, one easily finds that

$$\sup_{0 \leqslant t \leqslant T} \mathbb{E} \int_{\mathbb{R}^d} [\rho_{12}]_+(t,x)dx \leqslant \exp\{\|f'\|_{L^\infty([-a,a])}\|\mathrm{div}b\|_{L^\infty(\mathbb{R}^d)}T\} \int_{\mathbb{R}^d} [\rho_{12}]_+(0,x)dx. \tag{20}$$

Similar arguments imply that

$$\sup_{0 \leqslant t \leqslant T} \mathbb{E} \int_{\mathbb{R}^d} [\rho_{21}]_+(t,x)dx \leqslant \exp\{\|f'\|_{L^\infty([-a,a])}\|\mathrm{div}b\|_{L^\infty(\mathbb{R}^d)}T\} \int_{\mathbb{R}^d} [\rho_{21}]_+(0,x)dx. \tag{21}$$

Combining (20) and (21), we complete the proof. □

3. Proof of Theorem 2

(i) We prove the existence of stochastic strong entropy solutions for (1) by the method of vanishing viscosity, that is, we regard (1) as the $\varepsilon \downarrow 0$ limit of the viscosity equation

$$\begin{cases} d\rho^\varepsilon(t,x) + b(x) \cdot \nabla f(\rho^\varepsilon(t,x))dt = \varepsilon\Delta\rho^\varepsilon(t,x)dt + A(\rho^\varepsilon(t,x))dW_t, \ t > 0, \ x \in \mathbb{R}^d, \\ \rho^\varepsilon(t,x)|_{t=0} = \rho_0^\varepsilon(x), \ x \in \mathbb{R}^d, \end{cases} \tag{22}$$

where ρ_0^ε is an approximation to ρ_0.

We now divide the proof into three steps.

Step 1. Existence and uniqueness of mild solutions to the Cauchy problem (22).

Here, ρ^ε is said to be a mild solution of (22), if $\rho^\varepsilon(t)$ is an $\mathcal{F}_t$-adapted $L^2(\mathbb{R}^d)$-valued stochastic process and satisfies

$$\begin{aligned} \rho^\varepsilon(t,x) &= \int_{\mathbb{R}^d} G_\varepsilon(t,x-z)\rho_0^\varepsilon(z)dz + \int_0^t \int_{\mathbb{R}^d} \text{div}_z(G_\varepsilon(t-r,x-z)b(z))f(\rho^\varepsilon(r,z))dzdr \\ &+ \int_0^t dW_r \int_{\mathbb{R}^d} G_\varepsilon(t-r,x-z)A(\rho^\varepsilon(r,z))dz, \quad \mathbb{P}-a.s., \end{aligned} \tag{23}$$

for every $t \geqslant 0$, almost everywhere $x \in \mathbb{R}^d$, where the heat kernel $G_\varepsilon(t,x) = e^{-\frac{|x|^2}{4\varepsilon t}}/(4\pi\varepsilon t)^{d/2}$.

We choose $\rho_0^\varepsilon \in L^1 \cap L^\infty \cap H^1(\mathbb{R}^d)$ such that $\rho_0^\varepsilon \to \rho_0$ in $L^1 \cap L^2(\mathbb{R}^d)$ as $\varepsilon \downarrow 0$. For every fixed ε, every $p \in [1,\infty]$, $\|\rho_0^\varepsilon\|_{L^p(\mathbb{R}^d)} \leqslant \|\rho_0\|_{L^p(\mathbb{R}^d)}$. With the help of Banach contraction mapping principle, there is a unique mild solution ρ^ε to (22). Moreover, for every $T > 0$,

$$\rho^\varepsilon \in \mathcal{C}([0,T];L^2(\Omega;H^1(\mathbb{R}^d))) \cap L^2([0,T]\times\Omega;H^2(\mathbb{R}^d)) \cap L^\infty([0,T];L^p(\Omega\times\mathbb{R}^d)), \ \forall p \in [1,\infty).$$

Furthermore, for every $1 \leqslant p < \infty$, every $T > 0$, we have

$$\begin{aligned} &\sup_{0\leqslant t\leqslant T} \mathbb{E}\left[\|\rho^\varepsilon(t)\|_{L^p(\mathbb{R}^d)}^p\right] + \mathbb{E}\left[\varepsilon\int_0^T \|\nabla\rho^\varepsilon(t)\|_{L^2(\mathbb{R}^d)}^2 dt\right] \\ &\leqslant C(\|b\|_{L^\infty(\mathbb{R}^d)}, \|\text{div}b\|_{L^\infty(\mathbb{R}^d)}, \|f'\|_{L^\infty(\mathbb{R})}, T)\left[\|\rho_0^\varepsilon\|_{L^p(\mathbb{R}^d)}^p + \|\rho_0^\varepsilon\|_{L^2(\mathbb{R}^d)}^2\right] \\ &\leqslant C(\|b\|_{L^\infty(\mathbb{R}^d)}, \|\text{div}b\|_{L^\infty(\mathbb{R}^d)}, \|f'\|_{L^\infty(\mathbb{R})}, T)\left[\|\rho_0\|_{L^p(\mathbb{R}^d)}^p + \|\rho_0\|_{L^2(\mathbb{R}^d)}^2\right] \end{aligned} \tag{24}$$

and

$$\mathbb{E}\|\nabla^2\rho^\varepsilon\|_{L^2([0,T]\times\mathbb{R}^d)} \leqslant C(\|b\|_{W^{1,\infty}(\mathbb{R}^d)}, \|f'\|_{L^\infty(\mathbb{R})}, \varepsilon)\|\nabla\rho_0^\varepsilon\|_{L^2(\mathbb{R}^d)}. \tag{25}$$

We show that (4) holds for ρ^ε. Let η_θ be given by (15). $\forall M > 0$, define $\eta_\theta^M(r) = \eta_\theta(r-M)$, then

$$\eta_\theta^M(r) \to (r-M)_+ \text{ as } \theta \downarrow 0. \tag{26}$$

Let $\tilde{\varrho}$ be a d-dimensional standard mollifier, i.e.,

$$\tilde{\varrho}(x) = C_1 \exp(\frac{1}{|x|^2-1})1_{|x|<1}(x), \int_{\mathbb{R}^d} \tilde{\varrho}(x)dx = 1. \tag{27}$$

For $\delta > 0$, we define $\tilde{\varrho}_\delta(x) = \tilde{\varrho}(x/\delta)/\delta^d$. Let $\varphi(x) = C_1 e^{-|x|}$, with $C_1 = [\int_{\mathbb{R}^d} e^{-|x|}dx]^{-1}$ and for every given natural number $n \in \mathbb{N}$, we set $\varphi_\delta^n(x) = (\varphi 1_{|x|<n}(\cdot)) * \tilde{\varrho}_\delta(x)$.

By using Itô's formula and the integration by parts, then

$$\mathbb{E}\int_{\mathbb{R}^d}\varphi_\delta^\eta(x)\eta_\theta^M(\rho^\varepsilon(t,x))dx - \int_{\mathbb{R}^d}\varphi_\delta^\eta(x)\eta_\theta^M(\rho_0^\varepsilon(x))dx$$

$$\leqslant \mathbb{E}\int_0^t\int_{\mathbb{R}^d}\mathrm{div}(b(x)\varphi_\delta^\eta(x))q_M^\delta(\rho^\varepsilon(r,x))dxdr + \varepsilon\mathbb{E}\int_0^t\int_{\mathbb{R}^d}\eta_\theta^M(\rho^\varepsilon(r,x))\Delta\varphi_\delta^\eta(x)dxdr$$

$$+\frac{1}{2}\mathbb{E}\int_0^t\int_{\mathbb{R}^d}(\eta_\theta^M)''(\rho^\varepsilon(r,x))A^2(\rho^\varepsilon(r,x))\varphi_\delta^\eta(x)dxdr, \tag{28}$$

where in (28) we have used the fact

$$\Delta\eta_\theta^M(\rho^{\varepsilon_1}(t,x)) \geqslant (\eta_\theta^M)'(\rho^{\varepsilon_1}(t,x))\Delta\rho^{\varepsilon_1}(t,x). \tag{29}$$

For θ, δ, M and ε be fixed, if one lets n approach to infinity, (28) turns to

$$\mathbb{E}\int_{\mathbb{R}^d}\varphi_\delta(x)\eta_\theta^M(\rho^\varepsilon(t,x))dx - \int_{\mathbb{R}^d}\varphi_\delta(x)\eta_\theta^M(\rho_0^\varepsilon(x))dx$$

$$\leqslant \mathbb{E}\int_0^t\int_{\mathbb{R}^d}\mathrm{div}(b(x)\varphi_\delta(x))q_\theta^M(\rho^\varepsilon(r,x))dxdr + \varepsilon\mathbb{E}\int_0^t\int_{\mathbb{R}^d}\eta_\theta^M(\rho^\varepsilon(r,x))\Delta\varphi_\delta(x)dxdr$$

$$+\frac{1}{2}\mathbb{E}\int_0^t\int_{\mathbb{R}^d}(\eta_\theta^M)''(\rho^\varepsilon(r,x))A^2(\rho^\varepsilon(r,x))\varphi_\delta(x)dxdr$$

$$\leqslant \mathbb{E}\int_0^t\int_{\mathbb{R}^d}\mathrm{div}(b(x)\varphi_\delta(x))q_\theta^M(\rho^\varepsilon(r,x))dxdr + \varepsilon\mathbb{E}\int_0^t\int_{\mathbb{R}^d}\eta_\theta^M(\rho^\varepsilon(r,x))\varphi_\delta(x)dxdr$$

$$+C\mathbb{E}\int_0^t\int_{\mathbb{R}^d}\frac{1}{\theta}\mathbf{1}_{|\rho^\varepsilon(r,x)-M|\leqslant\theta}A^2(\rho^\varepsilon(r,x))\varphi_\delta(x)dxdr,$$

where $\varphi_\delta(x) = (\varphi * \tilde{\varrho}_\delta)(x)$ and in the last inequality we use the fact $\Delta\varphi_\delta(x) \leqslant \varphi_\delta(x)$. Then, taking $\delta \to 0$, we arrive at

$$\mathbb{E}\int_{\mathbb{R}^d}\varphi(x)\eta_\theta^M(\rho^\varepsilon(t,x))dx - \int_{\mathbb{R}^d}\varphi(x)\eta_\theta^M(\rho_0^\varepsilon(x))dx$$

$$\leqslant \mathbb{E}\int_0^t\int_{\mathbb{R}^d}\mathrm{div}(b(x)\varphi(x))q_\theta^M(\rho^\varepsilon(r,x))dxdr + \varepsilon\mathbb{E}\int_0^t\int_{\mathbb{R}^d}\eta_\theta^M(\rho^\varepsilon(r,x))\varphi(x)dxdr$$

$$+C\mathbb{E}\int_0^t\int_{\mathbb{R}^d}\frac{1}{\theta}\mathbf{1}_{|\rho^\varepsilon(r,x)-M|\leqslant\theta}A^2(\rho^\varepsilon(r,x))\varphi(x)dxdr. \tag{30}$$

Observing that f' is bounded, $(\eta_\theta^M)(M) = (\eta_\theta^M)'(M) = 0$ and $(\eta_\theta^M)'' \geqslant 0$, then

$$|q_\theta^M(\rho^\varepsilon)| = \left|\int_M^{\rho^\varepsilon}f'(v)(\eta_\theta^M)'(v)dv\right| \leqslant \|f'\|_{L^\infty(\mathbb{R})}\left|\int_M^{\rho^\varepsilon}(\eta_\theta^M)'(v)dv\right| = \|f'\|_{L^\infty(\mathbb{R})}\eta_\theta^M(\rho^\varepsilon). \tag{31}$$

By virtue of (11), taking $M > N$, from (30) and (31), we have

$$\mathbb{E}\int_{\mathbb{R}^d}\varphi(x)\eta_\theta^M(\rho^\varepsilon(t,x))dx - \int_{\mathbb{R}^d}\varphi(x)\eta_\theta^M(\rho_0^\varepsilon(x))dx$$

$$\leqslant [C(\|b\|_{W^{1,\infty}(\mathbb{R}^d)}, \|f'\|_{L^\infty(\mathbb{R})}, T) + \varepsilon]\mathbb{E}\int_0^t\int_{\mathbb{R}^d}\varphi(x)\eta_\theta^M(\rho^\varepsilon(r,x))dxdr + C\theta,$$

for all $0 \leqslant t \leqslant T$ ($T > 0$ is a given real number). Therefore,

$$\mathbb{E}\int_{\mathbb{R}^d}\varphi(x)\eta_\theta^M(\rho^\varepsilon(t,x))dx \leqslant C\int_{\mathbb{R}^d}\varphi(x)\eta_\theta^M(\rho_0^\varepsilon(x))dx + C\theta,$$

uniformly for $\varepsilon \leqslant 1$.

Due to (26), letting $\theta \downarrow 0$, for $M > \|\rho_0^\varepsilon\|_{L^\infty(\mathbb{R}^d)}$, we get

$$\mathbb{E}\int_{\mathbb{R}^d}\varphi(x)(\rho^\varepsilon(t,x) - M)_+ dx \leqslant C\int_{\mathbb{R}^d}\varphi(x)(\rho_0^\varepsilon(x) - M)_+ dx = 0. \tag{32}$$

Since ρ^ε is in $\mathcal{C}([0,T];L^2(\Omega \times \mathbb{R}^d))$, using the dominated convergence theorem, for $M > \|\rho_0^\varepsilon\|_{L^\infty(\mathbb{R}^d)}$, from (32), one has

$$\mathbb{E}\int_{\mathbb{R}^d}\varphi(x)(\rho^\varepsilon(t,x) - M)_+^2 dx = 0, \quad \forall\, t \in [0,T]. \tag{33}$$

By the convexity of η_θ^M, with the help of (28), (32) and (33), if $M > \max\{N, \|\rho_0^\varepsilon\|_{L^\infty(\mathbb{R}^d)}\}$, we have

$$
\begin{aligned}
&\mathbb{E}\sup_{0\leqslant t\leqslant T}\int_{\mathbb{R}^d}\varphi(x)(\rho^\varepsilon(t,x) - M)_+ dx\\
&\leqslant\ C\int_{\mathbb{R}^d}\varphi(x)(\rho_0^\varepsilon(x) - M)_+ dx + C\mathbb{E}\int_0^T\int_{\mathbb{R}^d}(\rho^\varepsilon(t,x) - M)_+\varphi(x)dxdt\\
&\quad + C\Big[\mathbb{E}\int_0^T\Big|\int_{\mathbb{R}^d}(\rho^\varepsilon(t,x) - M)_+\varphi(x)dx\Big|^2 dt\Big]^{\frac{1}{2}}\\
&\leqslant\ C\int_{\mathbb{R}^d}\varphi(x)(\rho_0^\varepsilon(x) - M)_+ dx + C\mathbb{E}\int_0^T\int_{\mathbb{R}^d}(\rho^\varepsilon(t,x) - M)_+\varphi(x)dxdt\\
&\quad + C\Big[\mathbb{E}\int_0^T\int_{\mathbb{R}^d}(\rho^\varepsilon(t,x) - M)_+^2\varphi(x)dxdt\Big]^{\frac{1}{2}} = 0.
\end{aligned}
$$

For the above calculations for η_θ^M adapted to $\zeta_\theta^M = \zeta_\theta(r + M)$, if $M > \max\{N, \|\rho_0^\varepsilon\|_{L^\infty(\mathbb{R}^d)}\}$, we have

$$\mathbb{E}\sup_{0\leqslant t\leqslant T}\int_{\mathbb{R}^d}\varphi(x)(\rho^\varepsilon(t,x) + M)_- dx \leqslant C\int_{\mathbb{R}^d}\varphi(x)(\rho_0^\varepsilon(x) + M)_- dx = 0,$$

where $\zeta_\theta(r) = \zeta(r/\theta)/\theta$, $\zeta : \mathbb{R} \to \mathbb{R}$ is a $\mathcal{C}^\infty$ convex function satisfying

$$\zeta(0) = 0, \quad \zeta'(r)\begin{cases} = 0, & \text{when } r > 0,\\ \in [-1,0], & \text{when } -2 \leqslant r \leqslant 0,\\ = -1, & \text{when } r < -2. \end{cases}$$

Therefore, (4) is true for ρ^ε, and

$$\sup_{0\leqslant t\leqslant T}\|\rho^\varepsilon(t)\|_{L^\infty(\Omega\times\mathbb{R}^d)} \leqslant \max\{N, \|\rho_0\|_{L^\infty(\mathbb{R}^d)}\}. \tag{34}$$

Step 2. Existence of the stochastic entropy solution to the Cauchy problem (1).

We choose ρ_0^ε as in **Step 1**, and when $\varepsilon = \varepsilon_i$ $(i = 1,2)$ in (22), we use the notation ρ^{ε_i} $(i = 1,2)$ to denote the unique stochastic entropy solution now. Suppose that η_θ is given by (15), then

$$\Delta_y\eta_\theta(\rho^{\varepsilon_1}(t,x) - \rho^{\varepsilon_2}(t,y)) \geqslant -\eta_\theta'(\rho^{\varepsilon_1}(t,x) - \rho^{\varepsilon_2}(t,y))\Delta_y\rho^{\varepsilon_2}(t,y). \tag{35}$$

Let $0 \leqslant J, \varphi \in \mathcal{C}_0^2(\mathbb{R}^d)$, such that

$$
\begin{cases} J(x) = 0, & \text{when } |x| \geqslant 1,\\ |\nabla_x J(x)| \leqslant CJ(x), & \text{when } x \in \mathbb{R}^d,\\ \int_{\mathbb{R}^d}J(x)dx = 1. \end{cases}
\qquad
\begin{cases} \varphi(x) = 1, & \text{when } |x| \leqslant 1,\\ |\nabla_x\varphi(x)| \leqslant C\varphi(x), & \text{when } x \in \mathbb{R}^d. \end{cases}
\tag{36}
$$

For any $\delta > 0$, we set

$$\psi_\delta(x,y) = J_\delta(x-y)\varphi(\frac{x+y}{2}) = \delta^{-d}J(\frac{x-y}{\delta})\varphi(\frac{x+y}{2}) \in C_0^2(\mathbb{R}^{2d}).$$

In view of (29) and (35), by using Itô's formula and the integration by parts,

$$\begin{aligned}
&\int_{\mathbb{R}^{2d}} \psi_\delta(x,y)\eta_\theta(\rho^{\varepsilon_1}(t,x) - \rho^{\varepsilon_2}(t,y))dxdy - \int_{\mathbb{R}^{2d}} \psi_\delta(x,y)\eta_\theta(\rho_0^{\varepsilon_1}(x) - \rho_0^{\varepsilon_2}(y))dxdy \\
&\leqslant \int_0^t \int_{\mathbb{R}^{2d}} [\operatorname{div}_x(b(x)\psi_\delta)q_\theta^{\varepsilon_1}(\rho^{\varepsilon_1}(r,x),\rho^{\varepsilon_2}(r,y)) + \operatorname{div}_y(b(y)\psi_\delta)\hat{q}_\theta^{\varepsilon_2}(\rho^{\varepsilon_1}(r,x),\rho^{\varepsilon_2}(r,y))]dxdydr \\
&\quad + \frac{1}{2}\int_0^t \int_{\mathbb{R}^{2d}} \psi_\delta(x,y)\eta_\theta''(\rho^{\varepsilon_1}(r,x) - \rho^{\varepsilon_2}(r,y))|A(\rho^{\varepsilon_1}(r,x)) - A(\rho^{\varepsilon_2}(r,y))|^2 dxdydr \\
&\quad + \int_0^t \int_{\mathbb{R}^{2d}} \eta_\theta(\rho^{\varepsilon_1}(r,x) - \rho^{\varepsilon_2}(r,y))[\varepsilon_1\Delta_x + \varepsilon_2\Delta_y]\psi_\delta(x,y)dxdydr \\
&\quad + \int_0^t dW_r \int_{\mathbb{R}^{2d}} \psi_\delta(x,y)\eta'(\rho^{\varepsilon_1}(r,x) - \rho^{\varepsilon_2}(r,y))[A(\rho^{\varepsilon_1}(r,x)) - A(\rho^{\varepsilon_2}(r,y))]dxdy \\
&= : H_1(t) + H_2(t) + H_3(t) + H_4(t),
\end{aligned} \tag{37}$$

where

$$\begin{aligned}
q_\theta^{\varepsilon_1}(\rho^{\varepsilon_1}(r,x),\rho^{\varepsilon_2}(r,y)) &= \int_{\rho^{\varepsilon_2}(r,y)}^{\rho^{\varepsilon_1}(r,x)} \eta_\theta'(v - \rho^{\varepsilon_2}(r,y))f'(v)dv, \\
\hat{q}_\theta^{\varepsilon_2}(\rho^{\varepsilon_1}(r,x),\rho^{\varepsilon_2}(r,y)) &= \int_{\rho^{\varepsilon_2}(r,y)}^{\rho^{\varepsilon_1}(r,x)} \eta_\theta'(\rho^{\varepsilon_1}(r,x) - v)f'(v)dv.
\end{aligned}$$

Clearly, $\mathbb{E}H_4(t) = 0$. For $\varepsilon_1, \varepsilon_2$ and δ are fixed, then

$$\begin{aligned}
\lim_{\theta\downarrow 0}\mathbb{E}H_1(t) &= \int_0^t \int_{\mathbb{R}^{2d}} [\operatorname{div}_x(b(x)\psi_\delta) + \operatorname{div}_y(b(y)\psi_\delta)]1_{[0,\infty)}(\rho^{\varepsilon_1}(r,x) - \rho^{\varepsilon_2}(r,y)) \\
&\quad \times [f(\rho^{\varepsilon_1}(r,x)) - f(\rho^{\varepsilon_2}(r,y))]dxdydr \\
&\leqslant C\int_0^t \int_{\mathbb{R}^{2d}} |\operatorname{div}_x(b(x)\psi_\delta) + \operatorname{div}_y(b(y)\psi_\delta)|[\rho^{\varepsilon_1}(r,x) - \rho^{\varepsilon_2}(r,y)]_+ dxdydr \\
&\leqslant C\|\operatorname{div}b\|_{L^\infty(\mathbb{R}^d)}\int_0^t \int_{\mathbb{R}^{2d}} \psi_\delta(x,y)[\rho^{\varepsilon_1}(r,x) - \rho^{\varepsilon_2}(r,y)]_+ dxdydr \\
&\quad + C\|\nabla b\|_{L^\infty(\mathbb{R}^d)}\int_0^t \int_{\mathbb{R}^{2d}} \varphi(\frac{x+y}{2})|\nabla_x J_\delta(x-y)|[\rho^{\varepsilon_1}(r,x) - \rho^{\varepsilon_2}(r,y)]_+ dxdydr \\
&\quad + C\|b\|_{L^\infty(\mathbb{R}^d)}\int_0^t \int_{\mathbb{R}^{2d}} |\nabla_x\varphi(\frac{x+y}{2})|J_\delta(x-y)[\rho^{\varepsilon_1}(r,x) - \rho^{\varepsilon_2}(r,y)]_+ dxdydr. \\
&\leqslant C\int_0^t \int_{\mathbb{R}^{2d}} \psi_\delta(x,y)[\rho^{\varepsilon_1}(r,x) - \rho^{\varepsilon_2}(r,y)]_+ dxdydr,
\end{aligned} \tag{38}$$

where in the last inequality we have used (36).

Moreover, $\lim_{\theta\downarrow 0}\mathbb{E}H_2(t) = 0$ and

$$\lim_{\theta\downarrow 0}\mathbb{E}H_3(t) = \int_0^t \int_{\mathbb{R}^{2d}} [\rho^{\varepsilon_1}(r,x) - \rho^{\varepsilon_2}(r,y)]_+[\varepsilon_1\Delta_x + \varepsilon_2\Delta_y]\psi_\delta(x,y)dxdydr. \tag{39}$$

For every $T > 0$, by (37)–(39), we obtain

$$\sup_{0 \leqslant t \leqslant T} \left[\mathbb{E} \int_{\mathbb{R}^{2d}} \psi_\delta(x,y) [\rho^{\varepsilon_1}(t,x) - \rho^{\varepsilon_2}(t,y)]_+ dxdy \right] - \int_{\mathbb{R}^{2d}} \psi_\delta(x,y) [\rho_0^{\varepsilon_1}(x) - \rho_0^{\varepsilon_2}(y)]_+ dxdy$$

$$\leqslant \sup_{0 \leqslant t \leqslant T} C \left[\mathbb{E} \int_0^t \int_{\mathbb{R}^{2d}} \psi_\delta(x,y) [\rho^{\varepsilon_1}(r,x) - \rho^{\varepsilon_2}(r,y)]_+ dxdydr \right]$$

$$+ \sup_{0 \leqslant t \leqslant T} \left[\mathbb{E} \int_0^t \int_{\mathbb{R}^{2d}} [\rho^{\varepsilon_1}(r,x) - \rho^{\varepsilon_2}(r,y)]_+ [\varepsilon_1 \Delta_x + \varepsilon_2 \Delta_y] \psi_\delta(x,y) dxdydr \right]. \tag{40}$$

Observing that

$$|[\varepsilon_1 \Delta_x + \varepsilon_2 \Delta_y] \psi_\delta(x,y)| \leqslant C \frac{\varepsilon_1 + \varepsilon_2}{\delta^2} \tilde{\psi}_\delta(x,y),$$

where

$$\tilde{\psi}_\delta(x,y) = \frac{1}{\delta^d} \tilde{J} \left(\frac{x-y}{\delta} \right) \tilde{\varphi} \left(\frac{x+y}{2} \right) \in \mathcal{C}_0(\mathbb{R}^{2d}), \quad \tilde{J}, \tilde{\varphi} \in \mathcal{C}_0(\mathbb{R}^d).$$

With the help of dominated convergence theorem, then

$$\lim_{\varepsilon_1 \downarrow 0, \varepsilon_2 \downarrow 0, \delta \downarrow 0, \frac{\varepsilon_1 + \varepsilon_2}{\delta^2} \to 0} \sup_{0 \leqslant t \leqslant T} \lim_{\theta \downarrow 0} \mathbb{E} H_3(t) = 0. \tag{41}$$

Combining (40), (41), and with the aid of Grönwall's inequality, then

$$\lim_{\varepsilon_1 \downarrow 0, \varepsilon_2 \downarrow 0, \delta \downarrow 0, \frac{\varepsilon_1 + \varepsilon_2}{\delta^2} \to 0} \sup_{0 \leqslant t \leqslant T} \mathbb{E} \int_{\mathbb{R}^{2d}} \psi_\delta(x,y) [\rho^{\varepsilon_1}(t,x) - \rho^{\varepsilon_2}(t,y)]_+ dxdy = 0.$$

Similar arguments also hint that

$$\lim_{\varepsilon_1 \downarrow 0, \varepsilon_2 \downarrow 0, \delta \downarrow 0, \frac{\varepsilon_1 + \varepsilon_2}{\delta^2} \to 0} \sup_{0 \leqslant t \leqslant T} \mathbb{E} \int_{\mathbb{R}^{2d}} \psi_\delta(x,y) [\rho^{\varepsilon_1}(t,x) - \rho^{\varepsilon_2}(t,y)]_- dxdy = 0.$$

Therefore,

$$\lim_{\varepsilon_1 \downarrow 0, \varepsilon_2 \downarrow 0, \delta \downarrow 0, \frac{\varepsilon_1 + \varepsilon_2}{\delta^2} \to 0} \sup_{0 \leqslant t \leqslant T} \mathbb{E} \int_{\mathbb{R}^{2d}} \psi_\delta(x,y) |\rho^{\varepsilon_1}(t,x) - \rho^{\varepsilon_2}(t,y)| dxdy = 0. \tag{42}$$

On the other hand, we have

$$\int_{\mathbb{R}^{2d}} \psi_\delta(x,y) |\rho^{\varepsilon_1}(t,x) - \rho^{\varepsilon_2}(t,y)| dxdy$$

$$= \int_{\mathbb{R}^{2d}} J(u) \varphi(v) |\rho^{\varepsilon_1}(t,v + \frac{\delta u}{2}) - \rho^{\varepsilon_2}(t,v - \frac{\delta u}{2})| dudv$$

$$= \int_{\mathbb{R}^{2d}} J(u) \varphi(v) |\rho^{\varepsilon_1}(t,v) - \rho^{\varepsilon_2}(t,v - \delta u)| dudv$$

$$+ \int_{\mathbb{R}^{2d}} J(u) [\varphi(v - \delta u) - \varphi(v)] |\rho^{\varepsilon_1}(t,v) - \rho^{\varepsilon_2}(t,v - \delta u)| dudv. \tag{43}$$

In view of (34),

$$\limsup_{\delta \downarrow 0} \sup_{\varepsilon_1, \varepsilon_2} \sup_{0 \leqslant t \leqslant T} \mathbb{E} \int_{\mathbb{R}^{2d}} J(u) |\varphi(v - \delta u) - \varphi(v)| |\rho^{\varepsilon_1}(t,v) - \rho^{\varepsilon_2}(t,v - \delta u)| dudv = 0. \tag{44}$$

By (42)–(44), then

$$\lim_{\varepsilon_1 \downarrow 0, \varepsilon_2 \downarrow 0, \delta \downarrow 0, \frac{\varepsilon_1 + \varepsilon_2}{\delta^2} \to 0} \sup_{0 \leqslant t \leqslant T} \mathbb{E} \int_{\mathbb{R}^{2d}} J(u) \varphi(v) |\rho^{\varepsilon_1}(t,v) - \rho^{\varepsilon_2}(t,v - \delta u)| dudv = 0. \tag{45}$$

Let J and φ be given in (36), then, for $\delta = (\varepsilon_1 \wedge \varepsilon_2)^{1/3}$, we have

$$\int_{\mathbb{R}^d} \varphi(v)|\rho^{\varepsilon_1}(t,v) - \rho^{\varepsilon_2}(t,v)|dv = \int_{\mathbb{R}^{2d}} J(u)\varphi(v)|\rho^{\varepsilon_1}(t,v) - \rho^{\varepsilon_2}(t,v)|dvdu$$

$$\leqslant \int_{\mathbb{R}^{2d}} J(u)\varphi(v)|\rho^{\varepsilon_1}(t,v) - \rho^{\varepsilon_2}(t,v-\delta u)|dvdu + \int_{\mathbb{R}^{2d}} J(u)\varphi(v)|\rho^{\varepsilon_2}(t,v) - \rho^{\varepsilon_2}(t,v-\delta u)|dvdu.$$

We conclude that

$$\lim_{\varepsilon_1 \downarrow 0, \varepsilon_2 \downarrow 0} \sup_{0 \leqslant t \leqslant T} \mathbb{E} \int_{\mathbb{R}^d} \varphi(v)|\rho^{\varepsilon_1}(t,v) - \rho^{\varepsilon_2}(t,v)|dv = 0. \tag{46}$$

Let $R > 0$ be a real number. If one takes $\varphi_R(x) = \varphi(x/R)$ instead of φ in the above calculations, then we get an analogue of (46),

$$\lim_{\varepsilon_1 \downarrow 0, \varepsilon_2 \downarrow 0} \sup_{0 \leqslant t \leqslant T} \mathbb{E} \int_{\mathbb{R}^d} \varphi_R(v)|\rho^{\varepsilon_1}(t,v) - \rho^{\varepsilon_2}(t,v)|dv = 0. \tag{47}$$

Thus, there is an $\mathcal{F}_t$-adapted L^1_{loc} valued random process $\rho(t)$, such that: $\rho \in C([0,T]; L^1(\Omega; L^1_{loc}(\mathbb{R}^d)))$ and $\rho^\varepsilon \to \rho$ in $C([0,T]; L^1(\Omega; L^1_{loc}(\mathbb{R}^d)))$. Moreover, by applying the estimates (24) and (34), (4) holds true.

On the other hand, for every entropy flux pair (η, q) ($\eta \in C^\infty(\mathbb{R})$, $\eta'' \geqslant 0$ and $q(v) = \int^v f'(s)\eta'(s)ds$ for every $0 \leqslant s < t < \infty$ and every $0 \leqslant \varphi \in C^2_0(\mathbb{R}^d)$,

$$\int_{\mathbb{R}^d} \varphi(x)\eta(\rho^\varepsilon(t,x))dx - \int_{\mathbb{R}^d} \varphi(x)\eta(\rho^\varepsilon(s,x))dx$$

$$\leqslant \int_s^t \int_{\mathbb{R}^d} \operatorname{div}(b(x)\varphi(x))q(\rho^\varepsilon(r,x))dxdr + \frac{1}{2}\int_s^t \int_{\mathbb{R}^d} \eta''(\rho^\varepsilon(r,x))A^2(\rho^\varepsilon)\varphi(x)dxdr$$

$$+ \int_s^t dW_r \int_{\mathbb{R}^d} \eta'(\rho^\varepsilon(r,x))A(\rho^\varepsilon)\varphi(x)dx + \varepsilon \int_s^t \int_{\mathbb{R}^d} \eta(\rho^\varepsilon(r,x))\Delta\varphi(x)dxdr, \quad \mathbb{P}-a.s. \tag{48}$$

Furthermore, if one approaches $\varepsilon \downarrow 0$ in (48), then (5) holds for $\rho(t,x)$. Thus, ρ is a stochastic entropy solution to (1).

Step 3. Existence of the stochastic strong entropy solution to the Cauchy problem (1).

For every $\{\mathcal{F}_t\}_{t \geqslant 0}$-adapted $L^2(\mathbb{R}^d)$-valued stochastic process $\tilde{\rho}(t,x,\omega)$ (meeting (3) and (4)), every given $\psi \in C^2_0(\mathbb{R}^{2d})$ and every given smooth convex function η, we set $\tilde{\eta}$ by (6) and

$$S(\eta, \psi)(s,t,v,y) = \int_s^t \int_{\mathbb{R}^d} \eta(\tilde{\rho}(r,x) - v)A(\tilde{\rho}(r,x))\psi(x,y)dxdW_r,$$

then

$$\int_{\mathbb{R}^d} \left[\int_s^t \tilde{\eta}(r,v,y)dW_r \right]_{v=\rho^\varepsilon(t,y)} dy = \int_{\mathbb{R}^d} S(\eta', \psi)(s,t,\rho^\varepsilon(t,y),y)dy,$$

where ρ^ε is the unique solution of (22).

Let ϱ be given in (14), and set $\varrho_\delta(\cdot) = \varrho(\cdot/\delta)/\delta$, then for almost all $\omega \in \Omega$, we have

$$\int_{\mathbb{R}^d} S(\eta', \psi)(s,t,\rho^\varepsilon(t,y),y)dy = \lim_{\delta \downarrow 0} \int_{\mathbb{R}^d} \int_{\mathbb{R}} S(\eta', v, \psi)(s,t,v,y)\varrho_\delta(v - \rho^\varepsilon(t,y))dvdy. \tag{49}$$

In view of the Itô formula for semi-martingales ($d(XY) = XdY + YdX + d[X,Y]$), (49) and the integration by parts, one derives that

$$\mathbb{E} \int_{\mathbb{R}^d} S(\eta', \psi)(s, t, \rho^\varepsilon(t, y), y) dy$$

$$= \mathbb{E} \int_{\mathbb{R}^d} \int_s^t S(\eta', \psi)(s, r, \rho^\varepsilon(r, y), y) dr dy$$

$$+ \mathbb{E} \int_{\mathbb{R}^d} \int_s^t S(\eta'', \psi)(s, r, \rho^\varepsilon(r, y), y)(-b(y) \cdot \nabla_y f(\rho^\varepsilon(r, y))) dr dy$$

$$+ \mathbb{E} \int_{\mathbb{R}^d} \int_s^t S(\eta'', \psi)(s, r, \rho^\varepsilon(r, y), y) \varepsilon \Delta_{yy} \rho^\varepsilon(r, y) dr) dy$$

$$+ \frac{1}{2} \mathbb{E} \int_{\mathbb{R}^d} \int_s^t S(\eta''', \psi)(s, r, \rho^\varepsilon(r, y), y) A^2(\rho^\varepsilon(r, y)) dr dy$$

$$- \mathbb{E} \int_{\mathbb{R}^d} \int_{\mathbb{R}^d} \int_s^t \eta''(\tilde{\rho}(r, x) - \rho^\varepsilon(r, y)) A(\tilde{\rho}(r, x)) A(\rho^\varepsilon(r, y)) \psi(x, y) dr dx dy$$

$$=: I_\varepsilon^1(s, t) + I_\varepsilon^2(s, t) + I_\varepsilon^3(s, t) + I_\varepsilon^4(s, t) + I_\varepsilon^5(s, t). \tag{50}$$

The calculations for $I_\varepsilon^i(s, t)$ ($i = 1, 2, 3, 4$) are similar, and we take $I_\varepsilon^2(s, t)$ for a typical example. Firstly, through integration by parts, it follows that

$$|I_\varepsilon^2(s, t)|$$

$$= \left| \mathbb{E} \int_{\mathbb{R}^d} \int_s^t \left[\int_s^r \int_{\mathbb{R}^d} \eta''(\tilde{\rho}(\tau, x) - v) A(\tilde{\rho}(\tau, x)) \mathrm{div}_y(\psi(x, y) b(y)) dx dW_\tau \right]_{v = \rho^\varepsilon(r, y)} f(\rho^\varepsilon(r, y)) dr dy \right| \tag{51}$$

$$\leqslant C \mathbb{E} \int_{\mathbb{R}^d} \int_s^t \sup_{|v| \leqslant N_1} \left| \int_s^r \int_{\mathbb{R}^d} \eta''(\tilde{\rho}(\tau, x) - v) A(\tilde{\rho}(\tau, x)) \mathrm{div}_y(\psi(x, y) b(y)) dx dW_\tau \right| dr dy,$$

where $N_1 = N \vee \|\rho_0\|_{L^\infty}$.

For $p > d \vee 2$, using the Sobolev embedding theorem $W^{1,p}(-N_1, N_1) \subset L^\infty(-N_1, N_1)$ and Hölder inequality, from (51), we have

$$\liminf_{\varepsilon \to 0} I_\varepsilon^2(s, t)$$

$$\leqslant C \int_{\mathbb{R}^d} \int_s^t \left(\int_{-N_1}^{N_1} \mathbb{E} \left| \int_s^r \int_{\mathbb{R}^d} \eta''(\tilde{\rho}(\tau, x) - v) A(\tilde{\rho}(\tau, x)) \mathrm{div}_y(\psi(x, y) b(y)) dx dW_\tau \right|^p dv \right)^{\frac{1}{p}} dr dy$$

$$+ C \int_{\mathbb{R}^d} \int_s^t \left(\int_{-N_1}^{N_1} \mathbb{E} \left| \int_s^r \int_{\mathbb{R}^d} \eta'''(\tilde{\rho}(\tau, x) - v) A(\tilde{\rho}(\tau, x)) \mathrm{div}_y(\psi(x, y) b(y)) dx dW_\tau \right|^p dv \right)^{\frac{1}{p}} dr dy \tag{52}$$

$$\leqslant C \int_{\mathbb{R}^d} \int_s^t \left[\int_{-N_1}^{N_1} \mathbb{E} \left(\int_s^r \left| \int_{\mathbb{R}^d} \eta''(\tilde{\rho}(\tau, x) - v) A(\tilde{\rho}(\tau, x)) \mathrm{div}_y(\psi(x, y) b(y)) dx \right|^2 d\tau \right)^{\frac{p}{2}} dv \right]^{\frac{1}{p}} dr dy$$

$$+ C \int_{\mathbb{R}^d} \int_s^t \left[\int_{-N_1}^{N_1} \mathbb{E} \left(\int_s^r \left| \int_{\mathbb{R}^d} \eta'''(\tilde{\rho}(\tau, x) - v) A(\tilde{\rho}(\tau, x)) \mathrm{div}_y(\psi(x, y) b(y)) dx \right|^2 d\tau \right)^{\frac{p}{2}} dv \right]^{\frac{1}{p}} dr dy$$

$$\leqslant C(N_1, T, \|b\|_{W^{1,\infty}}, \eta, \psi) \int_s^t (r - s)^{\frac{1}{2}} dr = C(N_1, T, \|b\|_{W^{1,\infty}}, \eta, \psi) |t - s|^{\frac{3}{2}} =: D(s, t),$$

where D is a deterministic function which meets the property (8).

By using dominated convergence theorem, we also have

$$\lim_{\varepsilon \to 0} I_\varepsilon^5(s, t) = - \int_{\mathbb{R}^d} \int_{\mathbb{R}^d} \int_s^t \eta''(\tilde{\rho}(r, x) - \rho(r, y)) A(\tilde{\rho}(r, x)) A(\rho(r, y)) \psi(x, y) dr dx dy \tag{53}$$

and

$$\lim_{\varepsilon \to 0} \mathbb{E} \int_{\mathbb{R}^d} S(\eta', \psi)(s, t, \rho^\varepsilon(t, y), y) dy = \mathbb{E} \int_{\mathbb{R}^d} S(\eta', \psi)(s, t, \rho(t, y), y) dy. \tag{54}$$

Combining (50) and (52)–(54), we know that (7) is true for ρ.

(ii) In this case, we choose $\rho_0^\varepsilon \in BV \cap L^\infty \cap H^1(\mathbb{R}^d)$ such that $\rho_0^\varepsilon \to \rho_0$ in $L^2 \cap BV(\mathbb{R}^d)$ as $\varepsilon \downarrow 0$. Let $\eta : \mathbb{R} \to \mathbb{R}$ be a $\mathcal{C}^\infty$ even function satisfying

$$\eta(0) = 0, \quad \eta'' \geqslant 0, \quad \eta'(r) = \begin{cases} -1, & \text{when } r < -1, \\ \in [-1,1], & \text{when } |r| \leqslant 1, \\ 1, & \text{when } r > 1. \end{cases}$$

For any $\delta > 0$, we define η_δ by $\eta_\delta(r) = \delta\eta(r/\delta)$. Then,

$$\eta_\delta(r) \to |r| \quad \text{as } \delta \downarrow 0. \tag{55}$$

Let $\varphi(x) = C_1 e^{-|x|}$, with $C_1 = [\int_{\mathbb{R}^d} e^{-|x|} dx]^{-1}$. Since $\rho^\varepsilon \in \mathcal{C}([0,T]; L^2(\Omega; H^2(\mathbb{R}^d)))$ for every $T > 0$, we can take the derivative of (22) with respect to x_i first, then by using the Itô formula to $\eta_\delta(\rho_{x_i}^\varepsilon(t,x))$,

$$
\begin{aligned}
&d\eta_\delta(\rho_{x_i}^\varepsilon(t,x)) + \eta_\delta'(\rho_{x_i}^\varepsilon(t,x))\partial_{x_i}(b(x) \cdot \nabla_x f(\rho^\varepsilon(t,x)))dt \\
={}& d\eta_\delta(\rho_{x_i}^\varepsilon(t,x)) + \eta_\delta'(\rho_{x_i}^\varepsilon(t,x))\partial_{x_i}b(x) \cdot \nabla_x f(\rho^\varepsilon(t,x))dt \\
&+ b(x) \cdot \nabla_x(\eta_\delta'(\rho_{x_i}^\varepsilon(t,x)))f'(\rho^\varepsilon(t,x))\partial_{x_i}\rho^\varepsilon(t,x)dt \\
&- \eta_\delta''(\rho_{x_i}^\varepsilon(t,x))f'(\rho^\varepsilon(t,x))\partial_{x_i}\rho^\varepsilon(t,x)b(x) \cdot \nabla_x\rho_{x_i}^\varepsilon(t,x)dt \\
={}& \varepsilon\eta_\delta'(\rho_{x_i}^\varepsilon(t,x))\Delta\rho_{x_i}^\varepsilon(t,x)dt + \eta_\delta'(\rho_{x_i}^\varepsilon(t,x))A'(\rho^\varepsilon(t,x))\rho_{x_i}^\varepsilon(t,x)dW_t \\
&+ \frac{1}{2}\eta_\delta''(\rho_{x_i}^\varepsilon(t,x))|A'(\rho^\varepsilon(t,x))\rho_{x_i}^\varepsilon(t,x)|^2 dt \\
={}& \varepsilon\Delta\eta_\delta(\rho_{x_i}^\varepsilon(t,x))dt + \eta_\delta'(\rho_{x_i}^\varepsilon(t,x))A'(\rho^\varepsilon(t,x))\rho_{x_i}^\varepsilon(t,x)dW_t \\
&+ \frac{1}{2}\eta_\delta''(\rho_{x_i}^\varepsilon(t,x))|A'(\rho^\varepsilon(t,x))\rho_{x_i}^\varepsilon(t,x)|^2 dt - \varepsilon\eta_\delta''(\rho_{x_i}^\varepsilon(t,x))|\nabla_x\rho_{x_i}^\varepsilon(t,x)|^2 dt \\
\leqslant{}& \varepsilon\Delta\eta_\delta(\rho_{x_i}^\varepsilon(t,x))dt + \eta_\delta'(\rho_{x_i}^\varepsilon(t,x))A'(\rho^\varepsilon(t,x))\rho_{x_i}^\varepsilon(t,x)dW_t \\
&+ \frac{1}{2}\eta_\delta''(\rho_{x_i}^\varepsilon(t,x))|A'(\rho^\varepsilon(t,x))\rho_{x_i}^\varepsilon(t,x)|^2 dt. \tag{56}
\end{aligned}
$$

Assume $R > 0$, we set $\varphi_R(\cdot) = \varphi(\cdot/R)$, then

$$
\begin{aligned}
&\mathbb{E}\int_{\mathbb{R}^d} \eta_\delta(\rho_{x_i}^\varepsilon(t,x))\varphi_R(x)dx - \int_{\mathbb{R}^d} \eta_\delta(\rho_{0,x_i}^\varepsilon(x))\varphi_R(x)dx \\
\leqslant{}& \frac{1}{2}\mathbb{E}\int_0^t \int_{\mathbb{R}^d} \eta_\delta''(\rho_{x_i}^\varepsilon(r,x))|A'(\rho^\varepsilon(r,x))\rho_{x_i}^\varepsilon(r,x)|^2\varphi_R(x)dxdr \\
&+ \frac{\varepsilon C}{R^2}\mathbb{E}\int_0^t \int_{\mathbb{R}^d} \eta_\delta(\rho_{x_i}^\varepsilon(r,x))\varphi_R(x)dxdr \\
&+ C(\|b\|_{W^{1,\infty}(\mathbb{R}^d)}, \|f'\|_{L^\infty(\mathbb{R})})\mathbb{E}\int_0^t \int_{\mathbb{R}^d} |\eta_\delta''(\rho_{x_i}^\varepsilon(r,x))||\rho_{x_i}^\varepsilon(r,x)||\nabla_x\rho_{x_i}^\varepsilon(r,x)|\varphi_R(x)dxdr \\
&+ C(\|b\|_{W^{1,\infty}(\mathbb{R}^d)}, \|f'\|_{L^\infty(\mathbb{R})})\mathbb{E}\int_0^t \int_{\mathbb{R}^d} |\nabla_x\rho^\varepsilon(r,x)|\varphi_R(x)dxdr \\
\leqslant{}& C\mathbb{E}\int_0^t \int_{\mathbb{R}^d} |\rho_{x_i}^\varepsilon(r,x)|1_{|\rho_{x_i}^\varepsilon(r,x)|\leqslant\delta}\varphi_R(x)dxdr + \frac{\varepsilon C}{R^2}\mathbb{E}\int_0^t \int_{\mathbb{R}^d} \eta_\delta(\rho_{x_i}^\varepsilon(r,x))\varphi_R(x)dxdr \\
&+ C\Big[\mathbb{E}\int_0^t \int_{\mathbb{R}^d} \frac{1}{\delta}|\rho_{x_i}^\varepsilon(r,x)|1_{|\rho_{x_i}^\varepsilon(r,x)|\leqslant\delta}|\nabla_x\rho_{x_i}^\varepsilon(r,x)|\varphi_R(x)dxdr\Big] \\
&+ C\mathbb{E}\int_0^t \int_{\mathbb{R}^d} |\nabla_x\rho^\varepsilon(r,x)|\varphi_R(x)dxdr, \tag{57}
\end{aligned}
$$

where, in the last inequality, we apply the fact $\eta_\delta''(\rho_{x_i}^\varepsilon(r,x)) \leqslant C1_{|\rho_{x_i}^\varepsilon(r,x)|\leqslant\delta}/\delta$.

Observing that, for almost everywhere, $(t,x) \in [0,T] \times \mathbb{R}^d$, $|\rho^{\varepsilon}_{x_i}|1_{|\rho^{\varepsilon}_{x_i}| \leqslant \delta}/\delta \to 0$ almost surely, as $\delta \downarrow 0$, from (57) by using dominated convergence theorem, if one lets $\delta \downarrow 0$ first and sums over i from 1 to d next,

$$\mathbb{E} \int_{\mathbb{R}^d} |\nabla \rho^{\varepsilon}(t,x)|\varphi_R(x)dx - \int_{\mathbb{R}^d} |\nabla \rho^{\varepsilon}_0(x)|\varphi_R(x)dx$$
$$\leqslant \frac{\varepsilon C}{R^2}\mathbb{E} \int_0^t \int_{\mathbb{R}^d} |\nabla \rho^{\varepsilon}(r,x)|\varphi_R(x)dxdr + C\mathbb{E} \int_0^t \int_{\mathbb{R}^d} |\nabla_x \rho^{\varepsilon}(r,x)|\varphi_R(x)dxdr.$$

Therefore,

$$\sup_{0 \leqslant t \leqslant T} \mathbb{E} \int_{\mathbb{R}^d} |\nabla \rho^{\varepsilon}(t,x)|dx \leqslant C(\|b\|_{W^{1,\infty}(\mathbb{R}^d)}, \|f'\|_{L^{\infty}(\mathbb{R})}, \varepsilon) \int_{\mathbb{R}^d} |\nabla \rho^{\varepsilon}_0(x)|dx. \tag{58}$$

Let η_δ be defined as before (meeting property (55)), and $\varphi(x) = 1$ when $|x| \leqslant 1$. We multiply φ_R on both sides of (56), in view of integration by parts, we derive that

$$\mathbb{E} \int_{\mathbb{R}^d} \eta_\delta(\rho^{\varepsilon}_{x_i}(t,x))dx - \int_{\mathbb{R}^d} \eta_\delta(\rho^{\varepsilon}_{0,x_i}(x))dx$$
$$\leqslant \frac{1}{2}\mathbb{E} \int_0^t \int_{\mathbb{R}^d} \eta''_\delta(\rho^{\varepsilon}_{x_i}(r,x))|A'(\rho^{\varepsilon}(r,x))\rho^{\varepsilon}_{x_i}(r,x)|^2\varphi_R(x)dxdr$$
$$+ \frac{\varepsilon}{R^2}\mathbb{E} \int_0^t \int_{\mathbb{R}^d} \Delta\varphi(\frac{x}{R})\eta_\delta(\rho^{\varepsilon}_{x_i}(r,x))dxdr$$
$$- \mathbb{E} \int_0^t \int_{\mathbb{R}^d} \eta'_\delta(\rho^{\varepsilon}_{x_i}(r,x))\partial_{x_i}b(x) \cdot \nabla_x f(\rho^{\varepsilon}(r,x))\varphi(\frac{x}{R})dxdr$$
$$+ \mathbb{E} \int_0^t \int_{\mathbb{R}^d} \eta'_\delta(\rho^{\varepsilon}_{x_i}(r,x))f'(\rho^{\varepsilon}(r,x))\partial_{x_i}\rho^{\varepsilon}(r,x)\mathrm{div}_x(b(x)\varphi(\frac{x}{R}))dxdr \tag{59}$$
$$- \mathbb{E} \int_0^t \int_{\mathbb{R}^d} \eta''_\delta(\rho^{\varepsilon}_{x_i}(r,x))f'(\rho^{\varepsilon}(r,x))\partial_{x_i}\rho^{\varepsilon}(r,x)b(x) \cdot \nabla_x \rho^{\varepsilon}_{x_i}(r,x)\varphi(\frac{x}{R})dxdr$$
$$\leqslant C\mathbb{E} \int_0^t \int_{\mathbb{R}^d} |\rho^{\varepsilon}_{x_i}(r,x)|1_{|\rho^{\varepsilon}_{x_i}(r,x)| \leqslant \delta}\varphi_R(x)dxdr + \frac{\varepsilon}{R^2}\mathbb{E} \int_0^t \int_{\mathbb{R}^d} \Delta\varphi(\frac{x}{R})\eta_\delta(\rho^{\varepsilon}_{x_i}(r,x))dxdr$$
$$+ C(\|b\|_{W^{1,\infty}(\mathbb{R}^d)}, \|f'\|_{L^{\infty}(\mathbb{R})})\mathbb{E} \int_0^t \int_{\mathbb{R}^d} |\nabla_x \rho^{\varepsilon}(r,x)|\Big[\varphi(\frac{x}{R}) + \frac{1}{R}|\nabla\varphi(\frac{x}{R})|\Big]dxdr$$
$$+ C(\|b\|_{L^{\infty}(\mathbb{R}^d)}, \|f'\|_{L^{\infty}(\mathbb{R})})\mathbb{E} \int_0^t \int_{\mathbb{R}^d} \frac{1}{\delta}|\rho^{\varepsilon}_{x_i}(r,x)|1_{|\rho^{\varepsilon}_{x_i}(r,x)| \leqslant \delta}|\nabla_x \rho^{\varepsilon}_{x_i}(r,x)|\varphi(\frac{x}{R})dxdr.$$

With the help of (25) and (58), from (59), by taking $\delta \downarrow 0$ first, $R \uparrow \infty$ next, then

$$\sup_{0 \leqslant t \leqslant T} \mathbb{E} \int_{\mathbb{R}^d} |\nabla \rho^{\varepsilon}(t,x)|dx \leqslant C(\|b\|_{W^{1,\infty}(\mathbb{R}^d)}, \|f'\|_{L^{\infty}(\mathbb{R})}) \int_{\mathbb{R}^d} |\nabla \rho^{\varepsilon}_0(x)|dx. \tag{60}$$

From (24) and (60), and noting that $\rho^{\varepsilon}_0 \to \rho_0$ in $L^2 \cap BV(\mathbb{R}^d)$, by letting $\varepsilon \downarrow 0$, (12) is true and we finish the proof. $\square$

4. Proof of Theorem 3

For $\varepsilon > 0$, we denote ρ^{ε} the unique solution of (22) with $\rho^{\varepsilon}_0 \in L^{\infty} \cap BV \cap H^1(\mathbb{R}^d)$ and $\rho^{\varepsilon}_0 \to \rho_0 \in L^2 \cap BV(\mathbb{R}^d)$, as $\varepsilon \downarrow 0$. Let $\tilde{\rho}^{\varepsilon}_0 \in L^1 \cap L^{\infty} \cap H^1(\mathbb{R}^d)$ and $\tilde{\rho}^{\varepsilon}_0 \to \tilde{\rho}_0$ in $L^1 \cap L^2(\mathbb{R}^d)$, as $\varepsilon \downarrow 0$. We assume $\tilde{\rho}^{\varepsilon}$ is the unique stochastic strong entropy solution of the following Cauchy problem:

$$\begin{cases} d\tilde{\rho}^{\varepsilon}(t,x) + \tilde{b}(x) \cdot \nabla_x \tilde{f}(\tilde{\rho}^{\varepsilon}(t,x))dt = \varepsilon\Delta\tilde{\rho}^{\varepsilon}(t,x)dt + A(\tilde{\rho}^{\varepsilon}(t,x))dW_t, & t > 0, \ x \in \mathbb{R}^d, \\ \tilde{\rho}^{\varepsilon}(t,x)|_{t=0} = \tilde{\rho}^{\varepsilon}_0(x), \ x \in \mathbb{R}^d. \end{cases} \tag{61}$$

Let η_δ be given by (56). We set the difference $\rho^{\varepsilon}(t,x) - \tilde{\rho}^{\varepsilon}(t,x)$ by $\xi^{\varepsilon}(t,x)$. Since $\rho^{\varepsilon}_0, \tilde{\rho}^{\varepsilon}_0 \in H^1(\mathbb{R}^d)$, $\xi^{\varepsilon} \in L^2([0,T] \times \Omega; H^2(\mathbb{R}^d))$. From (22) and (61) and by applying Itô's formula, then

$$
\begin{aligned}
d\eta_\delta(\xi^\varepsilon(t,x)) &= -\eta_\delta'(\xi^\varepsilon(t,x))[b(x)\cdot\nabla_x f(\rho^\varepsilon(t,x)) - \tilde{b}(x)\cdot\nabla\tilde{f}(\tilde\rho^\varepsilon(t,x))]dt \\
&\quad +\varepsilon\eta_\delta'(\xi^\varepsilon(t,x))\Delta\xi^\varepsilon(t,x)dt + \frac{1}{2}\eta_\delta''(\xi^\varepsilon(t,x))|A(\rho^\varepsilon(t,x)) - A(\tilde\rho^\varepsilon(t,x))|^2 dt \\
&\quad +\eta_\delta'(\xi^\varepsilon(t,x))[A(\rho^\varepsilon(t,x)) - A(\tilde\rho^\varepsilon(t,x))]dW_t \\
&\leqslant -\eta_\delta'(\xi^\varepsilon(t,x))[b(x) - \tilde{b}(x)]\cdot\nabla_x f(\rho^\varepsilon(t,x))dt \\
&\quad -\eta_\delta'(\xi^\varepsilon(t,x))\tilde{b}(x)\cdot\nabla[f(\rho^\varepsilon(t,x)) - \tilde{f}(\rho^\varepsilon(t,x))]dt \\
&\quad -\eta_\delta'(\xi^\varepsilon(t,x))\tilde{b}(x)\cdot\nabla[\tilde{f}(\rho^\varepsilon(t,x)) - \tilde{f}(\tilde\rho^\varepsilon(t,x))]dt \\
&\quad +\varepsilon\Delta\eta_\delta(\xi^\varepsilon(t,x))dt + \eta_\delta'(\xi^\varepsilon(t,x))[A(\rho^\varepsilon(t,x)) - A(\tilde\rho^\varepsilon(t,x))]dW_t \\
&\quad +\frac{1}{2}\eta_\delta''(\xi^\varepsilon(t,x))|A(\rho^\varepsilon(t,x)) - A(\tilde\rho^\varepsilon(t,x))|^2 dt.
\end{aligned}
\tag{62}
$$

Let φ_R be given in (57) and we integrate (62) against φ_R. By analogue calculations from (56) to (59), and then letting $\delta\downarrow 0$ first, $R\uparrow\infty$ next, it yields that

$$
\begin{aligned}
\mathbb{E}\int_{\mathbb{R}^d}|\xi^\varepsilon(t,x)|dx &\leqslant \int_{\mathbb{R}^d}|\xi^\varepsilon(0,x)|dx + C\|\mathrm{div}\tilde{b}\|_{L^\infty(\mathbb{R}^d)}\|\tilde{f}'\|_{L^\infty(\mathbb{R})}\mathbb{E}\int_0^t\int_{\mathbb{R}^d}|\xi^\varepsilon(r,x)|dxdr \\
&\quad +C\Big[\|b-\tilde{b}\|_{L^\infty(\mathbb{R}^d)}\|f'\|_{L^\infty(\mathbb{R})} + \|\tilde{b}\|_{L^\infty(\mathbb{R}^d)}\|f'-\tilde{f}'\|_{L^\infty(\mathbb{R})}\Big]\mathbb{E}\int_0^t\int_{\mathbb{R}^d}|\nabla\rho^\varepsilon(r,x)|dxdr.
\end{aligned}
$$

With the help of (60), then

$$
\begin{aligned}
\mathbb{E}\int_{\mathbb{R}^d}|\xi^\varepsilon(t,x)|dx &\leqslant \int_{\mathbb{R}^d}|\xi^\varepsilon(0,x)|dx + C\|\mathrm{div}\tilde{b}\|_{L^\infty(\mathbb{R}^d)}\|\tilde{f}'\|_{L^\infty(\mathbb{R})}\mathbb{E}\int_0^t\int_{\mathbb{R}^d}|\xi^\varepsilon(r,x)|dxdr \\
&\quad +C\Big(\|b\|_{W^{1,\infty}(\mathbb{R}^d)},\|f'\|_{L^\infty(\mathbb{R})},\|\tilde{b}\|_{L^\infty(\mathbb{R}^d)}\Big)\Big[\|b-\tilde{b}\|_{L^\infty(\mathbb{R}^d)} + \|f'-\tilde{f}'\|_{L^\infty(\mathbb{R})}\Big]\int_{\mathbb{R}^d}|\nabla\rho_0^\varepsilon|dx.
\end{aligned}
\tag{63}
$$

From (63), there is a constant $C>0$, which is dependent on $\|b\|_{W^{1,\infty}(\mathbb{R}^d)}$, $\|f'\|_{L^\infty(\mathbb{R})}$, $\|\tilde{f}'\|_{L^\infty(\mathbb{R})}$, $\|\mathrm{div}\tilde{b}\|_{L^\infty(\mathbb{R}^d)}$, $\|\tilde{b}\|_{L^\infty(\mathbb{R}^d)}$ and T, such that

$$
\mathbb{E}\int_{\mathbb{R}^d}|\xi^\varepsilon(t,x)|dx \leqslant \int_{\mathbb{R}^d}|\xi^\varepsilon(0,x)|dx + C\Big[\|b-\tilde{b}\|_{L^\infty(\mathbb{R}^d)} + \|f'-\tilde{f}'\|_{L^\infty(\mathbb{R})}\Big]\int_{\mathbb{R}^d}|\nabla\rho_0^\varepsilon(x)|dx.
\tag{64}
$$

From (64), by taking $\varepsilon\downarrow 0$, one ends up with the inequality (13). $\square$

5. Conclusions

In this paper, we have established three results on the existence and uniqueness of stochastic entropy solutions for a nonlinear transport equation by a stochastic perturbation, and the continuous dependence of stochastic strong entropy solutions on the coefficient b and the nonlinear function f. Compared with the results on uniqueness given in [11,17], Theorem 1 is new since the $1/2$-Hölder continuity of A is enough to ensure the uniqueness, and compared with the results on uniqueness for stochastic differential equations in [32], the hypotheses of $1/2$-Hölder continuity on A is optimal. Moreover, we develop a new method of parabolic approximation to obtain the existence of solutions, which sheds some new light on the method of vanishing viscosity put forth by Feng and Nualart [11].

Author Contributions: All authors carried out the proofs and conceived the study. All authors read and approved the final manuscript.

Funding: This work was supported by National Natural Science Foundation of China under grant No. 11471129.

Acknowledgments: The authors are grateful to the anonymous referees for helpful comments and suggestions that greatly improved the presentation of this paper.

Conflicts of Interest: The authors declare no conflict of interest.

References

1. Kang, J.; Tang, Y. Value function regularity in option pricing problems under a pure jump model. *Appl. Math. Optim.* **2017**, *76*, 303–321. [CrossRef]
2. Kang, J.; Tang, Y. Asymptotical behavior of partial integral-differential equation on nonsymmetric layered stable processes. *Asymptot. Anal.* **2017**, *102*, 55–70. [CrossRef]
3. Wang, G.; Tang, Y. Fractal dimension of a random invariant set and applications. *J. Appl. Math.* **2013**, *2013*, 415764. [CrossRef]
4. Wang, G.; Tang, Y. Random attractors for stochastic reaction-diffusion equations with multiplicative noise in H_0^1. *Math. Nachr.* **2014**, *287*, 1774–1791. [CrossRef]
5. Wu, E.; Tang, Y. Random perturbations of reaction-diffusion waves in biology. *Wave Motion* **2012**, *49*, 632–637. [CrossRef]
6. Glimm, J.; Darchesin, D.; Mcbryan, O. Statistical fluid dynamics: Unstable fingers. *Comm. Math. Phys.* **1980**, *74*, 1–13. [CrossRef]
7. Glimm, J.; Darchesin, D.; Mcbryan, O. A numerical method for two phase flow with an unstable interface. *J. Comput. Phys.* **1981**, *39*, 179–200. [CrossRef]
8. Glimm, J.; Sharp, D. Stochastic partial differential equations: Selected applications in continuum physics. In *Stochastic Partial Differential Equations: Six Perspectives, Mathematical Surveys Monographs*; American Mathematical Society: Providence, RI, USA, 1997; pp. 3–44.
9. Holden, H.; Risebro, N.H. Conservation laws with a random source. *Appl. Math. Opt.* **1997**, *36*, 229–241. [CrossRef]
10. Kim, J.U. On a stochastic scalar conservation law. *Indiana Univ. Math. J.* **2003**, *52*, 227–255. [CrossRef]
11. Feng, J.; Nualart, D. Stochastic scalar conservation laws. *J. Funct. Anal.* **2008**, *255*, 313–373. [CrossRef]
12. Bauzet, C.; Vallet, G.; Wittbold, P. The Cauchy problem for a conservation law with a multiplicative stochastic perturbation. *J. Hyperbolic Differ. Equ.* **2012**, *9*, 661–709. [CrossRef]
13. Biswas, I.H.; Karlsen, K.H.; Majee, A.K. Conservation laws driven by Lévy white noise. *J. Hyperbolic Differ. Equ.* **2015**, *2*, 518–654. [CrossRef]
14. Karlsen, K.H.; Storrøsten, E.B. On stochastic conservation laws and Malliavin calculus. *J. Funct. Anal.* **2017**, *272*, 421–497. [CrossRef]
15. Debussche, A.; Vovelle, J. Scalar conservation laws with stochastic forcing. *J. Funct. Anal.* **2010**, *259*, 1014–1042. [CrossRef]
16. Hofmanová, M. A Bhatnagar-Gross-Krook approximation to stochastic scalar conservation laws. *Ann. Inst. H. Poincaré Probab. Statist.* **2015**, *51*, 1500–1528. [CrossRef]
17. Chen, G.Q.; Ding, Q.; Karlsen, K.H. On nonlinear stochastic balance laws. *Arch. Ration. Mech. Anal.* **2012**, *204*, 707–743. [CrossRef]
18. Bauzet, C.; Vallet, G.; Wittbold, P. The Dirichlet problem for a conservation law with a multiplicative stochastic perturbation. *J. Funct. Anal.* **2014**, *266*, 2503–2545. [CrossRef]
19. Lv, G.; Wu, J. Renormalized entropy solutions of stochastic scalar conservation laws with boundary condition. *J. Funct. Anal.* **2016**, *271*, 2308–2338. [CrossRef]
20. Vallet, G.; Wittbold, P. On a stochastic first-order hyperbolic equation in a bounded domain. *Infin. Dimens. Anal. Qu.* **2009**, *12*, 613–651. [CrossRef]
21. Debussche, A.; Vovelle, J. Invariant measure of scalar first-order conservation laws with stochastic forcing. *Probab. Theory Relat. Fields* **2015**, *163*, 575–611. [CrossRef]
22. Weinan, E.; Khanin, K.; Mazel, A.; Sinai, Y. Invariant measures for Burgers equation with stochastic forcing. *Ann. Math.* **2000**, *151*, 877–960.
23. Biswas, I.H.; Majee, A.K. Stochastic conservation laws: Weak-in-time formulation and strong entropy condition. *J. Funct. Anal.* **2014**, *267*, 2199–2252. [CrossRef]
24. Dong, Z.; Xu, T.G. One-dimensional stochastic Burgers equation driven by Lévy processes. *J. Funct. Anal.* **2007**, *243*, 631–678. [CrossRef]
25. Jacob, N.; Potrykus, A.; Wu, J.L. Solving a non-linear stochastic pseudo-differential equation of Burgers type. *Stoch. Proc. Appl.* **2010**, *120*, 2447–2467. [CrossRef]
26. Wu, J.; Xie, B. On a Burgers type non-linear equation perturbed by a pure jump Lévy noise in $\mathbb{R}^d$. *Bull. Sci. Math.* **2012**, *136*, 484–506. [CrossRef]

27. Gess, B.; Souganidis, P.E. Long-time behavior and averaging lemmata for stochastic scalar conservation laws. *Comm. Pure Appl. Math.* **2016**, *70*, 1562–1597. [CrossRef]

28. Gess, B.; Souganidis, P.E. Scalar conservation laws with multiple rough fluxes. *Commun. Math. Sci.* **2015**, *13*, 1569–1597. [CrossRef]

29. Lions, P.L.; Perthame, B.; Souganidis, P.E. Scalar conservation laws with rough (stochastic) fluxes. *Stoch. Partial Differ. Equ. Anal. Comput.* **2013**, *1*, 664–686. [CrossRef]

30. Lions, P.L.; Perthame, B.; Souganidis, P.E. Scalar conservation laws with rough (stochastic) fluxes: The spatially dependent case. *Stoch. Partial Differ. Equ. Anal. Comput.* **2014**, *2*, 517–538. [CrossRef]

31. Mariani, M. Large deviations principles for stochastic scalar conservation laws. *Probab. Theory Related Fields* **2010**, *147*, 607–648. [CrossRef]

32. Yamada, T.; Watanabe, S. On the uniqueness of solutions of stochastic differential equations. *J. Math. Kyoto Univ.* **1971**, *11*, 155–167. [CrossRef]

33. Chen, G.Q.; Karlsen, K.H. Quasilinear anisotropic degenerate parabolic equations with time-space dependent diffusion coefficients. *Commun. Pure Appl. Anal.* **2005**, *4*, 241–266.

entropy

MDPI

Article

Multivariate Multiscale Complexity Analysis of Self-Reproducing Chaotic Systems

Shaobo He [1]![ORCID], **Chunbiao Li [2,3,]***, **Kehui Sun [1]** and **Sajad Jafari [4]**

[1]   School of Physics and Electronics, Central South University, Changsha 410083, China;
      heshaobo_123@163.com (S.H.); kehui@csu.edu.cn (K.S.)
[2]   Jiangsu Collaborative Innovation Center of Atmospheric Environment and Equipment Technology
      (CICAEET), Nanjing University of Information Science & Technology, Nanjing 210044, China
[3]   Jiangsu Key Laboratory of Meteorological Observation and Information Processing,
      Nanjing University of Information Science & Technology, Nanjing 210044, China
[4]   Department of Biomedical Engineering, Amirkabir University of Technology, 424 Hafez Ave.,
      Tehran 15875-4413, Iran; sajadjafari@aut.ac.ir
*   Correspondence: chunbiaolee@nuist.edu.cn

Received: 9 June 2018; Accepted: 24 July 2018; Published: 27 July 2018

Abstract: Designing a chaotic system with infinitely many attractors is a hot topic. In this paper, multiscale multivariate permutation entropy (MMPE) and multiscale multivariate Lempel–Ziv complexity (MMLZC) are employed to analyze the complexity of those self-reproducing chaotic systems with one-directional and two-directional infinitely many chaotic attractors. The analysis results show that complexity of this class of chaotic systems is determined by the initial conditions. Meanwhile, the values of MMPE are independent of the scale factor, which is different from the algorithm of MMLZC. The analysis proposed here is helpful as a reference for the application of the self-reproducing systems.

Keywords: multiscale multivariate entropy; multistability; self-reproducing system; chaos

---

1. Introduction

Since the behaviors of dynamic systems with coexisting attractors depend on the initial conditions, multistable systems have been extensively studied [1–5]. Multistability in circuit implementation [6], synchronization [7], image encryption [8] and neural networks [9] have also aroused much interest. Multistable systems can have a limited number of coexisting attractors [10] or even infinitely many attractors [11,12]. Specifically, Li et al. proposed a class of self-reproducing systems (one case of conditional symmetry) giving one-dimensional infinitely many attractors [11] and a unique case with a two-dimensional lattice of infinitely many strange attractors by introducing periodic trigonometric functions into a two-dimensional offset-boostablesystem [12].

However, the dynamics of these systems are mainly analyzed by means of a bifurcation diagram, Lyapunov exponents (LEs) and the phase trajectory analysis. Generally, higher complexity means that the time series is closer to noise and consequently leads to better security for real applications. Complexity measuring methods can be used to reflect the dynamics and complexity of time series in chaotic systems and provide an effective means for parameter selection of chaotic systems in real applications. From this point of view, much research on the complexity of chaotic systems has been carried out [13–18]. For example, Balasubramanian et al. [15] classified periodic, chaotic and random sequences based on approximate entropy and Lempel–Ziv complexity measures. He et al. [17,18] analyzed the complexity of multi-scroll chaotic systems and fractional-order chaotic systems. Complexity measuring of chaotic systems is an important issue in the nonlinear research community, among which designing multivariate complexity measures for a chaotic attractor is a hot topic.

In fact, phase space analysis is one of the most useful methods for the explanation of the long term dynamics of multivariate systems [19]. Phase-space reconstruction can reflect the asymptotic nature of the interconnected time series, which are responsible for the original dynamics. Most of the current multivariate complexity algorithms are designed based on this, such as multivariate sample entropy (MvSampEn) [20] and multivariate neighborhood sample entropy (MN-SampEn) [21]. However, it is difficult to choose the embedding dimension and the delay parameter. An alternative way to measure the complexity of the system is to resort to its phase space directly. Recently, He et al. [22] proposed multivariate permutation entropy and applied it to the complexity analysis of chaotic attractors. Meanwhile, since multiscale coarse-graining [23] on time series could lead to better complexity measuring results, multiscale complexity algorithms are applied to analyze the complexity of nonlinear time series [18,23–25]. In this paper, combining the process of multiscale coarse-graining, multivariate multiscale complexity measure algorithms are designed to analyze the complexity for self-reproducing chaotic systems. It is necessary to analyze the complexity of attractors of this class of chaotic systems under different initial conditions for its multistability. Multiscale multivariate complexity measuring algorithms are designed to achieve this goal. The rest of this article is organized as follows: In Section 2, MMPE and MMLZC are designed. In Section 3, the complexity of different kinds of self-reproducing chaotic systems is analyzed. Finally, concluding remarks is presented in Section 4.

2. Designing the Complexity Measuring Algorithms

Compared with other nonlinear time series analysis methods, complexity measuring methods just need a segment of time series or multiple time series from the system and are robust to the algorithm parameters. Meanwhile, it is more convenient to analyze the characteristics of time series by employing complexity measuring algorithms. In this section, the complexity of three-dimensional (3D) chaotic systems is analyzed with the main purpose to measure the complexity of the attractors. Thus, it is necessary to design multivariate complexity measuring algorithms.

2.1. Data Processing and Quantification

Step 1: Normalization of time series. For given time series $\{x_j(n), n = 1, 2, 3, \cdots, N, j = 1, 2, \cdots, d\}$, where d is the number of time series or the dimension of the chaotic system. Since amplitudes of different time series are different, normalization processing is necessary. The normalization function is given by:

$$\tilde{x}_j(n) = \frac{x_j(n) - \min(x_j)}{\max(x_j) - \min(x_j)}. \tag{1}$$

Step 2: Coarse graining. To design multiscale complexity measuring algorithms, the multiscale coarse-grained processing should be carried out firstly. For the j-th time series, its consecutive coarse-grained time series is constructed by [23]:

$$y_j^\tau(k) = \frac{1}{s} \sum_{i=(k-1)s+1}^{ks} \tilde{x}_j(i), \tag{2}$$

where $1 \leq k \leq \lfloor N/\tau \rfloor$ and τ is the scale factor that represents the length of the non-overlapping windows.

Step 3: Data quantification. For the given k and scale factor τ, $[y_1^\tau(k), y_2^\tau(k), \cdots, y_d^\tau(k)]$ can be modeled as a pattern by introducing the idea of the Bandt–Pompe pattern [26]. Obviously, there are $d!$ possible patterns. Let the pattern space be given by $\Lambda = \{\pi_1, \pi_2, \cdots, \pi_{d!}\}$, and thus, a pattern series $\{\Psi^\tau(k) : \Psi^\tau(k) \in \Lambda, k = 1, 2, \cdots, \lfloor N/\tau \rfloor\}$ can be obtained. Moreover, let $\pi_l = l$ ($l = 1, 2, \cdots, d!$); we can get a quantification pattern series, which is given by $\{\Phi^\tau(k) : \Phi^\tau(k) \in N, k = 1, 2, \cdots, \lfloor N/\tau \rfloor\}$.

To understand the above process better, here we illustrate how $\pi_s(s = 1, 2, \cdots, d!)$ are obtained firstly. Take a 3D chaotic system as an example, the parameter $d = 3$. Thus, there are six possible order patterns under the k-th point, and they are shown as below:

$$
\begin{cases}
\left\{ \pi_1 : y_1^\tau(k) \leq y_2^\tau(k) \leq y_3^\tau(k) \right\} \\
\left\{ \pi_2 : y_1^\tau(k) \leq y_3^\tau(k) \leq y_2^\tau(k) \right\} \\
\left\{ \pi_3 : y_2^\tau(k) \leq y_1^\tau(k) \leq y_3^\tau(k) \right\} \\
\left\{ \pi_4 : y_2^\tau(k) \leq y_3^\tau(k) \leq y_1^\tau(k) \right\} \\
\left\{ \pi_5 : y_3^\tau(k) \leq y_1^\tau(k) \leq y_2^\tau(k) \right\} \\
\left\{ \pi_6 : y_3^\tau(k) \leq y_2^\tau(k) \leq y_1^\tau(k) \right\}
\end{cases}
\tag{3}
$$

For example, let $\left[y_1^\tau, y_2^\tau, y_3^\tau \right] = [1, 3, 4]$; it can be defined as order pattern π_1; while if $\left[y_1^\tau, y_2^\tau, y_3^\tau \right] = [1, 4, 3]$, it belongs to another kind of pattern, which can be classified as π_2. Secondly, suppose that we have three short time series, namely $\left\{ y_1^\tau = 1, 2, 3, 4, 5 \right\}$, $\left\{ y_2^\tau = 3, 3, 1, 5, 4 \right\}$ and $\left\{ y_3^\tau = 5, 1, 4, 3, 2 \right\}$; the conducted pattern series $\left\{ \Psi^\tau(k) = \pi_1, \pi_5, \pi_3, \pi_5, \pi_6 \right\}$, and the quantification pattern series is $\left\{ \Phi^\tau(k) = 1, 5, 3, 5, 6 \right\}$.

2.2. Complexity Measuring Algorithms

Permutation entropy [26] can be calculated based on the comparison of neighboring values (quantification pattern series) by combing with the concept of Shannon entropy. It is particularly useful in the presence of dynamical or observational noise. However, Lempel–Ziv complexity (LZC) [27,28] is not based on the probabilistic of the symbols, but on the way that these symbols are repeated along the sequences. Based on the quantification pattern series, multiscale multivariate permutation entropy and multiscale multivariate Lempel–Ziv complexity can be designed.

2.2.1. Multiscale Multivariate Permutation Entropy

The probability distribution $P^\tau = p^\tau(\pi_l) | l = 1, 2, \cdots, d!$ associated with the quantification pattern series $\left\{ \Phi^\tau(k) : \Phi^\tau(k) \in \mathbb{N}, k = 1, 2, \cdots, floor(N/\tau) \right\}$ is defined by:

$$
p^\tau(\pi_l) = \frac{\#\{k | k \leq floor(N/\tau), \Phi^\tau(k) = l\}}{floor(N/\tau)},
\tag{4}
$$

where the symbol # stands for "number" and $l = 1, 2, 3, \cdots, d!$. According to the definition of Shannon entropy, MMPE is defined as:

$$
MMPE(x, \tau) = -\frac{1}{\ln(d!)} \sum_{s=1}^{d!} p^\tau(\pi_s) \ln(p^\tau(\pi_s)),
\tag{5}
$$

Obviously, larger MMPE values mean the time series is more complex.

2.2.2. Multiscale Multivariate Lempel–Ziv Complexity

First of all, Lempel–Ziv complexity [28] is described, and the steps are shown as follows.

Step 1: Suppose that the quantification pattern series is $\left\{ \Phi^\tau(k) = s_1, s_2, s_3, \cdots, s_N \right\}$. Let S and Q be two character strings.

Step 2: For the step n ($n = 1, 2, 3, \cdots, N$), let $S = (s_1, s_2, s_3, \cdots, s_n)$, and $Q = s_{n+1}$ or $Q = (s_{n+1}, s_{n+2}, \cdots, s_{n+k})$, then we get:

$$
SQ = (s_1, s_2, s_3, \cdots, s_n, s_{n+1}),
\tag{6}
$$

or:

$$
SQ = (s_1, s_2, s_3, \cdots, s_n, s_{n+1}, s_{n+2}, \cdots, s_{n+k}).
\tag{7}
$$

Define:

$$SQ_v = (s_1, s_2, s_3, \cdots, s_n). \tag{8}$$

or:

$$SQ_v = (s_1, s_2, s_3, \cdots, s_n, s_{n+1}, s_{n+2}, \cdots, s_{n+k-1}). \tag{9}$$

If there exist an i $(1 \leq i \leq n)$ and the following relationship is satisfied:

$$(s_{n+1}, s_{n+2}, \cdots, s_{n+k}) = (s_i, s_{i+1}, \cdots, s_{i+k-1}). \tag{10}$$

it means that Q is a duplicate of SQ_v. Then, the size of Q should increase by one, and the above operation is carried out again until $Q \notin SQ_v$. When Q does not belong to SQ_v, we call Q an "insertion". When an "insertion" is found, we place a "·" behind S. Repeat the above operations until $n = N$.

Step 3:   In Step 2, we obtained a series of dots; thus, we can calculate the number of dots and denote the complexity as $c(n)$.

Step 4:   According to [27], Lempel–Ziv complexity will reach a stable value, which is given by:

$$LZ_{Stable} = \lim_{n \to \infty} c(n) = \frac{n}{\log_2(n)}. \tag{11}$$

where LZ_{Stable} is the stable complexity measure value of a finite long time series. Thus, the normalized multiscale multivariate Lempel–Ziv complexity is defined as:

$$MMLZC = \frac{c(n)}{LZ_{stable}}, (0 \leq MMLZC \leq 1). \tag{12}$$

The scale factor is given by the quantized time series $\{\Phi^\tau(k) = s_1, s_2, s_3, \cdots\}$.

Here, an example is given to show the steps of the LZC algorithm. Suppose that the quantized time series with length $n = 7$ is $\{\Phi^\tau = 1, 2, 3, 2, 3, 2, 3\}$. Firstly, Φ^τ is converted to a string time series.

$n = 1$, $S = 1$, $Q = 2$, $SQ_v = 1$. Because $Q \notin SQ_v$, then Q is an insertion $S = 1 \cdot 2$.
$n = 2$, $S = 12$, $Q = 3$, $SQ_v = 12$. Because $Q \notin SQ_v$, then Q is an insertion $S = 1 \cdot 2 \cdot 3$.
$n = 3$, $S = 123$, $Q = 2$, $SQ_v = 123$. Because $Q \in SQ_v$, then Q is a copy of $S = 1 \cdot 2 \cdot 32$.
$n = 4$, $S = 1232$, $Q = 23$, $SQ_v = 123$. Because $Q \in SQ_v$, then Q is a copy of $S = 1 \cdot 2 \cdot 323$.

Continue the above steps until $n = 7$; we can get $S = 1 \cdot 2 \cdot 32323$. This means that the time series can be divided into three parts; thus, $c(n) = 3$. According to Equation (12), $LZC = 0.8159$. However, this measuring result is not satisfying since the length of this example is not suitable. To make the measuring results stable, the length of the sequence should be larger than 3600 [29].

2.2.3. Process for Complexity Measuring

Here, take a 3D chaotic system as an example, the process of the complexity measure is illustrated as follows. The steps to analyze the complexity of attractors and the representation of the result in the complexity vs. entropy map are described.

Step 1:   Figure 1a. Solve the chaotic system and observe the state of the system based on the phase diagrams, preliminarily.

Step 2:   Figure 1b. Cut three segments of chaotic time series, which are the three state variables of the 3D chaotic system. Data processing and coarse graining are carried out by employing the method given in the Section 2.1.

Step 3:   Figure 1c. Quantize the scaled time series using the Bandt–Pompe approach; thus, a symbol time series is obtained.

Step 4:   Figure 1d. Estimate the MMLZC and MMPE according to the obtained sequence, where the steps of MMPE and MMLZC are shown in Sections 2.2.1 and 2.2.2, respectively.

Step 5:   Figure 1e. Illustrate the complexity measuring results with different figures. Here, the two measures are shown in the MMPLC-MMPE plane.

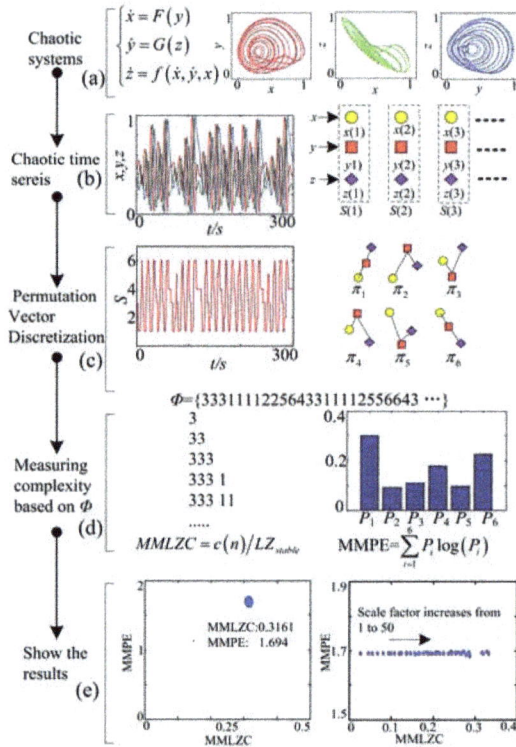

Figure 1. Steps to analyze the complexity of a chaotic system through the multiscale multivariate Lempel–Ziv complexity (MMLZC) and multiscale multivariate permutation entropy (MMPE) and algorithms. (**a**) Chaotic systems; (**b**) Chaotic time series; (**c**) Permutation Vector Discretization; (**d**) Measuring complexity based on Φ; (**e**) Show the results.

Note that we also illustrate the complexity with MMLZC and MMPE as shown by the curve and surfaces for comparison.

3. Complexity Analysis of Self-Reproducing Chaotic Systems

Multistable systems have multiple solutions under different initial conditions, and the self-reproducing system is a new kind of multistable system with infinitely many coexisting attractors by reproducing themselves along particular dimensions or directions. It should be pointed out that all of those coexisting attractors in a system share the same Lyapunov exponents, and the infinitely many attractors in self-reproducing chaotic systems are triggered by the initial condition [11,12]. Therefore, it is interesting to check whether those coexisting attractors have the same complexity. Here, the complexity of a one-directional self-reproducing system and two-directional self-reproducing system is analyzed by means of MMPE and MMLZC.

3.1. Case A: One-Directional Self-Reproducing System

For the self-reproducing system [10],

$$\begin{cases} \dot{x} = |y| - 1 \\ \dot{y} = z \\ \dot{z} = F(x) - by - az \end{cases} \qquad (13)$$

where x, y and z are the state variables, a and b are the bifurcation parameters and $F(x) = A\cos(x)$, $A = 1.55$. Fix $a = 0.6$, $b = 1$. Set the initial conditions $(x_0, y_0, z_0) = (2 - 2\pi, 0, -1)$, $(2 - \pi, 0, -1)$, $(2, 0, -1)$, $(2 + \pi, 0, -1)$ and $(2 + 2\pi, 0, -1)$; five attractors under different initial conditions are shown in Figure 2a. The red attractor is plotted with $(x_0, y_0, z_0) = (2 - 2\pi, 0, -1)$; the cyan attractor is plotted with $(x_0, y_0, z_0) = (2 - \pi, 0, -1)$; the blue attractor is plotted with $(x_0, y_0, z_0) = (2, 0, -1)$; the mauve attractor is plotted with $(x_0, y_0, z_0) = (2 + \pi, 0, -1)$; and the green attractor is plotted with $(x_0, y_0, z_0) = (2 + 2\pi, 0, -1)$. The MMLZC and MMPE analysis results are shown in Figure 2b,c, respectively. It is shown in Figure 2b that multivariate LZC increases with the scale factor τ. However, different complexities overlap with each other. This indicates that the Lempel–Ziv complexity of different attractors is of the same level. However, Figure 2c shows that the complexity of multivariate PE does not increase with the scale factor. This means that the complexity of an attractor does not increase with the scale factor in the sense of multivariate PE. Moreover, as shown in Figure 2c, the complexity of the attractors with $(x_0, y_0, z_0) = (2 - \pi, 0, -1)$ and $(2 + \pi, 0, -1)$ is lower than that of other cases. Here, let the scale factor τ for MMLZC be 100 and for MMPE be one. The MMLZC-MMPE plot in Figure 2d shows that the complexity of attractors with $(x_0, y_0, z_0) = (2 - \pi, 0, -1)$ and $(2 + \pi, 0, -1)$ is relatively lower than the other cases.

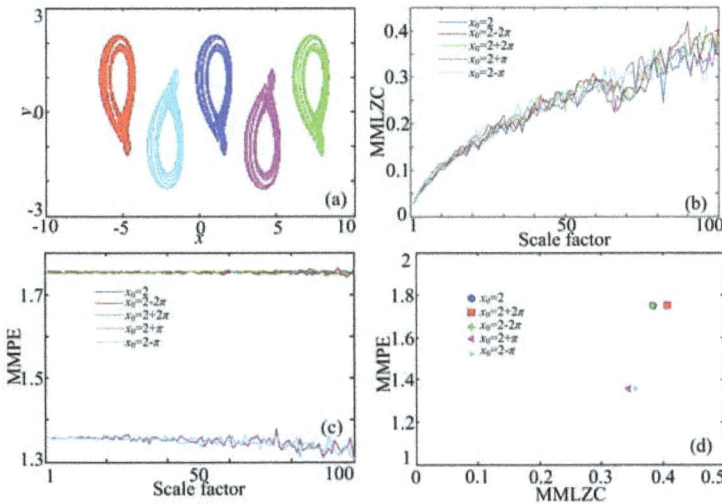

Figure 2. Coexisting attractors and complexity analysis results of System (13). (**a**) Coexisting attractors; (**b**) MMLZC; (**c**) MMPE; (**d**) MMLZC-MMPE plot.

Fix $y_0 = 0$ and $z_0 = -1$, and vary x_0 from $2 - 2\pi$ to $2 + 2\pi$ with a step size of 0.001. The complexity results are shown in Figure 3; MMPE of the chaotic system under different initial conditions varies with the values of x_0. Lines of MMPE and x smoothly evolve according to x_0. MMPE has the same variation tendency with the mean value of state variable x. Different from MMPE, MMLZC shows some robustness with the initial condition. As a matter of fact, it is indicated in [10] that different attractors have the same LEs (0.00285, 0, −0.6285). However, it should be noted that the complexity of the system still depends on the initial condition.

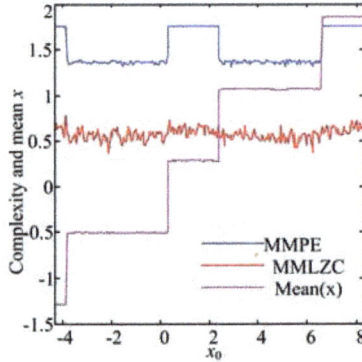

Figure 3. Complexity analysis result and mean value of *x* of System (13).

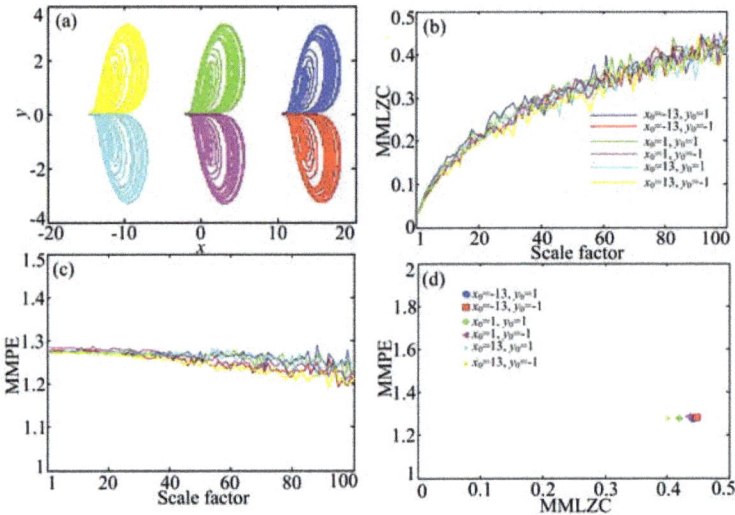

Figure 4. Coexisting attractors and complexity analysis results of System (14). (**a**) Coexisting attractors; (**b**) MMLZC; (**c**) MMPE; (**d**) MMLZC-MMPE plot.

Another one-directional self-reproducing chaotic system was defined by [10]:

$$\begin{cases} \dot{x} = az + y^2 - 1 \\ \dot{y} = byz \\ \dot{z} = -A\sin(Bx) - z \end{cases}, \tag{14}$$

where $a = 2.8$, $b = 4$, $A = 2.2$ and $B = 0.5$. When the initial conditions are given by $(x_0, y_0, z_0) = (-13, 1, 0), (-13, -1, 0), (1, 1, 0), (-1, 1, 0), (13, -1, 0)$ and $(-13, 1, 0)$, coexisting attractors are shown in Figure 4a. The position of chaotic attractors is decided by the values of x_0 and y_0. When $(x_0, y_0, z_0) = (-13, 1, 0)$ and $(-13, -1, 0)$, the yellow and the cyan attractors are plotted in the left part of Figure 4a. When $(x_0, y_0, z_0) = (1, 1, 0)$ and $(-1, 1, 0)$, the green and the mauve attractors are shown in the middle of Figure 4a, while when $(x_0, y_0, z_0) = (13, -1, 0)$ and $(-13, 1, 0)$, the red and the blue attractors are illustrated in the right part. The MMPE and MMLZC complexities of these attractors are calculated as shown in Figure 4b–d. According to Figure 4, we see that the multivariate LZC increases with the scale factor, but MMPE does not increase with the variation of the scale factor. According to the above analysis, the complexity of different attractors is at about the same level.

Let x_0 vary from 0 to 15 with a step size of 0.15 and y_0 vary from -1 to 1 with a step size of 0.02. MMLZC and MMPE of System (13) are computed, and the complexity of System (13) with simultaneous variations of x_0 and y_0 is analyzed. It is shown in Figure 5 that the complexity of System (14) does not change when y_0 takes different values and x_0 varies in the interval $x_0 \in [6.8183, 11.2122]$. As a matter of fact, when $x_0 \in [6.8183, 11.2122]$, the solution of the system goes to infinity , which means that the system is divergent.

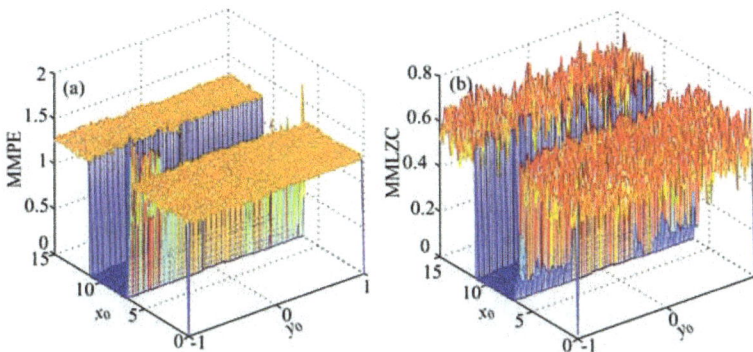

Figure 5. Complexity analysis results of System (14) with simultaneous variations of x_0 and y_0. (a) MMPE analysis result; (b) MMLZC analysis result.

3.2. Case B: Two-Directional Self-Reproducing System

Li et al. [12] designed a two-directional self-reproducing chaotic system,

$$\begin{cases} \dot{x} = \sin(y) \\ \dot{y} = a \sin(z) \\ \dot{z} = -\sin(y) - b \sin(z) - x + x^2 \end{cases}, \tag{15}$$

where x, y and z are the state variables and a and bare the system parameters. When $a = 1.05$ and $b = 0.5$, the system gives chaos with LEs $(0.0890, 0, -0.5808)$. Figure 6 illustrates the coexisting attractors of System (15) with initial conditions given by $(x_0, y_0, z_0) = (0, 0.1 - 2k\pi, 2l\pi \, (k, l = -1, 0, 1))$. It shows that attractors are distributed in a grid distribution according to the given initial conditions.

The complexity of System (15) is analyzed by varying system parameter b from 0.525 to 0.725 with a step size 0.002 and setting $y_0 = 0.1 + 2k\pi$, $z_0 = 0$ or $y_0 = 0.1$, $z_0 = 2k\pi$, $k \in [-25, 25]$, $k \in Z$. It is shown in Figure 7 that the system has lower complexity when parameter b takes values larger than 0.55. Additionally, in these cases, there are no significant changes of complexity with the variation of y_0 and z_0. However, when parameter b takes values between 0.525 and 0.55, some lower complexity analysis results are observed. This illustrates that the complexity of the system is determined by the initial conditions. Moreover, compared with MMPE, MMLZC obtains better analysis results when the system is in the route from period to chaos.

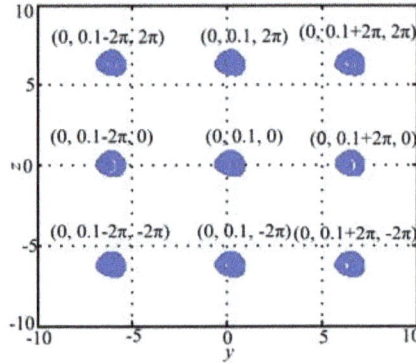

Figure 6. Coexisting attractors of System (15) with different initial conditions.

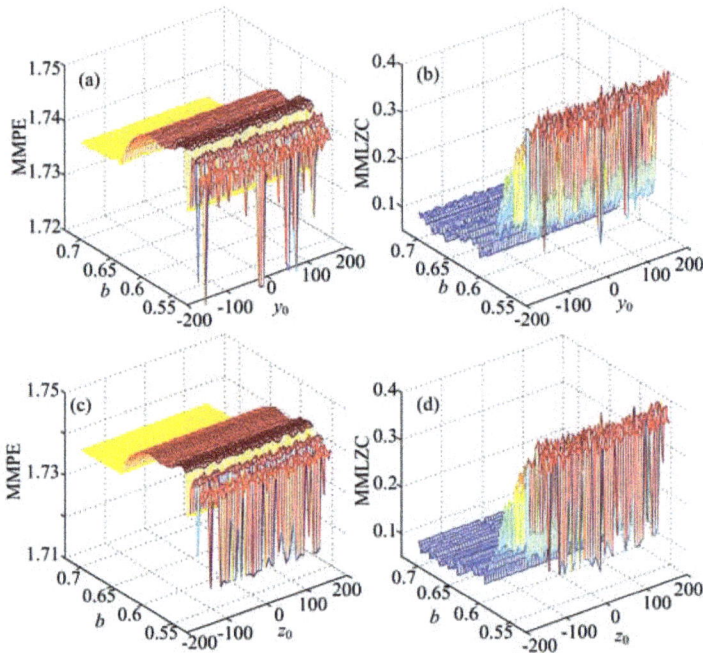

Figure 7. Complexity analysis results of System (15). (**a**) MMPE with y_0 and b; (**b**) MMPE with y_0 and b; (**c**) MMPE with y_0 and b; (**d**) MMLZC with z_0 and b.

Moreover, let $y_0 = 0.1 + 2k\pi$ and $z_0 = 2k\pi$, where $k \in [-50, 50]$ and $k \in Z$. The complexity analysis result of System (15) with different initial conditions is shown in Figure 8. It is shown that the system has high complexity in most cases. According to [12], when $y_0 = 0.1 + 2k\pi$ and $z_0 = 2k\pi$, chaotic attractors can be observed although in different positions. It is supposed to be a result that only high complexity could be found in this system. However, analysis results also indicate that the system could have a low complexity when y_0 and z_0 are varying. To explain this, phase portraits and time series of the system with $(x_0, y_0, z_0) = (0, 0.1 - 82\pi, -82\pi)$ are observed, and the results are shown in Figure 9. It is shown in Figure 9 that the system has chaotic attractors, but the fluctuation range of state variable z changes after a nonchaotic period $t \in (3000s, 4000s)$. In real practical applications, the nonchaotic period should be avoided, and complexity measure methods provide a method to fulfill this.

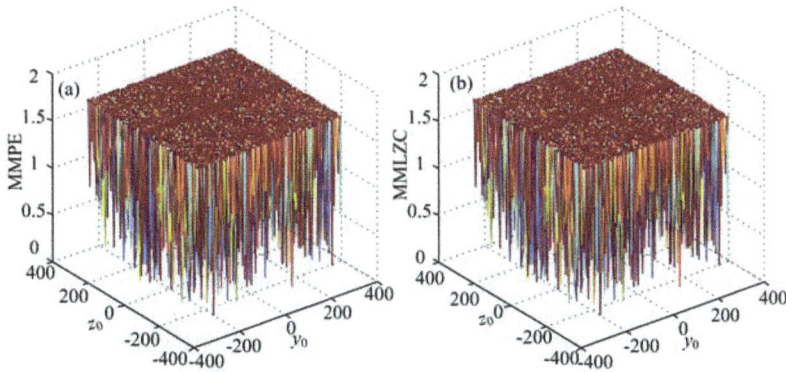

Figure 8. Complexity of System (15) with both y_0 and z_0 varying. (**a**) MMPE; (**b**) MMLZC.

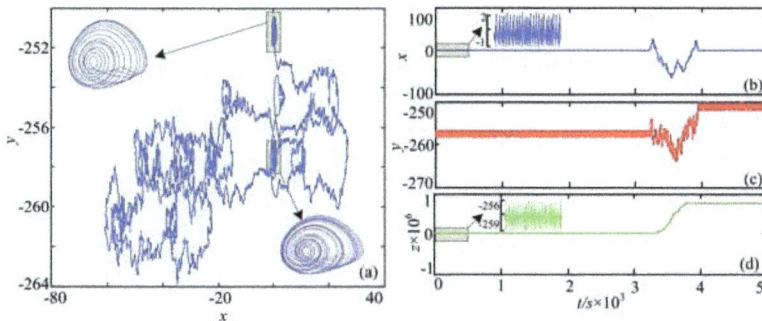

Figure 9. Dynamics of System (15) with $(x_0, y_0, z_0) = (0, 0.1 - 82\pi, -82\pi)$. (**a**) Phase diagram; (**b**) time series x; (**c**) time series y; (**d**) time series z.

4. Discussion

4.1. Comparison with the Corresponding Original Systems

Actually, self-reproducing chaotic systems can be constructed based on the existing systems. System (13) is designed based on the following system [30]:

$$\begin{cases} \dot{x} = |y| - 1 \\ \dot{y} = z \\ \dot{z} = x - by - az \end{cases}, \tag{16}$$

where $a = 0.6$ and $b = 1$. When solving this system, the initial condition is given by $(x_0, y_0, z_0) = (2, 0, -1)$. The LE calculation results are $(0.0363, 0, -0.6363)$. System (14) is designed based on the following equation [31]:

$$\begin{cases} \dot{x} = az + y^2 - 1 \\ \dot{y} = byz \\ \dot{z} = -x - z \end{cases}, \tag{17}$$

where $a = 2.8$ and $b = 4$. System (17) has chaotic attractors with Lyapunov exponents $(0.1149, 0, -1.1149)$ under the initial condition when $(x_0, y_0, z_0) = (1, 1, 0)$.

The two-directional self-reproducing chaotic system is designed based on the following [32]:

$$\begin{cases} \dot{x} = y \\ \dot{y} = z \\ \dot{z} = x^2 - x - 0.5\dot{y} - \dot{x} \end{cases}, \tag{18}$$

which gives offset boosting with an identical strange attractor with LEs $(0.0938, 0, -0.5938)$ under the initial condition $(x_0, y_0, z_0) = (13, -1, 0)$.

The complexities of the original systems and the corresponding self-reproducing versions are given in Table 1. These systems are solved in the time interval $[0, 200]$ with a step size of 0.01. The complexity of System (13) is calculated with initial condition $(x_0, y_0, z_0) = (2 + k\pi, 0, -1 \ (k = -5, -4, \cdots, 5))$, and its mean value of MMPE is 1.5579, while the mean value of MMLZC is 0.5376. Compared with the original System (16), System (13) with infinitely many attractors has a higher MMPE complexity, while a smaller maximum Lyapunov exponent (MLE). That is to say, the complexity of the modified self-reproducing system is almost the same as System (16). The initial conditions of System (14) are given in Figure 4. Table 1 shows that the modified system has higher complexity than its original system, which is given by Equation (17). Moreover, we compared the complexity of a two-directional self-reproducing chaotic system with its original system. It has higher MMLZC, but lower MLEs. In conclusion, our complexity analysis results illustrate that the new systems have relatively higher complexity. The effectiveness of the proposed methods and the potential application values in the fields including information encryption and secure communication of the proposed systems are shown.

Table 1. Complexity comparison in different systems.

(Original, New)	LE_{max}	MMPE	MMLZC
(Sys(16), Sys(13))	(0.0363, 0.0285)	(1.3805, 1.5579)	(0.5402, 0.5376)
(Sys(17), Sys(14))	(0.1149, 0.1101)	(1.1610, 1.2657)	(0.5402, 0.5907)
(Sys(18), Sys(15))	(0.0938, 0.0890)	(1.7435, 1.7414)	(0.6483, 0.6976)

LE_{max}: maximum Lyapunov exponent; MMPE: multiscale multivariate permutation entropy; MMLZC: multiscale multivariate Lempel–Ziv complexity; Sys: System.

4.2. Comparison of MMPE and MMLZC

It is shown in Figures 2 and 4 that measuring values of MMPE does not change with the scale factor, while MMLZC increases with the scale factor. Generally, the multiscale complexity of a continues chaotic system increases with the scale factor [18,23–25]. However, when multivariate states are considered, the situation is different. Firstly, we take System (15) as an example to show the effectiveness of the scale factor on multivariate measuring algorithms. The parameters of System (15) are $a = 1.05$ and $b = 0.525$. The chaotic attractor of System (15) is shown in Figure 10. Since it is plotted from the system directly, it shows the scale factor $\tau = 1$. Moreover, we also plot the phase diagrams, quantification pattern series and probability distribution when the scale factor is revised to be $\tau = 50$ and 100. It is shown in Figure 10 that although phase diagrams under a larger, different scale factor become rough, their shapes are similar. This means that the relationship between different variables is not changed.

Let function $\varphi(\cdot)$ represent the comparison of the processing of Equation (3), then the pattern series for one time series and multiple time series can be obtained by:

$$S_{One}^{\tau}(i) = \varphi\left(x_i^{\tau}, x_{i+1}^{\tau}, x_{i+2}^{\tau}\right), \tag{19}$$

and:

$$S_{Mul}^{\tau}(i) = \varphi\left(x_i^{\tau}, y_i^{\tau}, z_i^{\tau}\right), \tag{20}$$

respectively, where x^{τ}, y^{τ} and z^{τ} are the coarse-grained series under scale factor τ and $i = 1, 2, \cdots$. Obviously, these two processes are different. As for MMPE, it is calculated based on the probability distribution of different patterns. Because the relationship is not changed, the probability distribution (statistical result) will keep the same, which is verified in Figure 10. Therefore, the values of MMPE do not change with the scale factor. However, MMLZC calculates complexity based on the new reproduction in the pattern series. When the scale factor is larger, it is easier to find the insertion, and the complexity measuring result is larger. This gives an explanation, at least to a significant degree, why values of MMPE do not change with the increase of the scale factor.

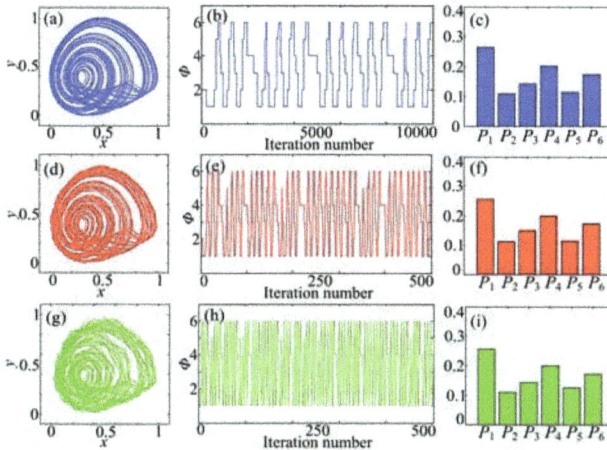

Figure 10. States of System (15) under different scale factors. (**a**) Phase diagram under $\tau = 1$; (**b**) quantification pattern series under $\tau = 1$; (**c**) probability distribution under $\tau = 1$; (**d**) phase diagram under $\tau = 50$; (**e**) quantification pattern series under $\tau = 50$; (**f**) probability distribution under $\tau = 50$; (**g**) phase diagram under $\tau = 100$; (**h**) quantification pattern series under $\tau = 100$; (**i**) probability distribution under $\tau = 100$.

Currently, multiscale processing is a hot topic, and it is widely used in many research works. Scientists found that it indeed makes the complexity measuring results better. However, when it comes to multiple time series, the situation is different according to our analysis. One should carefully introduce this multiscale process for complexity analysis of a multivariate system, since it sees that it is not necessary for entropy-based algorithms.

However, it should be pointed out that the two proposed complexity measures are reliable and effective for the complexity analysis of chaotic attractors in the following aspects. In the first place, the pattern series is obtained based on the time series from different dimensions of the system. It can reflect the nature of the complexity of attractors directly. Secondly, MMPE is designed by employing the concept of permutation entropy and Shannon entropy, while MMLZC is proposed based on the Lempel–Ziv complexity. As we all know, the PE algorithm and Lempel–Ziv algorithm are mature algorithms and have been widely used by many researchers in a vast quantity of literature. Moreover, it is shown in the literature and in the above analysis results that the multi-scale process could make the complexity measuring results better. Thus, the two proposed complexity measuring algorithms have potential application value in real applications.

5. Conclusions

In this paper, MMPE and MMLZC are employed to analyze the complexity of multivariate systems, and the MMPE vs. MMLZC map is introduced to demonstrate the complexity. How the complexity of the self-reproducing systems is determined by the initial condition is investigated. Moreover, we found that the multiscale coarse graining process does not affect the final result of MMPE, but a good MMLZC measuring result can be obtained by choosing a proper large-scale factor.

Since the self-reproducing chaotic systems can generate strange attractors in different positions under various initial conditions, multivariate complexities including MMPE and MMLZC of these systems with different initial conditions are analyzed. It is shown that the complexity of self-reproducing systems depends on the initial conditions. Especially, transition stages can be found in the two-directional self-reproducing chaotic systems since in this case the system has relatively lower complexity analysis results. Moreover, compared with their corresponding original systems, the newly-developed multistable systems have the same level or a higher level of complexity. The theoretical and practical significance of the multistable systems and complexity measuring algorithms is shown.

Author Contributions: Conceptualization, S.H. and C.L.; Methodology, C.L. and K.S.; Software, S.H.; Validation, S.H., C.L., K.S. and S.J.; Formal Analysis, S.J.; Writing-Original Draft Preparation, S.H.; Writing-Review & Editing, C.L., S.J. and K.S.; Funding Acquisition, C.L. and S.H.; All authors have read and approved the final manuscript.

Funding: This research was funded by [the Natural Science Foundation of Jiangsu Province] grant number [SBK2018021196], [the Postdoctoral Innovative Talents Support Program] grant number [BX20180386], [the Special Funds for Theoretical Physics of the National Natural Science Foundation of China] grant number [11747150], [the Natural Science Foundation of the Higher Education Institutions of Jiangsu Province] grant number [16KJB120004], [the Startup Foundation for Introducing Talent of NUIST] grant number [2016205] and a project funded by the Priority Academic Program Development of Jiangsu Higher Education Institutions.

Acknowledgments: The authors would like to thank the three anonymous reviewers for their constructive comments and insightful suggestions.

Conflicts of Interest: The authors declare no conflict of interest.

References

1. Feudel, U.; Kraut, S. Complex dynamics in multistable systems. *Int. J. Bifur. Chaos* **2008**, *18*, 1607–1626. [CrossRef]
2. Singh, J.P.; Roy, B.K. Multistability and hidden chaotic attractors in a new simple 4-D chaotic system with chaotic 2-torus behaviour. *Int. J. Dyn. Control* **2017**, *6*, 529–538. [CrossRef]
3. Li, C.B.; Hu, W.; Sprott, J.C.; Wang, X. Multistability in symmetric chaotic systems. *Eur. Phys. J. Spec. Top.* **2015**, *224*, 1493–1506. [CrossRef]
4. Liu, Y.; Chávez, J.P. Controlling multistability in a vibro-impact capsule system. *Nonlinear Dyn.* **2017**, *88*, 1289–1304. [CrossRef]
5. Zhusubaliyev, Z.T.; Mosekilde, E.; Rubanov, V.G.; Nabokov, R.A. Multistability and hidden attractors in a relay system with hysteresis. *Phys. D* **2015**, *306*, 6–15. [CrossRef]
6. Bao, B.C.; Li, Q.D.; Wang, N.; Xu, Q. Multistability in Chua's circuit with two stable node-foci. *Chaos* **2016**, *26*, 043111. [CrossRef] [PubMed]
7. Hu, A.; Xu, Z. Multi-stable chaotic attractors in generalized synchronization. *Commun. Nonl. Sci. Num. Simul.* **2011**, *16*, 3237–3244. [CrossRef]
8. Peng, G.; Min, F. Multistability analysis, circuit implementations and application in image encryption of a novel memristive chaotic circuit. *Nonlinear Dyn.* **2017**, *90*, 1607–1625. [CrossRef]
9. Shih, C.W. Multistability in recurrent neural networks. *SIAM J. Appl. Math.* **2006**, *66*, 1301–1320.
10. Lai, Q.; Chen, S. Research on a new 3D autonomous chaotic system with coexisting attractors. *Optik* **2016**, *127*, 3000–3004. [CrossRef]
11. Li, C.B.; Sprott, J.C.; Hu, W.; Xu, Y. Infinite multistability in a self-reproducing chaotic system. *Int. J. Bifur. Chaos* **2017**, *27*, 1750160. [CrossRef]

12. Li, C.B.; Sprott, J.C.; Mei, Y. An infinite 2-D lattice of strange attractors. *Nonlinear Dyn.* **2017**, *89*, 2629–2639. [CrossRef]
13. Kreinovich, V.; Kunin, I.A. Kolmogorov complexity and chaotic phenomena. *Int. J. Eng. Sci.* **2003**, *41*, 483–493. [CrossRef]
14. Micco, L.D.; Fernández, J.G.; Larrondo, H.A.; Plastino, A.; Rosso, O.A. Sampling period, statistical complexity, and chaotic attractors. *Phys. A* **2012**, *391*, 2564–2575. [CrossRef]
15. Balasubramanian, K.; Nair, S.S.; Nagaraj, N. Classification of periodic, chaotic and random sequences using approximate entropy and Lempel–Ziv complexity measures. *Pramana* **2015**, *84*, 365–372. [CrossRef]
16. Makark, S.A.; Starodubov, A.V.; Kalinin Yu. A. Application of permutation entropy method in the analysis of chaotic, noisy, and chaotic noisy series. *Tech. Phys.* **2017**, *62*, 1714–1719. [CrossRef]
17. He, S.B.; Sun, K.H.; Wang, H.H. Complexity Analysis and DSP Implementation of the Fractional-Order Lorenz Hyperchaotic System. *Entropy* **2015**, *17*, 8299–8311. [CrossRef]
18. He, S.B.; Sun, K.H.; Wang, H.H. Modified multiscale permutation entropy algorithm and its application for multiscroll chaotic systems. *Complexity* **2016**, *21*, 52–58.
19. Mukherjee, S.; Palit, S.K.; Banerjee, S.; Ariffin, M.R.K.; Rondoni, L.; Bhattacharya, D.K. Can complexity decrease in congestive heart failure? *Phys. A* **2015**, *439*, 93–102. [CrossRef]
20. Ahmed, M.U.; Mandic, D.P. Multivariate multiscale entropy: A tool for complexity analysis of multichannel data. *Phys. Rev. E* **2011**, *84*, 061918. [CrossRef] [PubMed]
21. Richman, J.S. Multivariate neighborhood sample entropy: A method for data reduction and prediction of complex data. *Meth. Enzymol.* **2011**, *487*, 397–408. [PubMed]
22. He, S.B.; Sun, K.; Wang, H. Multivariate permutation entropy and its application for complexity analysis of chaotic systems. *Phys. A* **2016**, *461*, 812–823. [CrossRef]
23. Costa, M.; Goldberger, A.L.; Peng, C.K. Multiscale entropy analysis of biological signals. *Phys. Rev. E* **2005**, *71*, 021906. [CrossRef] [PubMed]
24. Humeauheurtier, A. The Multiscale entropy algorithm and its variants: A review. *Entropy* **2015**, *17*, 3110–3123. [CrossRef]
25. Yu, S.N.; Lee, M.Y. Wavelet-Based Multiscale Sample Entropy and Chaotic Features for Congestive Heart Failure Recognition Using Heart Rate Variability. *J. Med. Biol. Eng.* **2015**, *35*, 338–347. [CrossRef]
26. Bandt, C.; Pompe, B. Permutation entropy: A natural complexity measure for time series. *Phys. Rev. Lett.* **2002**, *88*, 174102. [CrossRef] [PubMed]
27. Lempel, A.; Ziv, J. On the complexity of finite sequences. *IEEE Trans. Inform. Theory* **1976**, *22*, 75–81. [CrossRef]
28. Zozor, S.; Mateos, D.; Lamberti, P.W. Mixing Bandt-Pompe and Lempel–Ziv approaches: Another way to analyze the complexity of continuous-state sequences. *Eur. Phys. J. B* **2014**, *87*, 107. [CrossRef]
29. Hong, H.; Liang, M. Fault severity assessment for rolling element bearings using the Lempel–Ziv complexity and continuous wavelet transform. *J. Sound Vib.* **2009**, *320*, 452–468. [CrossRef]
30. Li, C.B.; Sprott, J.C.; Xing, H. Crisis in amplitude control hides in multistability. *Inter. J. Bifur. Chaos* **2017**, *26*, 1650233. [CrossRef]
31. Li, C.B.; Sprott, J.C.; Xing, H. Constructing chaotic systems with conditional symmetry. *Nonlinear Dyn.* **2016**, *87*, 1351–1358. [CrossRef]
32. Kengne, J.; Njitacke, Z.T.; Fotsin, H.B. Dynamical analysis of a simple autonomous jerk system with multiple attractors. *Nonlinear Dyn.* **2016**, *83*, 751–765. [CrossRef]

entropy

MDPI

Article

A New Fractional-Order Chaotic System with Different Families of Hidden and Self-Excited Attractors

Jesus M. Munoz-Pacheco [1,*] [iD], Ernesto Zambrano-Serrano [1] [iD], Christos Volos [2] [iD], Sajad Jafari [3] [iD], Jacques Kengne [4] and Karthikeyan Rajagopal [5]

[1] Faculty of Electronics Sciences, Autonomous University of Puebla, Puebla 72000, Mexico; erneszambrano@gmail.com
[2] Department of Physics, Aristotle University of Thessaloniki, 54124 Thessaloniki, Greece; volos@physics.auth.gr
[3] Department of Biomedical Engineering, Amirkabir University of Technology, Tehran 15875-4413, Iran; sajadjafari83@gmail.com
[4] Department of Electrical Engineering, University of Dschang, P.O. Box 134 Dschang, Cameroon; kengnemozart@yahoo.fr
[5] Center for Nonlinear Dynamics, Defence University, P.O. Box 1041 Bishoftu, Ethiopia; rkarthiekeyan@gmail.com
*   Correspondence: jesusm.pacheco@correo.buap.mx; Tel.: +52-222-229-5500

Received: 2 July 2018; Accepted: 25 July 2018; Published: 28 July 2018

Abstract: In this work, a new fractional-order chaotic system with a single parameter and four nonlinearities is introduced. One striking feature is that by varying the system parameter, the fractional-order system generates several complex dynamics: self-excited attractors, hidden attractors, and the coexistence of hidden attractors. In the family of self-excited chaotic attractors, the system has four spiral-saddle-type equilibrium points, or two nonhyperbolic equilibria. Besides, for a certain value of the parameter, a fractional-order no-equilibrium system is obtained. This no-equilibrium system presents a hidden chaotic attractor with a 'hurricane'-like shape in the phase space. Multistability is also observed, since a hidden chaotic attractor coexists with a periodic one. The chaos generation in the new fractional-order system is demonstrated by the Lyapunov exponents method and equilibrium stability. Moreover, the complexity of the self-excited and hidden chaotic attractors is analyzed by computing their spectral entropy and Brownian-like motions. Finally, a pseudo-random number generator is designed using the hidden dynamics.

Keywords: hidden attractor; self-excited attractor; fractional order; spectral entropy; coexistence; multistability

1. Introduction

Since Leonov et al. published their seminal paper [1], the attractors in dynamical systems have been categorized as self-excited attractors and hidden attractors. A self-excited attractor has a basin of attraction that is associated with an unstable equilibrium, the most of common examples of integer-order chaotic flows showing self-excited attractors are Lorenz, Chen, Rössler, and Lü systems, among many others [2–5]. Conversely, an attractor is called hidden if its basin of attraction does not intersect with small neighborhoods of the unstable equilibrium [6]. Additionally, the attractors in dynamical systems with no-equilibrium, with curves and surfaces of equilibria, and with stable equilibria also belong to the category of hidden attractors [1,6]. Hidden attractors are very important in engineering applications because they allow the study and understanding of the unexpected and potentially disastrous responses of the dynamical systems to perturbations, for instance, in mechanical structures, like a bridge or airplane wings [7–9], aircraft control systems [10], PLL circuits [1], drilling systems with induction motors [11], and secure communication schemes [1,12]. Hence, numerous integer-order chaotic flows with hidden attractors have been proposed [7,13–24].

However, it should be noted that most of the studies about hidden attractors have mainly concentrated on continuous-time dynamical systems of integer-order. In recent years, fractional calculus has received much attention due to fractional derivatives providing more accurate models than their integer-order counterparts. Many examples have been found in different interdisciplinary fields [25], ranging from the description of viscoelastic anomalous diffusion in complex liquids, D-decomposition technique for control problems, chaotic systems; to macroeconomic models with dynamic memory, forecast of the trend of complex systems, and so on [26–34]. Those works have demonstrated that fractional derivatives provide an excellent approach to describing the memory and hereditary properties of real physical phenomena.

Therefore, the research effort oriented to hidden attractors in fractional-order dynamical systems is vital to understand this exciting and still less-explored subject of importance. In the literature, few works have reported hidden attractors in fractional-order dynamical systems with one stable equilibrium [35,36], with no-equilibria [37–40], with a line or surfaces of equilibria [41,42], or even in fractional-order hyperchaotic systems [43,44]. However, those fractional order systems generate only one family of hidden attractors, i.e., line, surface, stable, and without equilibrium. A remaining research question is whether fractional-order dynamical systems whose dynamics can generate both self-excited and hidden attractors could exist. The first response was recently proposed by Rajagopal et al. [45] through a dynamical system and its fractional-order form, which changes from hidden to self-excited attractors and vice versa by modifying two system parameters.

Motivated by the aforementioned discussion, in this paper, we propose a new fractional-order dynamical system with four nonlinearities and a single system parameter. One salient feature of this fractional-order system is that it generates different families of self-excited and hidden attractors as a function of only one parameter. This parameter performs as a constant controller to select the required dynamics. More specifically, the proposed system exhibits a typical self-excited chaotic attractor with four equilibrium points of the type spiral saddle index 1 and index 2. Moreover, the proposed system has a self-excited chaotic attractor coexisting with two nonhyperbolic equilibrium points. A nonhyperbolic type of chaos is unusual because it does not satisfy the Shilnikov theorems.

Surprisingly, the proposed fractional-order system also has a hidden chaotic attractor without equilibria. Unlike other approaches, the resulting hidden attractor can be observed in a fractional order as low as 0.95. Finally, the multistability phenomenon was also found in the fractional-order no-equilibrium system. Multistability leads to different qualitative behavior in a given nonlinear dynamical system for the same parameter values. In the proposed system, a hidden chaotic attractor coexist with a periodic attractor. Since the system equations contain no unnecessary terms and the system parameter has a minimum of digits, the proposed fractional-order system can be considered elegant in the sense of Sprott [46]. Moreover, the criterion (*iii*) in [47] for reporting a new chaotic system is also satisfied. The multiple complex dynamics of the proposed system were studied by applying a numerical simulation approach to compute the Lyapunov exponents, basins of attraction, bifurcation diagrams, and phase portraits. Additionally, the 0–1 test was employed to detect a Brownian-like motion in the fractional-order system.

The complexity measure is an important property to characterize the dynamics of a chaotic system; it can also be used as the core in many applications of information security. The complexity is obtained using the spectral entropy for both self-excited and hidden attractors. From the spectral entropy analysis, the time series of the hidden attractor is used to design a pseudo-random number generator (PRNG).

The rest of this paper is organized as follows. Section 2 provides the mathematical background related to fractional calculus. Section 3 presents the new fractional-order system, along with the mechanism employed to get the hidden and self-excited attractors. Section 4 shows the results of the 0–1 test algorithm and spectral entropy. Section 5 reports the design of PRNG. Finally, Section 6 summarizes the conclusions.

2. Mathematical Background

In this section, we provide the background to support our main results. The integro-differential operator, denoted as $_aD_t^q$, is a combined differentiation and integration operator commonly used in fractional calculus. This operator is a notation for taking both the fractional derivative and the fractional integral of a function, combining them into a single expression that can be formally defined as

$$
_aD_t^q f = \begin{cases} \frac{d^q f}{dt^q}, & q > 0, \\ f, & q = 0, \\ \int_a^t f(d\tau)^q, & q < 0, \end{cases}
\tag{1}
$$

where f is a function of time, a and t are the limits of the operation, and $q \in R$ is the fractional order. As we know now, there are several different definitions for the fractional differential operator that can be adopted for (1). Hereafter, we consider the fractional derivative operator d^q/dt^q, with $m - 1 < q \le m \in N$, to be Caputo's derivative [48], with starting point $a = 0$, defined by

$$
D_t^q f(t) = \frac{1}{\Gamma(m - q)} \int_0^t \frac{f^{(m)}(\tau)}{(t - \tau)^{q+1-m}} d\tau,
\tag{2}
$$

where m is an integer number and $\Gamma(\cdot)$ is the gamma function. Caputo's derivative of order q is a formal generalization of the integer derivative under the Laplace transformation, and it is widely used in engineering [49].

2.1. Predictor–Corrector Scheme

The numerical method used in this work to compute the solution of the fractional-order system is the Adams–Bashforth–Moulton (ABM) predictor–corrector scheme, reported in [50–52]. The predictor–corrector scheme is based on the Caputo fractional differential operator (2), which allows us to specify both homogeneous and inhomogeneous initial conditions.

Consider the following fractional differential equation:

$$
\begin{aligned}
D^q y(t) &= f(t, y(t)), \quad 0 \le t \le T; \\
y^{(k)}(0) &= y_0^{(k)}, \qquad k = 0, 1, \dots, n - 1.
\end{aligned}
\tag{3}
$$

The solution of (3) is given by an integral equation of Volterra type as

$$
y(t) = \sum_{k=0}^{\lceil q \rceil - 1} y_0^k \frac{t^k}{k!} + \frac{1}{\Gamma(q)} \int_0^t (t - z)^{q-1} f(z, y(z)) dz.
\tag{4}
$$

As it is shown in [50], there is a unique solution of (3), within an interval $[0, T]$, thence we are interested in a numerical solution on the uniform grid $\{t_n = nh | n = 0, 1, \dots, N\}$ with an integer N and stepsize $h = T/N$. Then, (4) can be replaced by a discrete form to get the corrector, as follows

$$
\begin{aligned}
y_h(t_{n+1}) =\ & \sum_{k=0}^{\lceil q \rceil - 1} y_0^k \frac{t^k}{k!} + \frac{h^q}{\Gamma(q+2)} f\left(t_{n+1}, y_h^p(t_{n+1})\right) \\
& + \frac{h^q}{\Gamma(q+2)} \sum_{j=0}^n a_{j,n+1} f\left(t_j, y_h(t_j)\right),
\end{aligned}
\tag{5}
$$

where

$$
a_{j,n+1} = \begin{cases}
n^{q+1} - (n-q)(n+1)^q, & j = 0, \\
(n-j+2)^{q+1} + (n-j)^{q+1} \\
\quad -2(n-j+1)^{q+1}, & 1 \le j \le n, \\
1, & j = n+1,
\end{cases}
\tag{6}
$$

Moreover, the predictor has the following structure

$$
y_h^p(t_{n+1}) = \sum_{k=0}^{\lceil q \rceil - 1} y_0^k \frac{t^k}{k!} + \frac{1}{\Gamma(q)} \sum_{j=0}^{n} b_{j,n+1} f(t_j, y_h(t_j)),
\tag{7}
$$

with $b_{j,n+1}$ defined by

$$
b_{j,n+1} = \frac{h^q}{q}((n+1-j)^q - (n-j^q)).
\tag{8}
$$

The error of this approximation is given by

$$
\max_{j=0,1,\dots N} |y(t_j) - y_h(t_j))| = \mathcal{O}(h^P),
\tag{9}
$$

where $P = \min(2, 1+q)$.

2.2. Stability of Fractional-Order Systems

This subsection presents several definitions for the stability of fractional-order autonomous systems. Starting from Equations (1) and (2), it is possible to study the stability of fractional-order systems. A fractional-order differential equation with $0 < q < 1$ typically presents a stability region that is larger than that of the same equation with integer order $q = 1$.

Definition 1. *The roots of the equation $\mathbf{f}(\mathbf{x}) = 0$ are called the equilibria of the fractional-order differential system $D^q\mathbf{x} = \mathbf{f}(\mathbf{x})$, where $\mathbf{x} = (x_1, x_2, \dots, x_n)^T \in R$, $\mathbf{f}(\mathbf{x}) \in R$ and $D^q\mathbf{x} = (D^{q_1}x_1, D^{q_2}x_2, \dots, D^{q_n}x_n)^T$, $q_i \in R^+$, $i = 1, 2, \dots, n$.*

Theorem 1. *Consider a commensurate-order system described by*

$$
D^q\mathbf{x} = \mathbf{A}\mathbf{x}, \quad \mathbf{x}(0) = x_0
\tag{10}
$$

with $0 < q < 1$, $\mathbf{x} \in R^n$ and $\mathbf{A} \in R^{n \times n}$. It has been shown [53–58] that this fractional order system is asymptotically stable if and only if the following condition is satisfied

$$
|\arg(\lambda)| > q\pi/2,
\tag{11}
$$

where $|\arg(\lambda)|$ represents all eigenvalues of $\mathbf{A}$. Besides, the critical eigenvalues of $\mathbf{A}$ satisfying $|\arg(\lambda)| = q\pi/2$ must have a geometric multiplicity of one, which stands for the dimension of subspace of $\mathbf{v}$ for $\mathbf{A}\mathbf{v} = \lambda\mathbf{v}$.

Theorem 2. *Consider an incommensurate-order system described by*

$$
D^q\mathbf{x} = \mathbf{A}\mathbf{x}, \quad \mathbf{x}(0) = x_0
\tag{12}
$$

where $\mathbf{x} = (x_1, x_2, \dots, x_n)^T \in R$, $D^q\mathbf{x} = (D^{q_1}x_1, D^{q_2}x_2, \dots, D^{q_n}x_n)^T$, $q_i \in R^+$, $i = 1, 2, \dots, n$, $0 < q_i < 1$, and $\mathbf{A} = (a_{ij}) \in R^{n \times n}$, $i = 1, 2, \dots, n$, $j = 1, 2, \dots, n$. By assuming w as the lowest common multiple of the denominators u_i of q_i, where $q_i = v_i/u_i$, $(u_i, v_i) = 1$, $u_i, v_i \in Z^+$ for $i = 1, 2, \dots, n$, the characteristic matrix of (12) is defined by

$$\Delta(\lambda) = \begin{bmatrix} \lambda^{wq_1} - a_{11} & -a_{12} & \cdots & -a_{1n} \\ -a_{21} & \lambda^{wq_2} - a_{22} & \cdots & -a_{2n} \\ \vdots & \vdots & \ddots & \vdots \\ -a_{n1} & -a_{n2} & \cdots & \lambda^{wq_n} - a_{nn} \end{bmatrix}. \tag{13}$$

Then, the system (12) *is globally asymptotically stable in the Lyapunov sense if all roots* λ *of its characteristic polynomial, given by equation* $\det(\Delta(\lambda)) = 0$, *satisfy* $|\arg(\lambda)| > \pi/2w$ *[53–58].*

Theorem 3. *The equilibrium point* E_* *is asymptotically stable if and only if the instability measure*

$$\rho = (\pi/2w) - \min_i\{\arg(\lambda_i)\} \tag{14}$$

is strictly negative, where the λ_i *parameters are roots of equations:* $\det(diag([\lambda^{wq_1} \quad \lambda^{wq_2} \ldots \lambda^{wq_n}]) - \partial f/\partial x|_{x=E_*}) = 0$, $\forall E_* \in \Omega$ *[57,58]. If* $\rho \geq 0$ *and the critical eigenvalues satisfying* $\rho = 0$ *have the geometric multiplicity one, then* E_* *is stable.*

Remark 1. *If* ρ *is positive, then* E_* *is unstable and the system may exhibit chaotic behavior [57,58].*

3. A New Three-Dimensional Fractional-Order Chaotic System

Recently, Munoz-Pacheco et al. [59] proposed a fractional-order dynamical system with a line, lattice, and 3D grid of boostable variables. The chaotic attractors of that system are self-excited. Inspired from that work, we propose a new fractional-order chaotic system given by

$$\begin{aligned} D^{q_1}x &= yz + x(y - a), \\ D^{q_2}y &= 1 - |x|, \\ D^{q_3}z &= -xy - z, \end{aligned} \tag{15}$$

where a is a real parameter, $(q_1, q_2, q_3) \in [0, 1]$ are the fractional-order derivatives, and x, y, z are the states' variables. In the fractional-order system (15), the Caputo definition of fractional-order derivative (2) is used. The fractional-order system (15) presents a unique characteristic. The parameter a behaves as a controller of the diverse complex dynamics generated by the system, such as hidden and self-excited attractors. Therefore, the fractional-order system (15) belongs to different classes of dynamical systems, i.e., a new class of systems without equilibrium, a new class of systems with multistability, a subclass of systems with nonhyperbolic equilibria, and the well-known class of systems of the hyperbolic type. To the best knowledge of the authors, this is the first time reporting a fractional-order chaotic system that presents the unique characteristic of switching from self-excited chaotic attractors to hidden chaotic attractors, and the coexistence of hidden attractors which arise by varying just one single parameter. Also, the hidden chaotic attractor can be observed with a fractional order as low as $q = 0.95$.

In this manner, the study conducted herein could be straightforwardly expanded to find other fractional-order systems, with one single parameter generating different families of hidden and self-excited attractors, by applying a systematic computer search similar to [7,16].

3.1. Self-Excited Chaotic Attractor: Spiral Saddle Type of Equilibrium Points

In order to obtain the equilibrium points of the system (15), the left-hand side of the system is kept at zero, so the system's equations can be written as

$$\begin{aligned} 0 &= yz + x(y - a), \\ 0 &= 1 - |x|, \\ 0 &= -xy - z. \end{aligned} \tag{16}$$

The equilibria $E_* = (x_*, y_*, z_*)$ of the system (16) are

$$
\begin{aligned}
E_1 &= (1, (1 + \sqrt{1 - 4a})/2, -(1 + \sqrt{1 - 4a})/2), \\
E_2 &= (-1, (1 - \sqrt{1 - 4a})/2, (1 - \sqrt{1 - 4a})/2), \\
E_3 &= (1, (1 - \sqrt{1 - 4a})/2, -(1 - \sqrt{1 - 4a})/2), \\
E_4 &= (-1, (1 + \sqrt{1 - 4a})/2, (1 + \sqrt{1 - 4a})/2).
\end{aligned}
\tag{17}
$$

As can be seen from (17), the system parameter a is a controller for the kind of equilibria, i.e., the parameter a is also known as a bifurcation parameter. In this case, a self-excited attractor can be observed when $a < 1/4$. Let $a = -1$, then the equilibrium points E_* are as given in Table 1. For investigating the stability and type of these equilibrium points, the Jacobian matrix of system (16) is defined by

$$
J = \begin{bmatrix} y+1 & x+z & y \\ -sign(x) & 0 & 0 \\ -y & -x & -1 \end{bmatrix},
\tag{18}
$$

where the resulting eigenvalues evaluated at E_* are as shown in Table 1. Therefore, the fractional-order system (16) has four hyperbolic equilibrium points of the type spiral saddle index 1 and index 2, where the index is the number of eigenvalues with a positive real part, respectively. According to Theorem 1, the fractional-order system is asymptotically stable if $q < 0.9010$.

Lemma 1. *When $q = 0.93$ and $a = -1$, the system (15) exhibits a self-excited chaotic attractor.*

Proof. In order to generate a chaotic behavior in the system (15), the instability measure ρ defined in Theorem 3 must be positive. By selecting $q = 0.93$, $a = -1$, and $w = 100$, the characteristic equation of the equilibrium points E_1 and E_4 is

$$
\lambda^{279} - 1.6180\lambda^{186} - 0.6180\lambda^{93} - 2.2360,
\tag{19}
$$

with unstable root $\lambda = 1.0090$, while the characteristic equation at the equilibria E_2 and E_3 is

$$
\lambda^{279} + 0.6180\lambda^{186} + 1.6180\lambda^{93} + 2.2360,
\tag{20}
$$

with unstable roots $\lambda_{1,2} = 1.0039 \pm 0.0153i$. Then, the instability measure of the system is $\rho = (\pi/2m) - 0.0152 > 0$. Therefore, the fractional-order system (15) satisfies the necessary condition for exhibiting a self-excited chaotic attractor when $q = 0.93$ and $a = -1$. □

Numerical simulation results in Figure 1 illustrate the existence of a chaotic attractor for the given fractional order. All numerical analyses presented herein were obtained by the Adams–Bashforth–Moulton predictor–corrector scheme of Section 2.1, with $h = 0.01$.

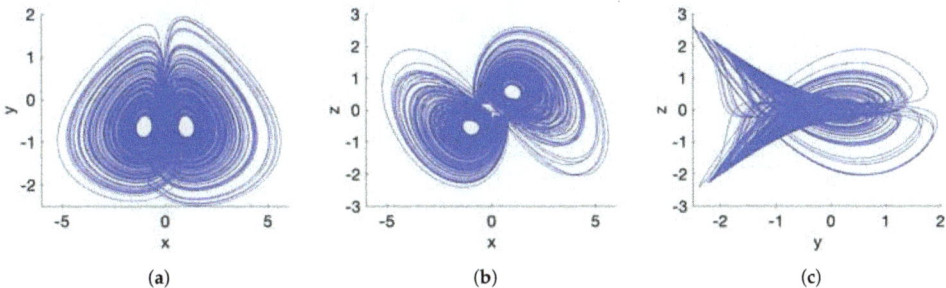

(a) (b) (c)

Figure 1. Self-excited attractor of the system (15) considering $a = -1$ and $q = 0.93$. (a) x–y plane; (b) x–z plane; (c) y–z plane.

To verify whether the system (15) is chaotic in the classical sense, its Lyapunov exponents are calculated. The Lyapunov exponents (LEs) are indicated by LE_1, LE_2, and LE_3 in Table 1. As is well known, a system is considered chaotic if $LE_1 > 0$, $LE_2 = 0$, $LE_3 < 0$ with $|LE_1| < |LE_3|$. Time series-based LEs calculation methods, like Wolf algorithm [60], Jacobian method [61], and neural network algorithm [62], are popular known ways of calculating Lyapunov exponents for integer and fractional-order systems. The Wolf algorithm [60] is used herein to calculate the LEs.

Table 1. Equilibria, eigenvalues, and Lyapunov exponents of the fractional-order chaotic system (15).

New System	Parameters	FO	Equilibria	Eigenvalues	x_0, y_0, z_0	LEs
Self-excited	$a = -1$;	$q = 0.93$	$(1, 1.6180, -1.6180)$ $(-1, -0.6180, -0.6180)$ $(1, -0.6180, 0.6180)$ $(-1, 1.6180, 1.6180)$	$2.3064, -0.3442 \pm 0.9225i$ $-1.0666, 0.2243 \pm 1.4304i$ $-1.0666, 0.2243 \pm 1.4304i$ $2.3064, -0.3442 \pm 0.9225i$	$(1, 1, 1)$	$LE_1 = 2.957$ $LE_2 = 0.01$ $LE_3 = -5.765$
Non-hyperbolic	$a = 0.25$;	$q = 0.99$	$(1, \frac{1}{2}, -\frac{1}{2})$ $(-1, \frac{1}{2}, \frac{1}{2})$	$0, -0.3750 + 0.5994i$ $0, -0.3750 + 0.5994i$	$(1, 1, 1)$	$LE_1 = 1.27$ $LE_2 = 0.010$ $LE_3 = -1.72$
Hidden	$a = 0.35$;	$q = 0.97$	no-equilibria		$(1, 1, 1)$	$LE_1 = 14.735$ $LE_2 = 0.010$ $LE_3 = -18.350$
Coexistence Chaotic	$a = 0.35$;	$q = 0.996$	no-equilibria		$(1, 1, 1)$	$LE_1 = 11.066$ $LE_2 = 0.080$ $LE_3 = -13.161$
Coexistence Periodic	$a = 0.35$;	$q = 0.996$	no-equilibria		$(0, 75, -50)$	$LE_1 = 0$ $LE_2 = -3.695$ $LE_3 = -3.705$

3.2. Degenerate Case: Self-Excited Chaotic Attractor with Nonhyperbolic Equilibria

A nonhyperbolic equilibrium point has one or more eigenvalues with a zero real part. In three-dimensional systems, 11 combinations can be determined [63]. Among them, six have only real eigenvalues, five present eigenvalues with a complex conjugate pair and one real part, and only two do not have nonzero real eigenvalues. Therefore, the stability of systems with nonhyperbolic equilibria cannot be obtained from their eigenvalues, because there is not an eigenvalue with a positive real part. Such systems can have neither homoclinic nor heteroclinic orbits, and thus the Shilnikov method cannot be used to verify the chaos [64]. Very few examples of fractional-order systems with nonhyperbolic equilibria have been previously reported.

As given in Table 1, the proposed fractional-order system (15) has two nonhyperbolic equilibrium points when parameter $a = 1/4$. The equilibria have a zero real eigenvalue and two complex conjugate eigenvalues with a negative real part. Therefore, the resulting self-excited attractor is of a nonhyperbolic type of chaos. Figure 2 shows the phase portraits.

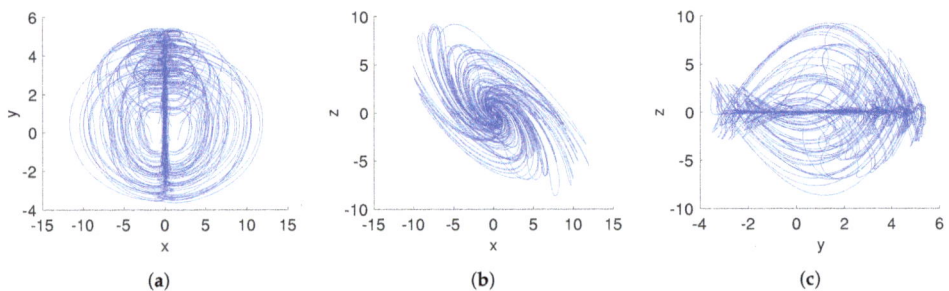

Figure 2. Chaotic attractor of the system (15) with nonhyperbolic equilibrium points, $a = 0.25$ and $q = 0.99$. (a) x–y plane; (b) x–z plane; (c) y–z plane.

Figure 3a shows the Lyapunov exponents spectrum when the fractional-order system (15) is nonhyperbolic. The positive Lyapunov exponent indicates a chaotic behavior. Additionally, the dynamical behavior of the system (15) can also be illustrated by the bifurcation diagram in Figure 3b. Due to system (15) having only one parameter, which must be $a = 1/4$ to present nonhyperbolic equilibrium points, it is interesting to analyze its dynamical behavior when a is fixed and the fractional-order q is varied. The bifurcation diagram in Figure 3b demonstrates a period-doubling route to chaos.

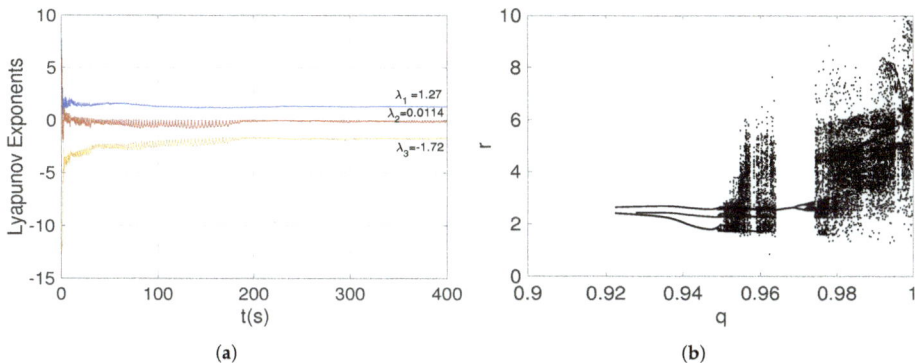

(a)   (b)

Figure 3. (**a**) Lyapunov exponents spectrum, and (**b**) bifurcation diagram of the fractional-order nonhyperbolic system (15) when $a = 1/4$.

3.3. Hidden Chaotic Attractor Localization in the Fractional-Order System without Equilibria

Most familiar examples of low-dimensional chaotic flows occur in systems having one or more saddle points. However, further studies showed that the self-excited periodic and chaotic oscillations did not give exhaustive information about the possible types of oscillations, i.e., "hidden oscillations" and "hidden attractors". So, this class of attractors should be introduced according to the following definition:

Definition 2. *An attractor is called a self-excited attractor if its basin of attraction intersects with any open neighborhood of an equilibrium, otherwise it is called a hidden attractor [1,6].*

With equilibrium, we are stating the equilibrium points of the state variables. Definition 2 also includes fractional-order dynamical systems with no-equilibria, line and surfaces of equilibria, and stable equilibria [35–42].

Similar to aforementioned scenarios, the parameter a is a controller of the dynamical behavior of the proposed system (15). In this case, if $a > 1/4$, a fractional-order system without equilibrium points is obtained. Hence, the resulting attractor is hidden using Definition 2. By selecting $a = 0.35$, and the fractional-order $q = 0.97$, the proposed system (15) generates the hidden chaotic attractor shown in Figure 4. It is important to note that the shape of the chaotic attractor in the x–z plane is similar to a hurricane. Moreover, the chaos generation is demonstrated by the Lyapunov exponents spectrum given in Figure 5a. As stated in Table 1, the largest Lyapunov exponent LE_1 is positive, and $|LE_1| < |LE_3|$, indicating a chaotic behavior.

By using the fractional-order q as bifurcation parameter, the bifurcation diagram of system (15) when it generates a hidden attractor ($a > 1/4$) is illustrated in Figure 5b. As can be seen from the bifurcation diagram, there are three regions where the chaotic behavior emerged, i.e., for $0.9285 < q < 0.931$, $0.962 < q < 0.973$, and $q > 0.9955$, a hidden attractor can be observed. This result indicates that the hidden chaotic attractor depends on the selected fractional order.

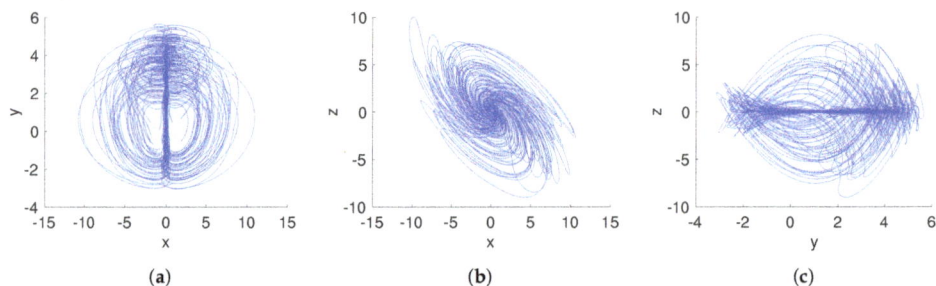

Figure 4. Hidden attractor of the system (15) considering $a = 0.35$, $q = 0.97$, and initial conditions $(x(0), y(0), z(0)) = (1, 1, 1)$. (**a**) x–y plane; (**b**) x–z plane; (**c**) y–z plane.

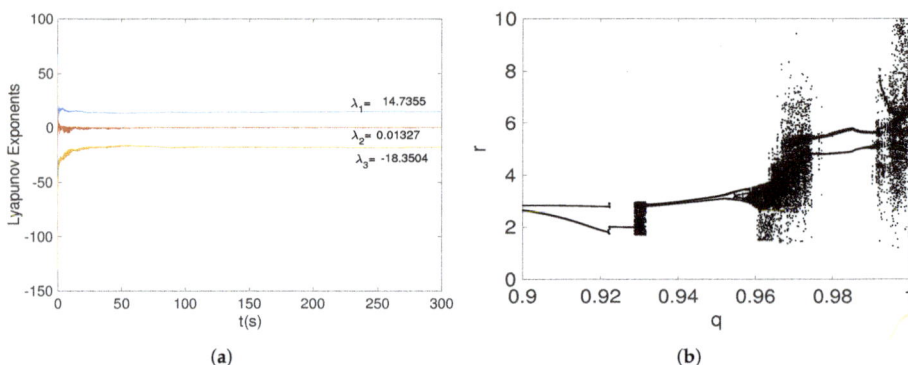

Figure 5. (**a**) Lyapunov exponents spectrum, and (**b**) bifurcation diagram of the fractional-order no-equilibrium system (15), when $a > 1/4$.

3.4. Coexistence of Hidden Attractors Regimes in the Fractional-Order System without Equilibria

The coexistence of attractors means that two or more different attractors are generated in a dynamical system from different initial conditions, which is an important and interesting nonlinear phenomenon [18,65]. In this subsection, we focus on studying the coexisting hidden attractors of the fractional-order no-equilibrium system (15). A necessary tool for analyzing the coexistence of attractors is the basin of attraction. All attractors, whether they be stable equilibria, limit cycles, attracting tori, or hidden strange attractors, are surrounded by a basin of attraction representing the set of initial conditions in the state space whose orbits approach and map out the attractor as time approaches infinity [66].

Figure 6 shows the basins of attraction of the system (15) for the cross-section in the y–z plane at $x = 0$ with $a = 0.35$ and $q = 0.996$. We found that the initial conditions inside of the yellow region converge to a hidden chaotic attractor, as shown in Figure 7, whereas the initial conditions belonging to the blue region lead to a hidden periodic attractor, as shown in Figure 8. This result confirms that there are two different hidden attractors coexisting in the proposed fractional-order chaotic system (15). Both coexisting attractors are also shown in Figure 7. Besides, this behavior also indicates multistability, because different initial conditions converge to different hidden attractors.

Table 1 gives the Lyapunov exponents spectrum for both hidden chaotic and periodic attractors, respectively. The positive, zero, and negative Lyapunov exponents of the hidden chaotic attractor indicate chaotic behavior, while a zero and two negative Lyapunov exponents point out a hidden periodic attractor.

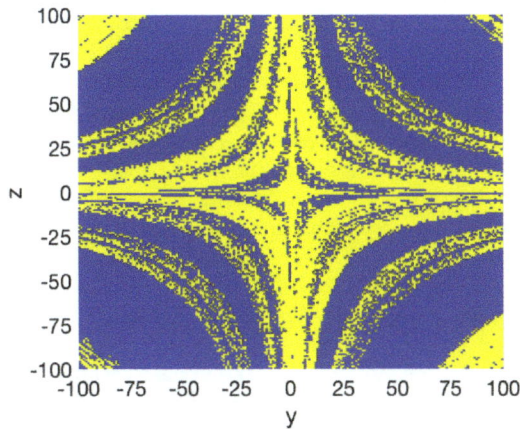

Figure 6. Cross-section of the basins of attraction of the two coexisting attractors in the y–z plane at $x = 0$ for the fractional-order chaotic system without equilibrium (15) when $a = 0.35$ and $q = 0.996$.

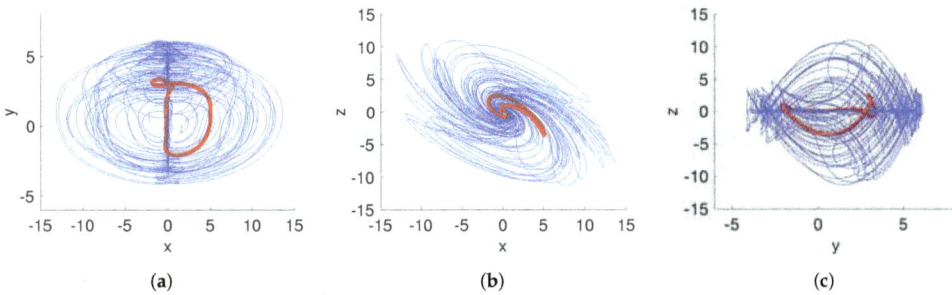

(a)   **(b)**   **(c)**

Figure 7. Coexistence of hidden chaotic and periodic attractors of the system (15) considering $a = 0.35$ and $q = 0.996$. (a) x–y plane; (b) x–z plane; (c) y–z plane.

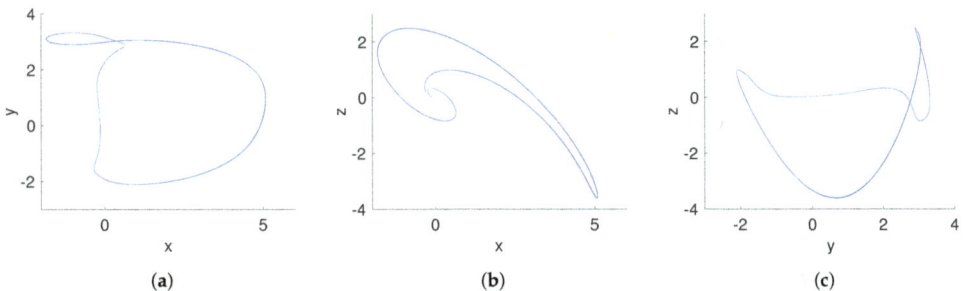

(a)   **(b)**   **(c)**

Figure 8. Hidden periodic attractor of the fractional-order system (15) with $a = 0.35$, $q = 0.996$, and initial conditions $(x(0), y(0), z(0)) = (0, 75, -50)$. (a) x–y plane; (b) x–z plane; (c) y–z plane.

3.5. Mechanism of the Different Dynamics

The mechanism of generating several types of equilibria in the proposed fractional-order system (15) is simple and intuitive. The basic idea consists of varying the single system parameter a in a range from negative to positive values, similar to the bifurcation analysis for integer-order systems.

By analyzing the symbolic equation of the equilibrium points (17), we realized that the number and stability of equilibria can be changed with the parameter a. One can easily see that system (15) has four unstable equilibrium points (spiral saddle index 1 and index 2) when $a < 1/4$. As a result, the fractional-order system (15) can be defined into a class of fractional-order chaotic systems with hyperbolic equilibrium points, which is the most typical form obtained for a chaotic attractor.

Next, with $a = 1/4$, the fractional-order system (15) degenerates, in the sense that their Jacobian eigenvalues at the equilibria consist of one zero eigenvalue and a complex conjugate pair with a negative real part. Clearly, the corresponding two equilibria are nonhyperbolic. Hence, the system (15) belongs to a subclass of fractional-order chaotic systems with nonhyperbolic equilibrium points.

Finally, the fractional-order system (15) has no-equilibrium points when $a > 1/4$. In this scenario, the resulting system can be categorized into a class of fractional-order no-equilibrium chaotic systems. It is interesting that if there are no-equilibrium points, the system (15) also presents multistability, since two distinct attractors are observed for different initial conditions. It is straightforward to observe that we added the simple constant control parameter a to the fractional-order chaotic system (15), trying to change the stability of its equilibria while preserving its chaotic dynamics. With the aim to analyze the relationship between the parameter a and the fractional-order q, we introduce the bi-dimensional map, that it is essentially a bifurcation diagram of two parameters, shown in Figure 9.

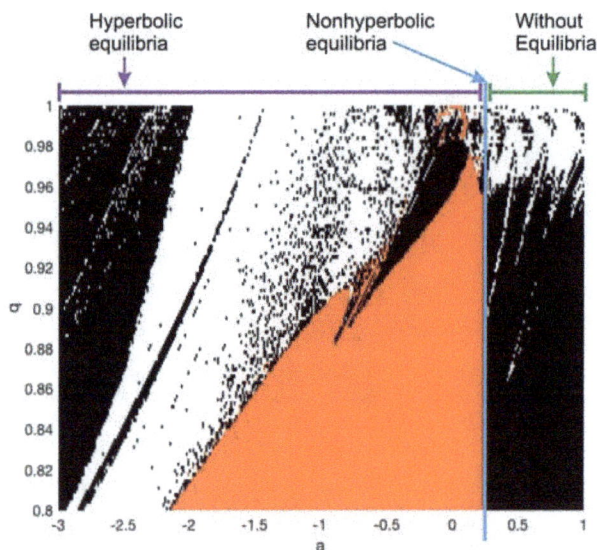

Figure 9. Bi-dimensional map for the different dynamical behaviors of the fractional-order system (15) as a function of the parameter a and order q. The white region leads to a chaotic attractor, the black region evolves to periodic attractors, and the orange region converges to unbounded orbits. Self-excited, nonhyperbolic, and hidden chaotic attractors for $a < 1/4$, $a = 1/4$, and $a > 1/4$, respectively.

This map indicates the type of equilibrium and the resulting dynamical behavior for a given a and q. The white, black, and orange regions evolve in chaotic, periodic, and unbounded behavior, respectively. From Figure 9, the minimal fractional order can be also determined. For instance, when $a = -1$, we observe that the chaotic attractor can appear for $q > 0.9010$, as was demonstrated in Section 3.1. However, unbounded trajectories are obtained if $q \leq 0.9010$, but a chaotic behavior can be detected for fractional orders as low as $q = 0.8$ when $a = -2.5$. Similarly, the chaotic attractor from nonhyperbolic equilibria is found for $q > 0.997$. For the case of a no-equilibria system ($a > 1/4$), we observed that the minimal fractional order wherein hidden chaotic attractors can emerge is about $q = 0.955$. For lower orders, the system only generates hidden periodic attractors.

To the best knowledge of the authors, this is the first time reporting a fractional-order chaotic system without equilibrium points and with the coexistence of hidden attractors.

4. Test 0–1 for Chaos

Gottwald and Melbourne [67] proposed a reliable and effective binary test method for testing whether a nonlinear system has chaotic behavior, which is called the "0–1 test". The test consists of creating a random dynamic process for the data and then studying how the scale of the stochastic process changes with time [67–69]. This test has been widely adopted as a suitable tool to confirm the chaotic behavior in fractional-order dynamical systems [26,33,40] because it is binary (minimizing issues of distinguishing small positive numbers from zero); the nature of the vector field, as well as its dimensionality, does not pose practical limitations; and it does not suffer from the difficulties associated with phase space reconstruction.

In this manner, the "0–1 test" is applied directly to the time series data of the fractional-order system (15). Since the test does not require phase space reconstruction, the dimension and origin of the system (15) are irrelevant. Let us consider a set of discrete data $\phi(n)$ with $n = 1, 2, \ldots, N$, representing a one-dimensional observable dataset obtained from the underlying dynamics of the system (15). For $c \in (0, \pi)$, we compute the translation variables $p_1(n) = \sum_{j=1}^{n} \phi(j) \cos(jc)$, and $p_2(n) = \sum_{j=1}^{n} \phi(j) \sin(jc)$. Next, the diffusive or non-diffusive behavior of p_1 and p_2 is obtained by the mean square displacement $M(n) = \lim_{N \to \infty} \frac{1}{N} \sum_{j=1}^{N} \left([p_1(j+n) - p_1(j)]^2 + [p_2(j+n) - p_2(j)]^2 \right)$, for $n \ll N$. Finally, the asymptotic growth rate K of $M(n)$ is given by

$$K = \lim_{n \to \infty} \frac{\log M(n)}{\log n}. \tag{21}$$

When $M(n)$ is bounded, the dynamics of the system (15) evolves in a periodic or quasi-periodic behavior. On the other hand, a chaotic behavior is detected if $M(n)$ grows linearly, similar to a Brownian motion. Moreover, a quantitative measure of the dynamics of the system (15) is given by K. For K close to 1, a chaotic behavior is observed, whereas for K close to 0, a regular behavior is obtained.

Detecting Chaos in the Proposed Fractional-Order System

In order to determine the chaotic and regular behaviors in the fractional-order system (15), we apply the "0–1 test" to the time series data obtained from the different scenarios in Section 3. The time series data were obtained by the ABM scheme with a time-step size $h = 0.01$.

Case 1: Self-excited attractor: When $q = 0.93$ and $a = -1$, the translation components (p_1, q_1) are as shown in Figure 10a. The unbounded behavior points out that the dynamics of the system (15) with unstable equilibria is chaotic. Also, the asymptotic growth rate K approaches one, with a value $K = 0.9988$, indicating the presence of chaotic dynamics. This result agrees with the self-excited chaotic attractor shown in Figure 1.

Case 2: Hidden chaotic attractor: When $q = 0.97$ and $a = 0.35$, a hidden chaotic attractor is localized, as shown by the phase portraits in Figure 4. In this case, the asymptotic growth rate of the time series of the system (15) with no-equilibrium is $K = 0.9985$. Additionally, the translation components (p_1, p_2) are shown in Figure 10b. The Brownian-like motion indicates chaotic behavior.

Case 3: Coexistence of hidden attractors: When $q = 0.996$, $a = 0.35$, and initial conditions $[1, 1, 1]^T$, we localize a hidden chaotic attractor, as shown in Figure 7. By applying the "0–1 test", $K = 0.9975$. Besides, the translation components (p_1, p_2), shown in Figure 11a, behave as Brownian-like motion. When the initial conditions are chosen as $[0, 75, -50]^T$, and the parameters a, q maintain the same value, the translation components (p_1, p_2) are now bounded, as shown in Figure 11b. Besides, the asymptotic growth rate is $K = 0.0364$. Therefore, the hidden attractor is periodic.

From Case 1 to Case 3, the "0–1 test" proved that three different dynamics can arise in the fractional-order system (15), i.e., a self-excited chaotic attractor, a hidden chaotic attractor, and the coexistence of hidden attractors.

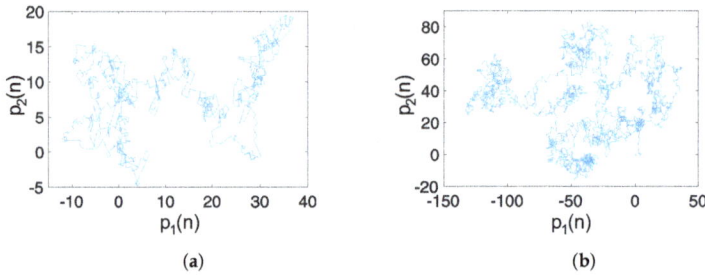

Figure 10. Dynamics of the translation components (p_1, p_2) of the fractional-order system (15): (a) Self-excited chaotic attractor ($q = 0.93$, $a = -1$) with an asymptotic growth rate $K = 0.9988$; (b) hidden chaotic attractor ($q = 0.97$, $a = 0.35$), with $K = 0.9985$.

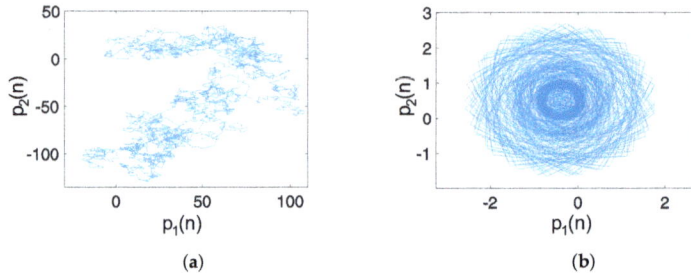

Figure 11. Dynamics of the translation components (p_1, p_2) of the fractional-order system (15): (a) Coexisting hidden chaotic attractor ($q = 0.996$, $a = 0.35$, $(x, y, z) = (1, 1, 1)$) with an asymptotic growth rate $K = 0.9975$; (b) coexisting hidden periodic attractor ($q = 0.996$, $a = 0.35$, $(x, y, z) = (0, 75, -50)$ with $K = 0.0364$.

5. Spectral Entropy Analysis

Complexity measures are an important way to characterize the complex behavior of a chaotic system. In information security, the complexity can reflect the security of a system [32]. Currently, there are several methods to measure the complexity of a time series [70]. In this sense, the complexity of chaotic sequences can be divided into behavior complexity and structural complexity. The former measures the size of the probability of a new pattern for a short-time window, while the latter is used to measure the complexity of a sequence by its frequency characteristic and energy spectrum in the transformation domain. Compared with the behavior complexity, the structural complexity has a global statistical significance, because it focuses on analyzing the energy characteristic based on all but the local sequence [70]. At present, the algorithms to evaluate structural complexity include spectral entropy (SE) and C_0 entropy.

Herein, we choose the spectral entropy algorithm to calculate the corresponding Shannon entropy value based on the Fourier transformation of the time series of the fractional-order system (15). By removing the direct-current, the steps are as follows. Given the time series $\{x^N(n), n = 0, 1, 2, \ldots, N - 1\}$ of the system (15) with length N, let $x(n) = x(n) - \bar{x}$, where $\bar{x}$ is the mean value of the time series, $\bar{x} = \frac{1}{N} \sum_{n=0}^{N-1} x(n)$. After that, the discrete Fourier transform (DFT) for the sequence $x(n)$ is computed with $X(k) = \sum_{n=0}^{N-1} x(n) e^{-j2\pi nk/N} = \sum_{n=0}^{N-1} x(n) W_N^{nk}$, where $k = 0, 1, 2, \ldots, N - 1$. Next, the relative power spectrum is derived with $P_k = \frac{|X(k)|^2}{\sum_{k=0}^{N/2-1} |X(k)|^2}$. By using $x(n)$, $X(k)$, and P_k, the spectral entropy of the time series of the system (15) for the scenarios in Section 3 can be determined by

$$SE = \frac{\sum_{k=0}^{N/2-1} |P_k \ln P_k|}{\ln(N/2)}, \tag{22}$$

where $\ln(N/2)$ is the entropy of a completely random signal.

5.1. Structural Complexity of the New Fractional-Order Chaotic System

The structural complexity of the self-excited and hidden attractors generated by the fractional-order system (15) is analyzed by Equation (22). The SE is computed from the time series $x(n)$ of the system (15) with length $N = 4.5 \times 10^4$. Figure 12a shows the SE for the case of the self-excited attractor, whereas Figure 12b displays the SE of the hidden attractor. The complexity of the self-excited attractor is almost constant in the interval $q \in [0.9, 1]$. On the other hand, the SE of the hidden attractor, as a function of fractional order, presents regions where the complexity is close to $SE = 0.6$, but other regions have a low SE. Therefore, we must be aware of the selected fractional order in the hidden attractor in order to have a relatively high structural complexity.

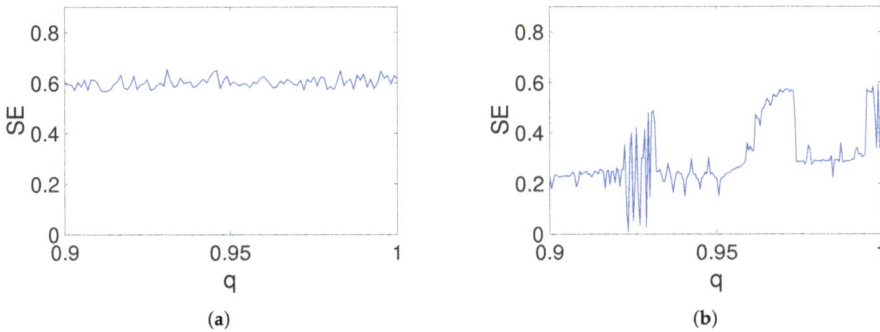

(a)

(b)

Figure 12. Spectral entropy versus fractional-order q for the system (15): (**a**) Structural complexity of the self-excited attractor in Figure 1 ($a = -1$); (**b**) structural complexity of the hidden chaotic attractor in Figure 4 ($a = 0.35$).

5.2. Design of a PRNG Using Hidden Attractors

By considering the results of the structural complexity, we select the hidden attractor of the system (15) to design a pseudo-random number generator (PRNG). More specifically, the chaotic signals obtained from the system (15) with $a = 0.35$ and $q = 0.97$ are used to generate a bitstream using the approach in [33,71]. In this manner, the chaotic signal $x(t)$ of the system (15) is sampled randomly to get samples γ_i with a suitable sample space. A ceil function is required to convert the real value into an integer value. Next, from each sampled value, we obtain $g_e(o)$ of 4-bit resolution composed of the four least-significant bits. As a post-processing operation, the output bits $g(o)$ are obtained XORing two consecutive $g_e(o)$.

The performance of the PRNG designed with the hidden dynamics is characterized by using the NIST SP 800-22 battery of statistical tests [72]. By selecting a confidence level $\alpha = 0.01$, the p-values are determined for sequences of 1 Mbit. As is well known, a p-value ≥ 0.01 means that the sequence is considered to be random with a confidence of 99%. Table 2 summarizes the results. As can be seen, the resulting PRNG using the hidden attractor of the system (15) satisfies all statistical tests.

Table 2. Results of NIST statistical tests for the bit sequences based on the system (15) when it presents a hidden chaotic attractor.

Statistical Test	*p*-Value	Results
Frequency	0.654721	success
Block Frequency	0.420199	success
Cusum-Forward	0.600222	success
Cusum-Reverse	0.446686	success
Runs	0.220773	success
Long Runs of Ones	0.012522	success
Rank	0.254592	success
Spectral DFT	0.538167	success
Non-Overlapping Templates	0.615839	success
Overlapping Templates	0.102065	success
Universal	0.830304	success
Approximate Entropy	0.635119	success
Random Excursions	0.407574	success
Random Excursions Variant	0.444982	success
Linear Complexity	0.634990	success
Serial	0.301388	success

6. Conclusions

In this paper, a fractional-order dynamical system with different families of hidden and self-excited attractors is introduced. As a function of only one parameter, the fractional-order system can be defined without equilibrium points, with nonhyperbolic equilibria, and with hyperbolic equilibria. A hidden chaotic attractor was identified in the proposed fractional-order system when it has no-equilibrium points. Additionally, it was found that two different attractors coexist for a determined fractional order, indicating multistability. Not only hidden dynamics were generated by the new system, but also two distinct self-excited chaotic attractors were obtained. Lyapunov exponents and the Brownian-like motion approach demonstrated the chaotic behavior of the system for each scenario. Finally, the structural complexity of the hidden and self-excited dynamics were evaluated using the spectral entropy. As an application, a PRNG with a suitable performance was designed with the time series of the hidden chaotic attractor.

As consequence, a contribution to this new phenomenon and little-explored area is the description of a fractional-order chaotic no-equilibrium system, along with the coexistence of hidden attractors. Such nonlinear systems without equilibrium and with multistability are appropriate for practical applications. Moreover, the fractional order is an extra parameter that permits the study of dynamical behaviors with more accuracy.

Author Contributions: Conceptualization, J.M.M.-P.; Formal analysis, J.M.M.-P., E.Z.-S. and C.V.; Funding acquisition, J.M.M.-P.; Methodology, J.M.M.-P.; Software, E.Z.-S., C.V., S.J. and K.R.; Validation, C.V., S.J. and J.K.; Visualization, J.K. and K.R.; Writing—original draft, J.M.M.-P., E.Z.-S., C.V. and S.J.

Funding: This research was funded by Consejo Nacional de Ciencia y Tecnología (CONACyT/MEXICO) grant number 258880.

Acknowledgments: J.M. Muñoz-Pacheco thanks CONACyT/MEXICO for the financial support to through Project No. 258880 (Proyecto Apoyado por el Fondo Sectorial de Investigación para la Educación). E. Zambrano-Serrano acknowledges SEP-PRODEP/MEXICO (Grant No. 511-6/18-1242) for the financial support to complete a postdoctoral visit to the research group "Sistemas Fotónicos y Nanoóptica" at BUAP.

Conflicts of Interest: The authors declare no conflict of interest.

References

1. Leonov, G.A.; Kuznetsov, N.V. Hidden attractors in dynamical systems. From hidden oscillations in Hilbert-Kolmogorov, Aizerman, and Kalman problems to hidden chaotic attractor in chua circuits. *Int. J. Bifurc. Chaos* **2013**, *23*, 1330002. [CrossRef]

2. Lorenz, E.N. Deterministic nonperiodic flow. *J. Atmos. Sci.* **1963**, *20*, 130–141. [CrossRef]
3. Chen, G.R.; Ueta, T. Yet another chaotic attractor. *Int. J. Bifurc. Chaos* **1999**, *9*, 1465–1466. [CrossRef]
4. Rössler, O. An equation for continuous chaos. *Phys. Lett. A* **1976**, *57*, 397–398. [CrossRef]
5. Lü, J.; Chen, G.R. Generating multiscroll chaotic attractors: Theories, methods and applications. *Int. J. Bifurc. Chaos* **2006**, *16*, 775–858. [CrossRef]
6. Leonov, G.A.; Kuznetsov, N.V.; Vagaitsev, V.I. Localization of hidden Chua's attractors. *Phys. Lett. A* **2011**, *375*, 2230–2233. [CrossRef]
7. Jafari, S.; Sprott, J.C. Simple chaotic flows with a line equilibrium. *Chaos Solitons Fractals* **2013**, *57*, 79–84. [CrossRef]
8. Dudkowski, D.; Jafari, S.; Kapitaniak, T.; Kuznetsov, N.V.; Leonov, G.A.; Prasad, A. Hidden attractors in dynamical systems. *Phys. Rep.* **2016**, *637*, 1–50. [CrossRef]
9. Leonov, G.A.; Kuznetsov, N.V.; Mokaev, T.N. Homoclinic orbits, and self-excited and hidden attractors in a Lorenz-like system describing convective fluid motion. *Eur. Phys. J. Spec. Top.* **2015**, *224*, 1421–1458. [CrossRef]
10. Andrievsky, B.R.; Kuznetsov, N.V.; Leonov, G.A.; Pogromsky, A.Y. Hidden oscillations in aircraft flight control system with input saturation. *IFAC Proc. Vol.* **2013**, *46*, 75–79. [CrossRef]
11. Leonov, G.A.; Kuznetsov, N.V.; Kiselev, M.A.; Solovyeva, E.P.; Zaretskiy, A.M. Hidden oscillations in mathematical model of drilling system actuated by induction motor with a wound rotor. *Nonlinear Dyn.* **2014**, *77*, 277–288. [CrossRef]
12. Yang, T. A Survey of Chaotic Secure Communication Systems. *Int. J. Comput. Cognit.* **2004**, *2*, 81–130.
13. Molaie, M.; Jafari, S.; Sprott, J.C.; Golpayegani, S.M.R.H. Simple chaotic flow with one stable equilibrium. *Int. J. Bifurc. Chaos* **2013**, *23*, 1350188. [CrossRef]
14. Kingni, S.; Jafari, S.; Pham, V.-T.; Wuafo, P. Constructing and analyzing of a unique three-dimensional chaotic autonomous system exhibiting three families of hidden attractors. *Math. Comput. Simul.* **2017**, *132*, 172–182. [CrossRef]
15. Lai, Q.; Akgul, A.; Li, C.; Xu, G.; Cavusoglu, Ü. A new chaotic system with multiple attractors: Dynamics analysis, circuit realization and s-box design. *Entropy* **2018**, *20*, 12. [CrossRef]
16. Jafari, S.; Sprott, J.C.; Golpayegani, S.M.R.H. Elementary quadratic chaotic flows with no equilibria. *Phys. Lett. A* **2013**, *377*, 699–702. [CrossRef]
17. Escalante-Gonzalez, R.J.; Campos-Canton, E.; Nicol, M. Generation of multi-scroll attractors without equilibria via piecewise linear systems. *Chaos* **2017**, *27*, 053109. [CrossRef] [PubMed]
18. Pham, V.-T.; Volos, C.; Jafari, S.; Kapitaniak, T. Coexistence of hidden chaotic attractors in a novel no-equilibrium system. *Nonlinear Dyn.* **2017**, *87*, 2001–2010. [CrossRef]
19. Messias, M.; Reinol, A.C. On the formation of hidden chaotic attractors and nested invariant tori in the Sprott A system. *Nonlinear Dyn.* **2017**, *88*, 807–821. [CrossRef]
20. Pham. V.-T.; Jafari, S.; Volos, C.; Gotthan, T.; Wang, X.; Hoang, D.V. A chaotic system with rounded square equilibrium and with no-equilibrium. *Optik* **2017**, *130*, 365–371.
21. Jafari, S.; Sprott, J.; Molaie, M. A simple chaotic flow with a plane of equilibria. *Int. J. Bifurc. Chaos* **2016**, *26*, 1650098. [CrossRef]
22. Jafari, M.A.; Mliki, E.; Akgul, A.; Pham, V.-T.; Kingni, S.T.; Wang, X.; Jafari, S. Chameleon: The most hidden chaotic flow. *Nonlinear Dyn.* **2017**, *88*, 2303–2317. [CrossRef]
23. Wang, C.; Ding, Q. A new two-dimensional map with hidden attractors. *Entropy* **2018**, *20*, 322. [CrossRef]
24. Gotthans, T.; Petrzela, J. New class of chaotic systems with circular equilibrium. *Nonlinear Dyn.* **2015**, *81*, 1143–1149. [CrossRef]
25. Sun, H.; Zhang, Y.; Baleanu, D.; Chen, W.; Chen, Y. A new collection of real world applications of fractional calculus in science and engineering. *Commun. Nonlinear Sci. Numer. Simul.* **2018**, *64*, 213–231. [CrossRef]
26. Tenreiro Machado, J.A.; Lopes, A.M. Complex and Fractional Dynamics. *Entropy* **2017**, *19*, 62. [CrossRef]
27. Chen, W.C. Nonlinear dynamics and chaos in a fractional-order financial system. *Chaos Soliton Fractals* **2008**, *36*, 1305–1314. [CrossRef]
28. Machado, J.; Mata, M.E.; Lopes, A.M. Fractional State Space Analysis of Economic Systems. *Entropy* **2015**, *17*, 5402–5421. [CrossRef]
29. Zambrano-Serrano, E.; Campos-Canton, E.; Munoz-Pacheco, J.M. Strange attractors generated by a fractional order switching system and its topological horseshoe. *Nonlinear Dyn.* **2016**, *83*, 1629–1641. [CrossRef]

30. Petras, I. *Fractional-Order Nonlinear Systems, Modeling, Analysis and Simulation*; Higher Education Press: Beijing, China; Springer: Berlin, Germany, 2011.
31. Alkahtani, B.S.T.; Atangana, A. Chaos on the Vallis Model for El Niño with Fractional Operators. *Entropy* **2016**, *18*, 100. [CrossRef]
32. He, S.; Sun, K.; Wang, H. Complexity Analysis and DSP Implementation of the Fractional-Order Lorenz Hyperchaotic System. *Entropy* **2015**, *17*, 8299–8311. [CrossRef]
33. Zambrano-Serrano, E.; Munoz-Pacheco, J.M.; Campos-Canton, E. Chaos generation in fractional-order switched systems and its digital implementation. *Int. J. Electron. Commun. (AEÜ)* **2017**, *79*, 43–52. [CrossRef]
34. Tacha, O.I.; Munoz-Pacheco, J.M.; Zambrano-Serrano, E.; Stouboulos, I.N.; Pham, V.-T. Determining the chaotic behavior in a fractional-order finance system with negative parameters. *Nonlinear Dyn.* **2018**, in press. [CrossRef]
35. Kingni, S.T.; Jafari, S.; Simo, H.; Woafo, P. Three-dimensional chaotic autonomous system with only one stable equilibrium: Analysis, circuit design, parameter estimation, control, synchronization and its fractional-order form. *Eur. Phys. J. Plus* **2014**, *129*, 76. [CrossRef]
36. Wang, X.; Ouannas, A.; Pham, V.-T.; Abdolmohammadi, H.R. A fractional-order form of a system with stable equilibria and its synchronization. *Adv. Differ. Equ.* **2018**, *20*, 1–13.
37. Pham, V.-T.; Ouannas, A.; Volos, C.; Kapitaniak, T. A simple fractional-order chaotic system without equilibrium and its synchronization. *Int. J. Electron. Commun. (AEÜ)* **2018**, *86*, 69–76. [CrossRef]
38. Pham, V.-T.; Kingni, S.T.; Volos, C.; Jafari, S.; Kapitaniak, T. A simple three-dimensional fractional-order chaotic system without equilibrium: Dynamics, circuitry implementation, chaos control and synchronization. *Int. J. Electron. Commun. (AEÜ)* **2017**, *78*, 220–227. [CrossRef]
39. Cafagna, D.; Grassi, G. Elegant chaos in fractional-order system without equilibria. *Math. Probl. Eng.* **2013**, *2013*, 380436. [CrossRef]
40. Cafagna, D.; Grassi, G. Chaos in a new fractional-order system without equilibrium points. *Commun. Nonlinear Sci. Numer. Simul.* **2014**, *19*, 2919–2927. [CrossRef]
41. Kingni, S.T.; Pham, V.T.; Jafari, S.; Kol, G.R.; Woafo, P. Three-dimensional chaotic autonomous system with a circular equilibrium: Analysis, circuit implementation and Its fractional-order form. *Circuits Syst. Signal Process* **2016**, *35*, 1933–1948. [CrossRef]
42. Kingni, S.T.; Pham, V.-T.; Jafari, S. A chaotic system with an infinite number of equilibrium points located on a line and on a hyperbola and its fractional-order form. *Chaos Soliton Fractals* **2017**, *99*, 209–218. [CrossRef]
43. Hoang, D.V.; Kingni, S.T.; Pham, V.T. A No-equilibrium hyperchaotic system and its fractional-order form. *Math. Probl. Eng.* **2017**, *2017*, 3927184. [CrossRef]
44. Volos, C.; Pham, V.T.; Zambrano-Serrano, E.; Munoz-Pacheco, J.M.; Vaidyanathan, S.; Tlelo-Cuautle, E. Analysis of a 4-D hyperchaotic fractional-order memristive system with hidden attractor. In *Advances in Memristors, Memristive Devices and Systems*; Vaidyanathan, S., Volos, C., Eds.; Springer: Cham, Switzerland, 2017; ISBN 978-3-31-951723-0.
45. Rajagopal, K.; Akgul, A.; Jafari, S.; Karthikeyan, A.; Koyuncu, I. Chaotic chameleon: Dynamic analyses, circuit implementation, FPGA design and fractional-order form with basic analyses. *Chaos Soliton Fractals* **2017**, *103*, 476–487. [CrossRef]
46. Sprott, J.C. *Elegant Chaos: Algebraically Simple Chaotic Flows*; World Scientific: Singapore, 2010.
47. Sprott, J.C. A proposed standard for the publication a new chaotic systems. *Int. J. Bifurc. Chaos* **2011**, *21*, 2391–2394. [CrossRef]
48. Caputo, M. Linear Models of Dissipation whose Q is almost Frequency Independent-II. *Geophys. J.* **1967**, *13*, 529–539. [CrossRef]
49. Deng, W.; Lü, J. Design of multidirectional multiscroll chaotic attractors based on fractional differential systems via switching control. *Chaos* **2006**, *16*, 043120. [CrossRef] [PubMed]
50. Diethelm, K.; Ford, N.J.; Freed, A.D. Detailed error analysis for a fractional Adams method. *Numer. Algorithms* **2004**, *36*, 31–52. [CrossRef]
51. Diethelm, K.; Ford, N.J. Analysis of fractional differential equations. *J. Math. Anal. Appl.* **2002**, *265*, 229–248. [CrossRef]
52. Garrapa, R. Numerical solution of fractional differential equations: A survey and a software tutorial. *Mathemathics* **2017**, *2*, 16. [CrossRef]

Entropy **2018**, *20*, 564

53. Odibat, Z.; Corson, N.; Aziz-Alaoui, M.A.; Alsaedi, A. Chaos in fractional order cubic Chua system and synchronization. *Int. J. Bifurc. Chaos* **2017**, *27*, 1750161. [CrossRef]

54. Ahmed, E.; El-Sayed, A.M.A.; El-Saka, H.A.A. Equilibrium points, stability and numerical solutions of fractional-order predator-prey and rabies models. *J. Math. Anal. Appl.* **2007**, *325*, 542–553. [CrossRef]

55. Tavazoei, M.S.; Haeri, M. Unreliability of frequency-domain approximation in recognising chaos in fractional-order systems. *IET Signal Proc.* **2007**, *1*, 171–181. [CrossRef]

56. Tavazoei, M.S.; Haeri, M. A necessary condition for double scroll attractor existence in fractional-order systems. *Phys. Lett. A* **2007**, *367*, 102–113. [CrossRef]

57. Danca, M. Hidden chaotic attractors in fractional-order systems. *Nonlinear Dyn.* **2017**, *89*, 577–586. [CrossRef]

58. Tavazoei, M.S.; Haeri, M. Chaotic attractors in incommensurate fractional order systems. *Phys. D* **2008**, *237*, 2628–2637. [CrossRef]

59. Munoz-Pacheco, J.M.; Zambrano-Serrano, E.; Volos, C.; Tacha, O.I.; Stouboulos, I.N.; Pham, V.-T. A fractional order chaotic system with a 3D grid of variable attractors. *Chaos Soliton Fractals* **2018**, *113*, 69–78. [CrossRef]

60. Wolf, A.; Swift, J.B.; Swinney, H.L.; Vastano, J.A. Determining Lyapunov exponents from a time series. *Physica D* **1985**, *16*, 285–317. [CrossRef]

61. Ellner, S.; Gallant, A.R.; McCaffrey, D.; Nychka, D. Convergence rates and data requirements for Jacobian-based estimates of Lyapunov exponents from data. *Phys. Lett. A* **1991**, *153*, 357–363. [CrossRef]

62. Maus, A.; Sprott, J.C. Evaluating lyapunov exponent spectra with neural networks. *Chaos Solitons Fractals* **2013**, *51*, 13–21. [CrossRef]

63. Wei, Z.; Sprott, J.C.; Chen, H. Elementary quadratic chaotic flows with a single non-hyperbolic equilibrium. *Phys. Lett. A* **2015**, *379*, 2184–2187. [CrossRef]

64. Sprott, J.C. Strange attractors with various equilibrium types. *Eur. Phys. J. Spec. Top.* **2015**, *224*, 1409–1419. [CrossRef]

65. Petras, I. Comments on "Coexistence of hidden chaotic attractors in a novel no-equilibrium system" (Nonlinear Dyn, doi:10.1007/s11071-016-3170-x). *Nonlinear Dyn.* **2017**, *90*, 749–754. [CrossRef] [PubMed]

66. Sprott, J.C.; Xiong, A. Classifying and quantifying basins of attraction. *Chaos* **2015**, *25*, 083101. [CrossRef] [PubMed]

67. Gottwald, G.A.; Melbourne, I. On the implementation of the 0–1 test for chaos. *SIAM J. Appl. Dyn. Syst.* **2009**, *8*, 129–145. [CrossRef]

68. Gottwald, G.A.; Melbourne, I. A new test for chaos in deterministic systems. *Proc. R. Soc. Lond. Ser. A Math. Phys. Sci.* **2004**, *460*, 603–611. [CrossRef]

69. Gottwald, G.A.; Melbourne, I. On the validity of the 0–1 test for chaos. *Nonlinearity* **2009**, *22*, 1367–1382. [CrossRef]

70. Sun, K. *Chaotic Secure Communication*; Walter de Gruyter GmbH: Berlin, Germany; Boston, MA, USA, 2016.

71. Wang, Q.; Yu, S.; Li, C.; Lü, J.; Fang, X.; Guyeux, C.; Bahi, J.M. Theoretical design and FPGA-based implementation of higher-dimensional digital chaotic systems. *IEEE Trans. Circuits Syst. I* **2016**, *63*, 401–412. [CrossRef]

72. Rukhin, A.; Soto, J.; Nechvatal, J.; Smid, M.; Barker, E. *A Statistical Test Suite for Random and Pseudorandom Number Generators for Cryptographic Applications*; Booz-Allen and Hamilton Inc.: Mclean, VA, USA, 2010.

entropy

MDPI

Article

A New Chaotic System with Stable Equilibrium: Entropy Analysis, Parameter Estimation, and Circuit Design

Tomasz Kapitaniak [1], S. Alireza Mohammadi [2], Saad Mekhilef [2] [iD], Fawaz E. Alsaadi [3], Tasawar Hayat [4,5] and Viet-Thanh Pham [6,*]

[1] Division of Dynamics, Lodz University of Technology, Stefanowskiego 1/15, 90-924 Lodz, Poland; tomasz.kapitaniak@p.lodz.pl
[2] Power Electronics and Renewable Energy Research Laboratory (PEARL), Department of Electrical Engineering, Faculty of Engineering, University of Malaya, Kuala Lumpur 50603, Malaysia; salirezam@siswa.um.edu.my (S.A.M.); saad@um.edu.my (S.M.)
[3] Department of Information Technology, Faculty of Computing and IT, King Abdulaziz University, Jeddah 21589, Saudi Arabia; fawazkau@gmail.com
[4] Department of Mathematics, Quaid-I-Azam University 45320, Islamabad 44000, Pakistan; fmgpak@gmail.com
[5] NAAM Research Group, King Abdulaziz University, Jeddah 21589, Saudi Arabia
[6] Modeling Evolutionary Algorithms Simulation and Artificial Intelligence, Faculty of Electrical & Electronics Engineering, Ton Duc Thang University, Ho Chi Minh City, Vietnam
* Correspondence: phamvietthanh@tdt.edu.vn

Received: 12 July 2018; Accepted: 2 August 2018; Published: 5 September 2018

Abstract: In this paper, we introduce a new, three-dimensional chaotic system with one stable equilibrium. This system is a multistable dynamic system in which the strange attractor is hidden. We investigate its dynamic properties through equilibrium analysis, a bifurcation diagram and Lyapunov exponents. Such multistable systems are important in engineering. We perform an entropy analysis, parameter estimation and circuit design using this new system to show its feasibility and ability to be used in engineering applications.

Keywords: chaotic flow; hidden attractor; multistable; entropy

1. Introduction

Chaotic systems are very important in nonlinear dynamics. Many researchers are investigating the reason for the existence of chaotic attractors. For many years, researchers thought that the existence of a saddle equilibrium [1,2] is a necessary condition for strange attractors. However, in recent years many chaotic systems with no saddle point have been proposed. For example, we note systems with chaotic attractors and without any equilibria [3,4], with stable equilibria [5,6], with a line of equilibria [7,8], with a curve of equilibria [9,10], with circle and square equilibria [11], with a circular equilibria [12], with ellipsoid equilibria [13] and with a plane or surface of equilibria [14–17].

Leonov and Kusnetsov have introduced a new topic in nonlinear dynamics that has been called hidden attractors [18–20]. Hidden attractors are attractors in which the basin of attraction does not intersect with any equilibrium point [21–23]. The opposite side of this definition is self-excited attractors. A self-excited attractor has a basin of attraction that is associated with at least one unstable equilibrium [24–26]. Many unusual chaotic systems that have been proposed recently are systems with hidden attractors [27]. Hidden attractors in fractional order systems are also studied in [28–30].

Multistability is another important phenomenon that can be observed in dynamic systems [31–33]. In multistable systems, the final state of the system is dependent on the initial conditions [34,35]. Chaotic systems with stable equilibria are examples of multistable systems [36,37].

The quantification of chaotic attractors is a challenging topic in nonlinear dynamics. There are many measures that are used in this area. The main such measure is the Lyapunov exponent [38]. Entropy is another measure that determines the unpredictability of complex dynamics [39]. Entropy can be helpful in short time series [40], while the Lyapunov exponent is not suitable for them.

In this paper, a new three-dimensional chaotic flow with one stable equilibrium is proposed. The chaotic attractor of this system is hidden since it cannot be found using the stable equilibrium point. The rest of the paper is organized as follows:

The new chaotic system is proposed in Section 2. Some of its dynamic properties are investigated in Section 3. Section 4 discuses the complexity of the system's attractors. Section 5 is devoted to the parameter estimation of the proposed system. The circuit implementation of the system is carried out in Section 6. Finally, the paper is concluded in Section 7.

2. System Description

In this paper, we are going investigate the dynamic properties of the following system:

$$\dot{x} = z$$
$$\dot{y} = -x - z \tag{1}$$
$$\dot{z} = 0.1x + 5y - z + xy - 0.3xz + a$$

where parameter $a = 1$. System (1) is a three-dimensional chaotic flow that can have a chaotic attractor. This system has been designed based on the method proposed in [41]. In the first step of investigating its dynamic properties, the equilibrium points of the system were calculated. By setting zero at the right hand side of this equation we obtain:

$$z = 0$$
$$x = 0 \tag{2}$$
$$y = -\tfrac{1}{5}$$

Thus, the system has one equilibrium point in $(0, -0.2, 0)$. A stability analysis of this equilibrium point can be carried out using the following Jacobian matrix at the equilibrium:

$$J = \begin{bmatrix} 0 & 0 & 1 \\ -1 & 0 & -1 \\ -0.1 & 5 & -1 \end{bmatrix} \tag{3}$$

By solving the equality $\det(\lambda I - J) = 0$, the characteristic equation of System (1) is determined as follows:

$$\lambda^3 + \lambda^2 + 5.1\lambda + 5 = 0 \tag{4}$$

Solving Equation (4), we find that System (1) has three eigenvalues ($\lambda_1 = -0.9835, \lambda_{2,3} = -0.0082 \pm 2.2547i$) for the equilibrium $(0, -0.2, 0)$. Thus, it is a stable equilibrium point. Every other possible attractor of this system coexists with this stable equilibrium point. The system shows a chaotic attractor if we choose initial conditions $(x_0, y_0, z_0) = (5.4, -1.8, 3.3)$. The chaotic attractor cannot be found using any equilibrium points of the system since the system has only one stable equilibrium point. Thus the strange attractor is hidden [27]. The time series of three states of System (1) for $a = 1$ are shown in Figure 1. Three projections of the chaotic attractor and its three-dimensional attractor are presented in Figure 2 and its Poincaré map is shown in Figure 3. In this plot, we use the peak values of *x* variable as the Poincaré map.

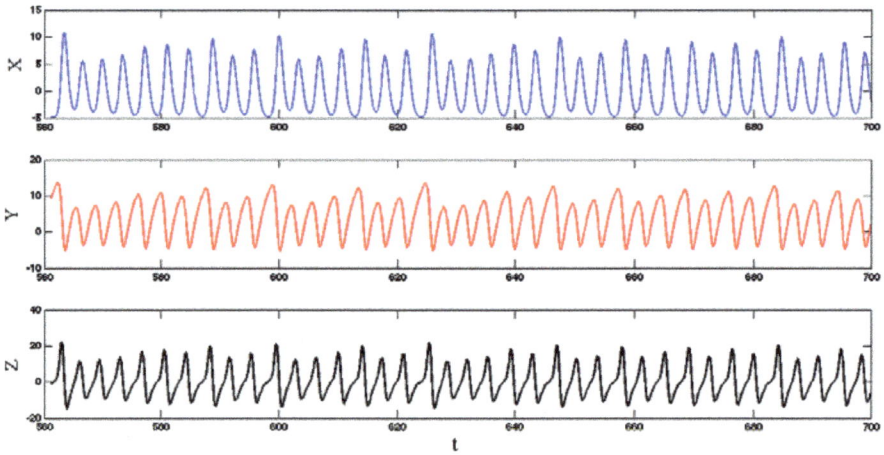

Figure 1. Time series of System (1) with parameter $a = 1$ and initial conditions $(x_0, y_0, z_0) = (5.4, -1.8, 3.3)$.

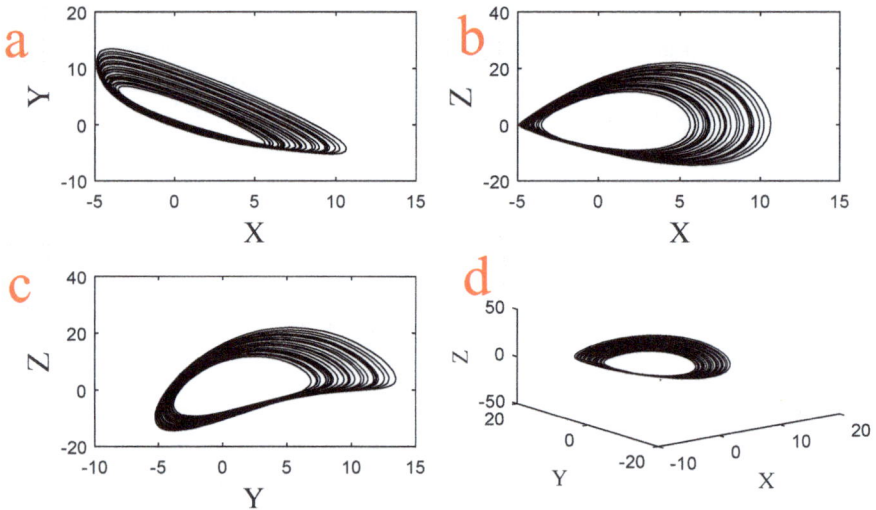

Figure 2. Three projections of the chaotic attractor of System (1) with parameter $a = 1$ and initial conditions $(x_0, y_0, z_0) = (5.4, -1.8, 3.3)$ in (**a**) *X-Y* plane. (**b**) *X-Z* plane. (**c**) *Y-Z* plane and (**d**) 3-D chaotic attractor.

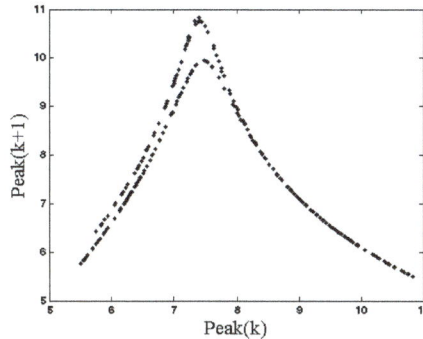

Figure 3. Poincaré map (peaks of x variable) of System (1) with parameter $a = 1$ and initial conditions $(x_0, y_0, z_0) = (5.4, -1.8, 3.3)$.

3. Bifurcation Analysis

In order to show the different dynamic behaviors of System (1), its bifurcation diagram was investigated. Figure 4a shows a bifurcation diagram of the system with respect to the changing parameter a. The system has an inverse route of period doubling after its chaotic behavior. The dynamic of the system also suddenly changes in $a = 1.46$ from a stable limit cycle to a stable equilibrium. In other words, System (1) has a chaotic attractor in parameter $a = 1$ and initial conditions $(x_0, y_0, z_0) = (5.4, -1.8, 3.3)$. The system also has a stable equilibrium point in $\left(0, -\frac{a}{5}, 0\right) = \left(0, -\frac{1}{5}, 0\right)$. Thus, there are some initial conditions in the vicinity of this stable equilibrium that are attracted to it. In the bifurcation diagram of Figure 4, we used the initial conditions $(x_0, y_0, z_0) = (5.4, -1.8, 3.3)$ for parameter $a = 1$ and applied the forward continuation method for the higher values of parameter a. In other words, in the higher values of parameter a, we used initial conditions from the end of trajectory in the previous parameter with forward changing. Thus, the trajectory of the system traps into one attractor that is chaotic in parameter $a = 1$ and bifurcates with an inverse route of period doubling to chaos. In parameter $a = 1.46$ the previous attractor becomes unstable and the system jumps from a stable limit cycle to the stable equilibrium point $\left(0, -\frac{a}{5}, 0\right)$. To be sure about the chaotic and other types of attractors of the system, it was necessary to calculate the Lyapunov exponents (Figure 4b). In the smaller values of the parameter a the system has a chaotic attractor (one positive, one zero, and one negative LE). It then shows the limit cycle, since its largest Lyapunov exponent is zero and the other two LEs are negative. After that, in the higher values of parameter a, the attractor changed to a stable equilibrium that has three negative Lyapunov exponents.

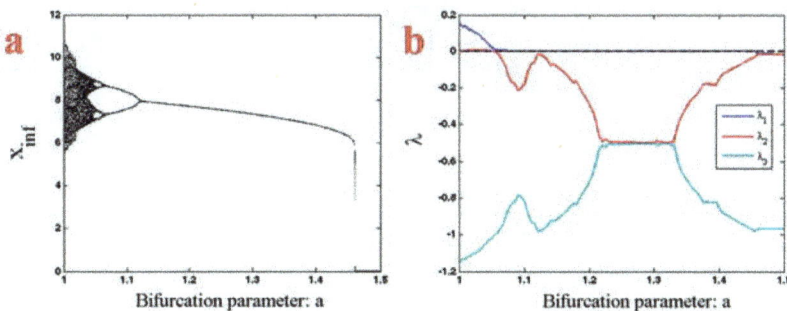

Figure 4. (**a**) Bifurcation diagram of System (1) with respect to the changing parameter a in the interval [1, 1.5] and forward continuation. (**b**) Lyapunov exponents of System (1) with respect to the changing parameter a in the interval [1, 1.5] and forward continuation.

4. Entropy Analysis

Entropy is a measure of unpredictability. Shannon has proposed a formulation for calculating entropy [42]. Since chaotic attractors have an infinite number of states, another type of entropy is needed to calculate their unpredictability. This entropy is called Kolmogorov–Sinai (H_{ks}) [40,43] and its formulation is shown in Equation (5).

$$H_{ks}(\beta[\varepsilon]) = \frac{1}{\tau_{min}(\beta[\varepsilon])} \sum_{\tau} \rho(\tau, \beta[\varepsilon]) log\left(\frac{1}{\rho(\tau, \beta[\varepsilon])}\right) \tag{5}$$

It is defined using the first Poincaré recurrence times (FPRs) denoted by τ_i. β is a D-dimensional box in the state space with side ε_1 where the FPRs are observed. $\rho(\tau, \beta)$ is the probability distribution of τ_i. For a smooth chaotic system H_{ks} is equal to the sum of all positive Lyapunov exponents [44,45]. The Kolmogorov–Sinai entropy of System (1) with respect to the changing parameter a is shown in Figure 5. Near a bifurcation point, the system's state becomes slower. In other words, the transient time increases near a bifurcation point [46,47]. In order to use Kolmogorov–Sinai entropy to anticipate a bifurcation point, we calculated it without removing the transient time of the trajectory. If we remove the transient time, then the estimated Kolmogorov–Sinai entropy became zero in regular dynamics and it changed through variations in the final state of the system. By applying Kolmogorov–Sinai entropy to the system's state without removing transient time, we were able to see complexity of transient parts as well as final state of the system. As Figure 5 shows, in small values of parameter a the system has a chaotic attractor and its unpredictability is high. By increasing parameter a, the system changes its dynamic to a regular dynamic and thus its entropy decreases. However, in the bifurcation points the system becomes slower and its transient time increases. That is the reason for the increasing entropy in the bifurcation points.

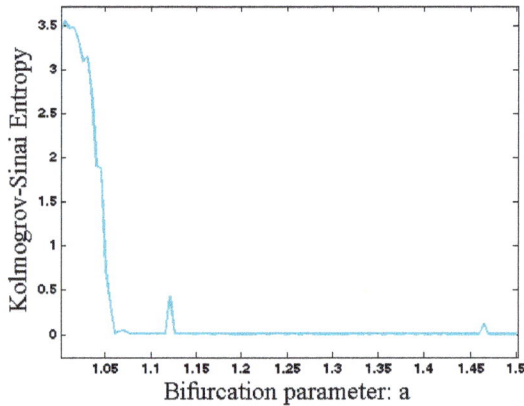

Figure 5. Kolmogorov–Sinai entropy of System (1) with respect to changing parameter a.

5. Parameter Estimation

There are various methods for parameter estimation in dynamic systems that are based on optimization methods [48–50]. The basis of these methods is a cost function associated with the differences between the time series obtained from a real system and the time series obtained from a known model with unknown parameters. However, these approaches are not appropriate for chaotic systems due to the butterfly effect [51–53]. Therefore an alternative method is proposed in [54,55]. This new method changes the analyzing domain of the chaotic system from time space to the state space. In the other words, this new model compares the topology and structure of the points in the state space. To this end, the algorithm searches the space of the parameter to find the most similar point from the model to the point of the data. Whenever the structure of the points in the state space gets close enough to the structure of the real data, the optimum parameter is found. For more complete details, see [55,56]. In this paper, we used this useful cost function along with WOA (whale optimization algorithm [57,58]) for the parameter estimation method. Figure 6 shows the result of the cost function with respect to changing the parameter a. As is shown in the figure, the global minimum is located exactly in the main parameter $a = 1$. Figure 7 shows the result of the cost function with respect to changing the parameters a and b (consider b as the coefficient of x in the third equation of Equation (1)). As is shown in the figure, the global minimum is located in the main parameters $a = 1$ & $b = 0.1$.

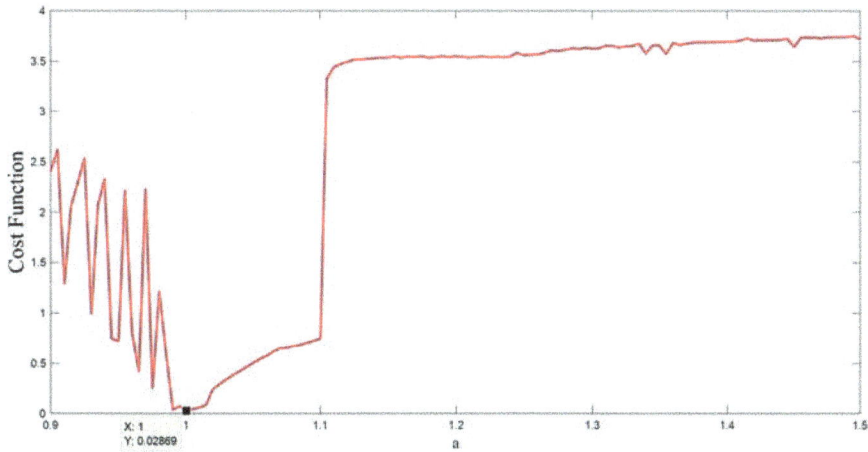

Figure 6. The value of the cost function with respect to changing the parameter a.

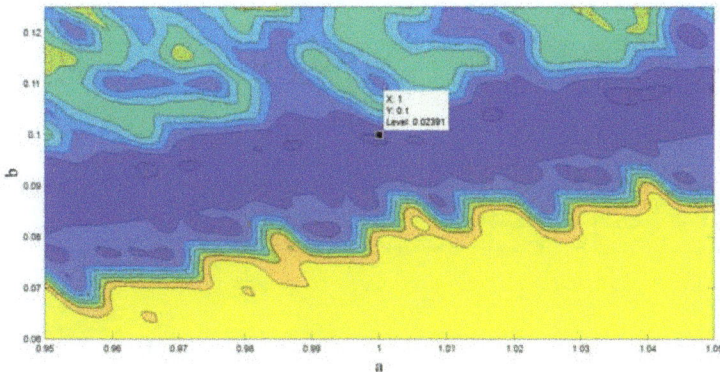

Figure 7. The value of the cost function with respect to changing the parameter a & b.

One of the most efficient categories of the optimization methods is meta-heuristic methods, which cover a wide range of problems, especially in engineering applications [50,59–61]. Most of them are inspired by nature. Humpback whales' hunting behavior in sea form the basis of the WOA (whale optimization algorithm) meta-heuristic algorithm [57,58]. The hunting behavior of Humpback whales, who encircle the recognized location of prey, has become the basis of the WOA algorithm. In this algorithm the target prey is the current best candidate or close to the optimum solution and the attacking strategy is a bubble-net strategy. By considering all these together, the WOA optimization method can be explained through three steps: Finding the prey, encircling the prey, and the bubble-net attacking behavior of humpback whales.

At first, the algorithm determines the best candidate solution. Then it updates the position of the other points in order to get closer to the best agent. The second step is about the attack strategy, which can be divided into two approaches. The first is a shrinking encircling mechanism and the other is a spiral updating position. For complete details see [57]. Figure 8 represents the result of the WOA for the 30 searching agents and 40 iterations.

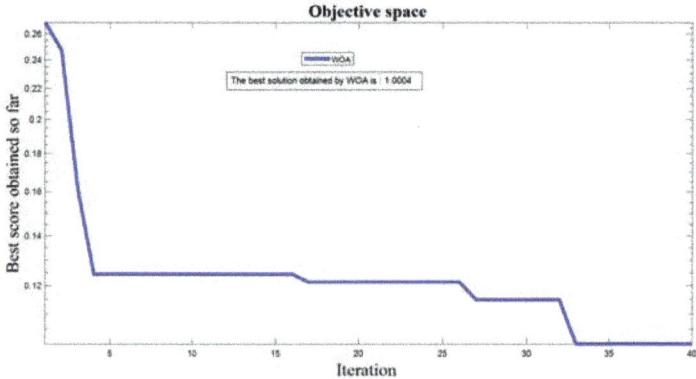

Figure 8. The result of WOA for the 30 searching agents and 40 iterations.

6. Circuit Design

This section presents a circuit implementation for the three-dimensional flow (1) (see Figure 9). The circuit implementation in Figure 9 was constructed using six operational amplifiers ($U_1 - U_6$) and electronic elements [62–66]. We used TL084 operational amplifiers and AD633 analog multipliers. Taking the voltages of three operational amplifiers (U_1, U_2, U_3) as X, Y, Z, it confirms that the circuit in Equation (6) corresponds to the flow (1):

$$\dot{X} = \frac{1}{R_1 C_1} Z$$
$$\dot{Y} = -\frac{1}{R_2 C_2} X - \frac{1}{R_3 C_2} Z$$
$$\dot{Z} = \frac{1}{R_4 C_3} X + \frac{1}{R_5 C_3} Y - \frac{1}{R_6 C_3} Z + \frac{1}{R_7 C_3 10V} XY - \frac{1}{R_8 C_3 10V} XZ - \frac{1}{R_9 C_3} V_a$$

(6)

where V_a is a DC voltage source.

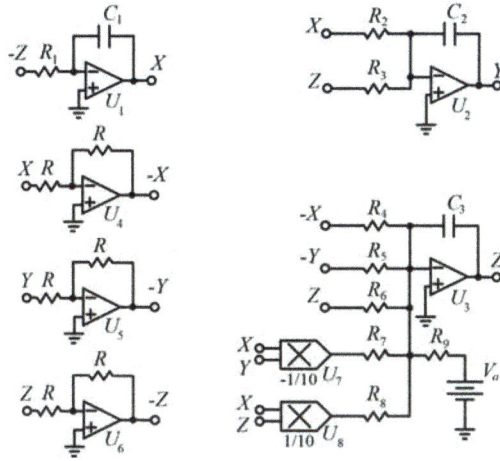

Figure 9. The circuit constructed by six operational amplifiers $(U_1 - U_6)$ and electronic elements.

The circuit generates chaos as illustrated in Figure 10 for the following set of components: $C_1 = C_2 = C_3 = 20$ nF, $R_1 = R_2 = R_3 = R_6 = R = 40$ kΩ, $R_4 = 400$ kΩ, $R_5 = 8$ kΩ, $R_7 = 1$ kΩ, $R_8 = 3.333$ kΩ, $R_9 = 160$ kΩ, and $V_a = -1V_{DC}$.

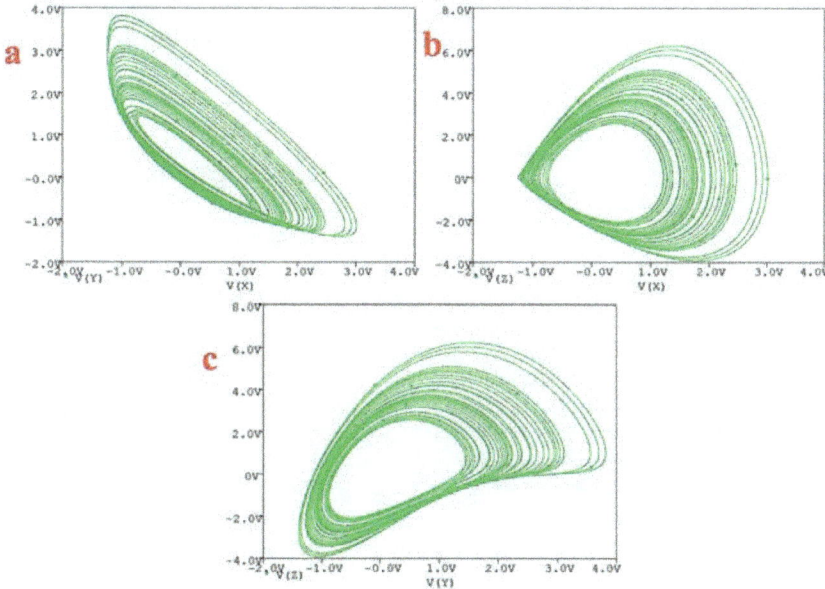

Figure 10. Generated attractors in PSpice of the circuit: (**a**) *X-Y* plane, (**b**) *X-Z* plane, (**c**) *Y-Z* plane.

7. Conclusions

A new three-dimensional chaotic system with one stable equilibrium was proposed in this paper. A bifurcation analysis of the system showed an inverse period doubling route to chaos with respect to increasing parameter *a*. The unpredictability of its dynamic was discussed using Kolmogorov–Sinai entropy. Parameter estimation of the system was carried out and circuit implementation of the system confirmed its feasibility. It is noted that the real practical realization and real laboratory measurements of the circuit should be carried out. Therefore, practical results will be reported in our next works.

Author Contributions: Data curation, S.A.M., S.M., F.E.A. and T.H.; Formal analysis, F.E.A., T.H. and V.-T.P.; Investigation, T.K., S.A.M. and S.M.; Supervision, T.K.; Validation, T.K., S.A.M., S.M., F.E.A., T.H. and V.-T.P.; Writing—original draft, T.H. and V.-T.P.; Writing—review & editing, T.K., S.A.M., S.M., F.E.A., T.H. and V.-T.P.

Funding: TK was supported by the Polish National Science Centre, MAESTRO Programme-Project No. 2013/327 08/A/ST8/00/780.

Conflicts of Interest: The authors declare no conflict of interest.

References

1. Lorenz, E.N. Deterministic nonperiodic flow. *J. Atmos. Sci.* **1963**, *20*, 130–141. [CrossRef]
2. Chen, G.; Ueta, T. Yet another chaotic attractor. *Int. J. Bifurc. Chaos* **1999**, *9*, 1465–1466. [CrossRef]
3. Jafari, S.; Sprott, J.; Golpayegani, S.M.R.H. Elementary quadratic chaotic flows with no equilibria. *Phys. Lett. A* **2013**, *377*, 699–702. [CrossRef]
4. Wei, Z. Dynamical behaviors of a chaotic system with no equilibria. *Phys. Lett. A* **2011**, *376*, 102–108. [CrossRef]
5. Lao, S.-K.; Shekofteh, Y.; Jafari, S.; Sprott, J.C. Cost function based on Gaussian mixture model for parameter estimation of a chaotic circuit with a hidden attractor. *Int. J. Bifurc. Chaos* **2014**, *24*, 1450010. [CrossRef]
6. Wang, X.; Chen, G. A chaotic system with only one stable equilibrium. *Commun. Nonlinear Sci. Numer. Simul.* **2012**, *17*, 1264–1272. [CrossRef]
7. Jafari, S.; Sprott, J. Simple chaotic flows with a line equilibrium. *Chaos Solitons Fractals* **2013**, *57*, 79–84. [CrossRef]
8. Rajagopal, K.; Karthikeyan, A.; Srinivasan, A. Bifurcation and chaos in time delayed fractional order chaotic memfractor oscillator and its sliding mode synchronization with uncertainties. *Chaos Solitons Fractals* **2017**, *103*, 347–356. [CrossRef]
9. Pham, V.-T.; Jafari, S.; Volos, C.; Vaidyanathan, S.; Kapitaniak, T. A chaotic system with infinite equilibria located on a piecewise linear curve. *Opt.-Int. J. Light Electron Opt.* **2016**, *127*, 9111–9117. [CrossRef]
10. Barati, K.; Jafari, S.; Sprott, J.C.; Pham, V.-T. Simple Chaotic Flows with a Curve of Equilibria. *Int. J. Bifurc. Chaos* **2016**, *26*, 1630034. [CrossRef]
11. Tolba, M.F.; Said, L.A.; Madian, A.H.; Radwan, A.G. FPGA implementation of fractional-order integrator and differentiator based on Grünwald Letnikov's definition. In Proceedings of the 2017 29th International Conference on Microelectronics (ICM), Beirut, Lebanon, 10–13 December 2017; pp. 1–4.
12. Kingni, S.T.; Pham, V.-T.; Jafari, S.; Kol, G.R.; Woafo, P. Three-Dimensional Chaotic Autonomous System with a Circular Equilibrium: Analysis, Circuit Implementation and Its Fractional-Order Form. *Circuits Syst. Signal Process.* **2016**, *35*, 1933–1948. [CrossRef]
13. Ismail, S.M.; Said, L.A.; Rezk, A.A.; Radwan, A.G.; Madian, A.H.; Abu-Elyazeed, M.F.; Soliman, A.M. Generalized fractional logistic map encryption system based on FPGA. *AEU-Int. J. Electron. Commun.* **2017**, *80*, 114–126. [CrossRef]
14. Jafari, S.; Sprott, J.C.; Molaie, M. A simple chaotic flow with a plane of equilibria. *Int. J. Bifurc. Chaos* **2016**, *26*, 1650098. [CrossRef]
15. Bao, B.; Jiang, T.; Wang, G.; Jin, P.; Bao, H.; Chen, M. Two-memristor-based Chua's hyperchaotic circuit with plane equilibrium and its extreme multistability. *Nonlinear Dyn.* **2017**, *89*, 1157–1171. [CrossRef]
16. Jafari, S.; Sprott, J.C.; Pham, V.-T.; Volos, C.; Li, C. Simple chaotic 3D flows with surfaces of equilibria. *Nonlinear Dyn.* **2016**, *86*, 1349–1358. [CrossRef]
17. Sun, K.; Sprott, J. A simple jerk system with piecewise exponential nonlinearity. *Int. J. Nonlinear Sci. Numer. Simul.* **2009**, *10*, 1443–1450. [CrossRef]

18. Leonov, G.; Kuznetsov, N.; Vagaitsev, V. Localization of hidden Chua's attractors. *Phys. Lett. A* **2011**, *375*, 2230–2233. [CrossRef]

19. Leonov, G.; Kuznetsov, N.; Vagaitsev, V. Hidden attractor in smooth Chua systems. *Phys. D Nonlinear Phenom.* **2012**, *241*, 1482–1486. [CrossRef]

20. Leonov, N.V.; Kuznetsov, G.A. Hidden attractors in dynamical systems. From hidden oscillations in Hilbert–Kolmogorov, Aizerman, Kalman problems to hidden chaotic attractor in Chua circuits. *Int. J. Bifurc. Chaos* **2013**, *23*, 1330002. [CrossRef]

21. Leonov, G.; Kuznetsov, N.; Kiseleva, M.; Solovyeva, E.; Zaretskiy, A. Hidden oscillations in mathematical model of drilling system actuated by induction motor with a wound rotor. *Nonlinear Dyn.* **2014**, *77*, 277–288. [CrossRef]

22. Leonov, G.; Kuznetsov, N.; Mokaev, T. Hidden attractor and homoclinic orbit in Lorenz-like system describing convective fluid motion in rotating cavity. *Commun. Nonlinear Sci. Numer. Simul.* **2015**, *28*, 166–174. [CrossRef]

23. Leonov, G.A.; Kuznetsov, N.V.; Mokaev, T.N. Homoclinic orbits, and self-excited and hidden attractors in a Lorenz-like system describing convective fluid motion. *Eur. Phys. J. Spec. Top.* **2015**, *224*, 1421–1458. [CrossRef]

24. Danca, M.-F.; Kuznetsov, N. Hidden chaotic sets in a Hopfield neural system. *Chaos Solitons Fractals* **2017**, *103*, 144–150. [CrossRef]

25. Danca, M.-F.; Kuznetsov, N.; Chen, G. Unusual dynamics and hidden attractors of the Rabinovich–Fabrikant system. *Nonlinear Dyn.* **2017**, *88*, 791–805. [CrossRef]

26. Kuznetsov, N.; Leonov, G.; Yuldashev, M.; Yuldashev, R. Hidden attractors in dynamical models of phase-locked loop circuits: Limitations of simulation in MATLAB and SPICE. *Commun. Nonlinear Sci. Numer. Simul.* **2017**, *51*, 39–49.

27. Dudkowski, D.; Jafari, S.; Kapitaniak, T.; Kuznetsov, N.V.; Leonov, G.A.; Prasad, A. Hidden attractors in dynamical systems. *Phys. Rep.* **2016**, *637*, 1–50. [CrossRef]

28. Hassard, B.D.; Kazarinoff, N.D.; Wan, Y.-H. *Theory and Applications of Hopf Bifurcation*; CUP Archive: Cambridge, UK, 1981; Volume 41.

29. Stankevich, N.V.; Dvorak, A.; Astakhov, V.; Jaros, P.; Kapitaniak, M.; Perlikowski, P.; Kapitaniak, T. Chaos and Hyperchaos in Coupled Antiphase Driven Toda Oscillators. *Regul. Chaotic Dyn.* **2018**, *23*, 120–126. [CrossRef]

30. Jaros, P.; Brezetsky, S.; Levchenko, R.; Dudkowski, D.; Kapitaniak, T.; Maistrenko, Y. Solitary states for coupled oscillators with inertia. *Chaos Interdiscip. J. Nonlinear Sci.* **2018**, *28*, 011103. [CrossRef] [PubMed]

31. Kapitaniak, T.; Leonov, G.A. Multistability: Uncovering hidden attractors. *Eur. Phys. J. Spec. Top.* **2015**, *224*, 1405–1408. [CrossRef]

32. Lai, Q.; Chen, S. Generating multiple chaotic attractors from Sprott B system. *Int. J. Bifurc. Chaos* **2016**, *26*, 1650177. [CrossRef]

33. Lai, Q.; Wang, L. Chaos, bifurcation, coexisting attractors and circuit design of a three-dimensional continuous autonomous system. *Opt.-Int. J. Light Electron Opt.* **2016**, *127*, 5400–5406. [CrossRef]

34. Kengne, J.; Njitacke, Z.; Fotsin, H. Dynamical analysis of a simple autonomous jerk system with multiple attractors. *Nonlinear Dyn.* **2016**, *83*, 751–765. [CrossRef]

35. Ma, J.; Wu, F.; Ren, G.; Tang, J. A class of initials-dependent dynamical systems. *Appl. Math. Comput.* **2017**, *298*, 65–76. [CrossRef]

36. Wei, Z.; Moroz, I.; Liu, A. Degenerate Hopf bifurcations, hidden attractors, and control in the extended Sprott E system with only one stable equilibrium. *Turk. J. Math.* **2014**, *38*, 672–687. [CrossRef]

37. Wei, Z.; Zhang, W. Hidden hyperchaotic attractors in a modified Lorenz–Stenflo system with only one stable equilibrium. *Int. J. Bifurc. Chaos* **2014**, *24*, 1450127. [CrossRef]

38. Hilborn, R.C. *Chaos and Nonlinear Dynamics: An Introduction for Scientists and Engineers*; Oxford University Press: Oxford, UK, 2000.

39. Vaidyanathan, S.; Volos, C.; Pham, V.-T.; Madhavan, K.; Idowu, B.A. Adaptive backstepping control, synchronization and circuit simulation of a 3-D novel jerk chaotic system with two hyperbolic sinusoidal nonlinearities. *Arch. Control Sci.* **2014**, *24*, 375–403. [CrossRef]

40. Tam, W.M.; Lau, F.C.; Chi, K.T. *Digital Communications with Chaos: Multiple Access Techniques and Performance*; Elsevier: New York, NY, USA, 2010.

41. Molaie, M.; Jafari, S.; Sprott, J.C.; Golpayegani, S.M.R.H. Simple chaotic flows with one stable equilibrium. *Int. J. Bifurc. Chaos* **2013**, *23*, 1350188. [CrossRef]

42. Kant, N.A.; Dar, M.R.; Khanday, F.A.; Psychalinos, C. Ultra-low-Voltage Integrable Electronic Realization of Integer-and Fractional-Order Liao's Chaotic Delayed Neuron Model. *Circuits Syst. Signal Process.* **2017**, *36*, 4844–4868. [CrossRef]

43. Tsirimokou, G.; Psychalinos, C.; Elwakil, A.S.; Salama, K.N. Electronically tunable fully integrated fractional-order resonator. *IEEE Trans. Circuits Syst. II Express Br.* **2018**, *65*, 166–170. [CrossRef]

44. Bertsias, P.; Psychalinos, C.; Radwan, A.G.; Elwakil, A.S. High-Frequency Capacitorless Fractional-Order CPE and FI Emulator. *Circuits Syst. Signal Process.* **2017**, *37*, 2694–2713. [CrossRef]

45. Tang, Y.X.; Khalaf, A.J.M.; Rajagopal, K.; Pham, V.T.; Jafari, S.; Tian, Y. A new nonlinear oscillator with infinite number of coexisting hidden and self-excited attractors. *Chin. Phys. B* **2018**, *27*, 040502. [CrossRef]

46. Alghassab, M.; Mahmoud, A.; Zohdy, M.A. Nonlinear Control of Chaotic Forced Duffing and Van der Pol Oscillators. *Int. J. Mod. Nonlinear Theory Appl.* **2017**, *6*, 26. [CrossRef]

47. Njah, A. Synchronization via active control of parametrically and externally excited Φ6 Van der Pol and Duffing oscillators and application to secure communications. *J. Vib. Control* **2011**, *17*, 493–504. [CrossRef]

48. He, Q.; Wang, L.; Liu, B. Parameter estimation for chaotic systems by particle swarm optimization. *Chaos Solitons Fractals* **2007**, *34*, 654–661. [CrossRef]

49. Tang, Y.; Guan, X. Parameter estimation for time-delay chaotic system by particle swarm optimization. *Chaos Solitons Fractals* **2009**, *40*, 1391–1398. [CrossRef]

50. Wang, L.; Xu, Y. An effective hybrid biogeography-based optimization algorithm for parameter estimation of chaotic systems. *Expert Syst. Appl.* **2011**, *38*, 15103–15109. [CrossRef]

51. Jafari, S.; Hashemi Golpayegani, S.M.R.; Jafari, A.H.; Gharibzadeh, S. Some remarks on chaotic systems. *Int. J. Gen. Syst.* **2012**, *41*, 329–330. [CrossRef]

52. Jafari, S.; Hashemi Golpayegani, S.M.R.; Daliri, A. Parameters identification of chaotic systems by quantum-behaved particle swarm optimization. *Int. J. Comput. Math.* **2009**, *86*, 2225–2235; Comment on *Int. J. Comput. Math.* **2013**, *90*, 903–905.

53. Jafari, S.; Hashemi Golpayegani, S.M.R.; Rasoulzadeh Darabad, M. Parameter identification and synchronization of fractional-order chaotic systems. *Commun. Nonlinear Sci. Numer. Simul.* **2012**, *17*, 305–316, Comment on *Commun. Nonlinear Sci. Numer. Simul.* **2013**, *18*, 811–814.

54. Shekofteh, Y.; Jafari, S.; Sprott, J.C.; Hashemi Golpayegani, S.M.R.; Almasganj, F. A gaussian mixture model based cost function for parameter estimation of chaotic biological systems. *Commun. Nonlinear Sci. Numer. Simul.* **2015**, *20*, 469–481. [CrossRef]

55. Jafari, S.; Sprott, J.C.; Pham, V.-T.; Hashemi Golpayegani, S.M.R.; Jafari, A.H. A New Cost Function for Parameter Estimation of Chaotic Systems Using Return Maps as Fingerprints. *Int. J. Bifurc. Chaos* **2014**, *24*, 1450134. [CrossRef]

56. Kingni, S.T.; Jafari, S.; Simo, H.; Woafo, P. Three-dimensional chaotic autonomous system with only one stable equilibrium: Analysis, circuit design, parameter estimation, control, synchronization and its fractional-order form. *Eur. Phys. J. Plus* **2014**, *129*, 76. [CrossRef]

57. Wang, Z.; Abdolmohammadi, H.R.; Alsaadi, F.E.; Hayat, T.; Pham, V.-T. A new oscillator with infinite coexisting asymmetric attractors. *Chaos Solitons Fractals* **2018**, *110*, 252–258. [CrossRef]

58. Hu, X.; Liu, C.; Liu, L.; Ni, J.; Yao, Y. Chaotic dynamics in a neural network under electromagnetic radiation. *Nonlinear Dyn.* **2018**, *91*, 1541–1554. [CrossRef]

59. Hu, X.; Liu, C.; Liu, L.; Ni, J.; Li, S. An electronic implementation for Morris–Lecar neuron model. *Nonlinear Dyn.* **2016**, *84*, 2317–2332. [CrossRef]

60. Lee, K.S.; Geem, Z.W. A new meta-heuristic algorithm for continuous engineering optimization: Harmony search theorypractice. *Comput. Methods Appl. Mech. Eng.* **2005**, *194*, 3902–3933. [CrossRef]

61. Hu, X.; Liu, C.; Liu, L.; Yao, Y.; Zheng, G. Multi-scroll hidden attractors and multi-wing hidden attractors in a 5-dimensional memristive system. *Chin. Phys. B* **2017**, *26*, 110502. [CrossRef]

62. Hu, X.; Liu, C.; Liu, L.; Ni, J.; Li, S. Multi-scroll hidden attractors in improved Sprott A system. *Nonlinear Dyn.* **2016**, *86*, 1725–1734. [CrossRef]

63. Majhi, S.; Perc, M.; Ghosh, D. Chimera states in a multilayer network of coupled and uncoupled neurons. *Chaos Interdiscip. J. Nonlinear Sci.* **2017**, *27*, 073109. [CrossRef] [PubMed]

64. Namazi, H.; Daneshi, A.; Azarnoush, H.; Jafari, S.; Towhidkhah, F. Fractal based analysis of the influence of auditory stimuli on eye movements. *Fractals* **2018**, *26*, 1850040. [CrossRef]

65. Gosak, M.; Stožer, A.; Markovič, R.; Dolenšek, J.; Marhl, M.; Slak Rupnik, M.; Perc, M. The relationship between node degree and dissipation rate in networks of diffusively coupled oscillators and its significance for pancreatic beta cells. *Chaos Interdiscip. J. Nonlinear Sci.* **2015**, *25*, 073115. [CrossRef] [PubMed]
66. Smirnova, R.; Zakrzhevsky, M.; Schukin, I. Global analysis of the nonlinear Duffing-van der Pol type equation by a bifurcation theory and complete bifurcation groups method. *Vibroeng. Procedia* **2014**, *3*, 139–143.

entropy

MDPI

Article

Optimization of Thurston's Core Entropy Algorithm for Polynomials with a Critical Point of Maximal Order

Gamaliel Blé * [ID] **and Domingo González** [ID]

División Académica de Ciencias Básicas, Universidad Juárez Autónoma de Tabasco,
Carretera Cunduacán-Jalpa Km 1, Cunduacán Tabasco 86690, México; domingo.gonzalez@ujat.mx
* Correspondence: gble@ujat.mx

Received: 1 August 2018; Accepted: 5 September 2018; Published: 11 September 2018

Abstract: This paper discusses some properties of the topological entropy systems generated by polynomials of degree d in their Hubbard tree. An optimization of Thurston's core entropy algorithm is developed for a family of polynomials of degree d.

Keywords: core entropy; Thurston's algorithm; Hubbard tree; external rays

1. Introduction

The topological entropy of a polynomial P, denoted by P allows us to measure the complexity of the orbits of the dynamical system generated by P. This concept has been used to classify the dynamics in different polynomial families, for example, in the case of real one-parameter families of polynomials of degree 2, it has been shown that the entropy behaves monotonically [1,2]. For real cubic maps, it was shown that each locus of constant topological entropy is a connected set [3]. Later, this result was shown for a quartic polynomial family and for real multimodal maps [4,5]. In the complex polynomials family, the entropy is concentrated in the Julia set; it is constant and only depends on the degree of the polynomial family [1,6]. In order to study the dynamics of a polynomial with a finite postcritical set, Douady and Hubbard introduced the Hubbard tree; the theory of admissible Hubbard trees and critical portraits was later studied by Poirier [7]. Afterwards Thurston proposed to study the entropy, restricted to its *Hubbard tree*, of a polynomial with finite postcritical set, which, in this setting, is called the *core entropy*. He showed that the core entropy generalizes the concept defined for an invariant interval in the real case [8]. Furthermore, Thurston proposed an algorithm in order to calculate the core entropy. It is based on a linear transformation A (defined in terms of the external arguments of the postcritical set) whose spectral radius coincides with the core entropy [9].

In the case of the quadratic family, Li proved that the core entropy grows through the veins of the Mandelbrot set. Later Tiozzo proved, for the same family, that the core entropy can be extended as a continuous function of the external argument on the boundary of the Mandelbrot set [10,11]. He generalizes this result for polynomials of higher degrees [12].

In this article, we show a simplification of Thurston's algorithm for a family of polynomials of degree $d \geq 3$ with one free critical point and one fixed critical point of maximum multiplicity. We always assume that the free critical point is either periodic or eventually periodic. According to [13], this family is conjugated to

$$P_a(z) = z^{d-1}\left(z + \frac{da}{d-1}\right). \tag{1}$$

The polynomial function $P_a(z)$ has two critical points: zero which is the fixed critical point of maximal multiplicity and $-a$ which is the free critical point. The parameter space of this polynomial family has been studied by Milnor [14,15] in the cubic case, and by Roesch [13], who studied the topological properties of the hyperbolic components in the case of degree $d \geq 3$.

To simplify Thurston's algorithm, we construct a linear transformation A' with a definition based on the external arguments of the orbit of the critical point $(-a)$. As we will show, this linear transformation is defined in a space with smaller dimension than the one proposed by Thurston. Consequently, the spectral radius is easier to calculate. Here is the main result of this paper.

Main Theorem. *Let P_a be a postcritically finite polynomial of the family* (1). *If A denotes the matrix obtained via Thurston's algorithm, then A and A' have equal spectral radii ρ.*

In order to prove the Main Theorem, we use of the concept of external rays, the Thurston algorithm, and some properties of the entropy and non-negative matrices [16–20].

2. Thurston's Algorithm

The algorithm proposed by Thurston allows us to compute the core entropy of a polynom of degree d. With the purpose of defining this algorithm, we present some needed concepts which can be found in the work of Gao, [9].

2.1. The Algorithm of Thurston for Polynomials of Degree d

Let $P(z)$ be a postcritically finite polynomial of degree d. Thus, $P(z)$ has exactly $d - 1$ critical points, say, $c_1, ..., c_n$ (counting multiplicities). Each c_i is either in the Julia set, J_p, or is the center of a Fatou component. Furthermore, J_p is locally connected [18,21]. The algorithm is based on the analysis of the external rays that land either on the critical points or on the boundaries of Fatou components that contain the critical points.

Definition 1. *We say that an external ray $R(\theta)$ supports a bounded Fatou component U if:*

(1) The ray lands on a point q at the boundary of U.

(2) There exists a sector based at q, delimited by $R(\theta)$ and the internal ray of U that lands at q, such that the sector does not contain any other external ray that lands on q.

Given a postcritically finite polynomial of degree d and a critical Fatou component U, that is, a Fatou component containing a critical point, let $\delta = deg(P|_U)$. We define the set $\Theta(U)$ as follows:

(1) If U is periodic with orbit

$$U \to P(U) \to \cdots \to P^n(U) = U,$$

we build $\Theta(U', z', \theta)$ for all U' in this orbit simultaneously.

Using the Böttcher coordinates in U, we can find $z \in \partial U$ with internal argument 0. This z is a root of U, which depends on the choice of the coordinates. This means that z is a periodic point of minimal period on the boundary of U. This choice determines a root for each Fatou component ($P^k(U)$, for $k = 1, 2 \ldots, n - 1$). We call this root a *preferred root of $P^k(U)$*. If U' is any component in the cycle and z' is its preferred root, consider a ray ($R(\theta)$) which supports U' at z'. Define $\Theta(U', z', \theta)$ to be the set consisting of $\delta_{U'}$ arguments of the support rays for the component U' that are the inverse image of $P(R(\theta))$.

(2) If the Fatou component U is strictly preperiodic, take n as the smallest number for which $P^n(U)$ is a critical Fatou component. Let $z \in \partial U$ be such that $P^n(z) = \gamma(\alpha)$, where $\gamma(\alpha)$ is the point where $R(\alpha)$ lands on $\partial P^n(U)$ and $\alpha \in \Theta(P^n(U), \gamma(\alpha), \alpha)$. Consider a ray ($R(\theta)$) that supports component U which contains z. Define $\Theta(U, z, \theta)$ as the set of the δ_U arguments of the supporting rays of U that, under P^n, go to $R(\alpha)$.

Remark 1. *For each critical Fatou component U, there exists, at most, a finite number of sets $(\Theta(U', z', \theta))$, each one dependent on the choice of the root (z) in U and the argument (θ). We can choose any of them and denote it by $\Theta(U)$.*

Definition 2. *Let P be a polynomial with finite postcritical set. Let $U_1, ..., U_n$ be the pairwise disjoint critical Fatou components, and let $c_1, ..., c_m$ be the critical points in the Julia set ($m + n$ is the number of different critical points of P). The finite collection of subsets of the circle*

$$\Theta_P = \{\Theta(c_1), ..., \Theta(c_m), \Theta(U_1), ..., \Theta(U_n)\}$$

is called the critical marking of P, if each of the $\Theta(U_i)$ is chosen as in Remark 1 and each $\Theta(c_j)$ consists purely of the angles of the external rays that lands on c_j.

Let $\Theta = \{\Theta_1, \Theta_2 ..., \Theta_l\}$ be the critical marking of a polynomial P of degree d. We define the critical and postcritical sets of Θ as

$$crit(\Theta) = \bigcup_{k=1}^{l} \Theta_k \ \text{ and } \ post(\Theta) = \bigcup_{n \geq 1} \tau^n crit(\Theta),$$

respectively, where $\tau : \mathbb{T} \to \mathbb{T}$ is the function given by $\tau(\theta) = d\theta \mod 1$. From the definition of critical marking, it is easy to see that the following holds:

1. Each $\tau\Theta_i$, $i \in \{1, 2, ..., l\}$, consists of a unique angle.
2. The convex hulls of Θ_i and Θ_j in the unit disk intersect each other in, at most, one point of $\mathbb{T}$, for any $i \neq j$ in the set $\{1, 2, ..., l\}$.
3. For each i, $\#\Theta_i \geq 2$ and $\sum_{i=1}^{l}(\#\Theta_i - 1) = d - 1$.

Let $\mathbb{D}$ be the unit disk endowed with the hyperbolic metric. We identify any point in $\partial\mathbb{D}$ with the argument in $\mathbb{T}$. By doing this, each angle in the circle is considered to be mod 1. A leaf is either a point in $\mathbb{T}$ or the closure in $\mathbb{D}$ of a hyperbolic chord (non-trivial). Indeed, from now on, each time we mention chord or hull in the disk, it will be in the hyperbolic sense. For each set ($S \subset \mathbb{T}$), we denote the convex hull of S as a subset of $\mathbb{D}$ by $hull(S)$.

A *critical portrait* of degree d is a finite collection of finite subsets of the circumference, $\Theta = \{\Theta_1, \Theta_2 ..., \Theta_k\}$ satisfying properties **1**, **2**, **3**.

Notice that any critical marking of a postcritically finite polynomial seen in the unit disk is a critical portrait.

Definition 3. *Let $\Theta = \{\Theta_1, \Theta_2 ..., \Theta_l\}$ be a critical portrait. Given any two angles $x, y \in \mathbb{T}$ that are not necessarily different, and an element Θ of Θ, we say that the chord $\overline{xy}$ crosses the convex hull, $hull(\Theta)$ if $x, y \notin \Theta$, and $\overline{xy} \cap hull(\Theta) \neq \emptyset$. In this setting, we also say that x, y are separated by Θ.*

Definition 4. *Given any pair of angles $x, y \in \mathbb{T}$, the separation set relative to Θ is the set $\{k_1, ..., k_p\}$ where the chord $\overline{xy}$ successively crosses $hull(\Theta_{k_1}), ..., hull(\Theta_{k_p})$, $\Theta_{k_j} \in \Theta$, and no other element of Θ separates the angles x, y. We say that the angles x, y are not separated by Θ if its separation set relative to Θ is empty.*

If P_d is a polynomial with finite postcritical set, then each element of its critical portrait Θ is rational and $post(\Theta)$ is a finite set. Hence, it is possible to define the finite set S consisting of pairs (not ordered) of $\{x, y\}$ with $x \neq y \in post(\Theta)$ as long as $card(post(\Theta)) \geq 2$. In the case of $post(\Theta) = \{x\}$, S has only one element and in this case, x is a fixed point of τ.

Once we have defined the set S, the entropy of P_d restricted to the Hubbard tree ($H(P_d)$) is given by the Algorithm 1.

Algorithm 1 Thurston's Algorithm

- Let $\mathcal{V}$ be the vector space over $\mathbb{R}$ which has the elements of S as a basis.
- Let $\mathcal{A} : \mathcal{V} \rightarrow \mathcal{V}$ be the linear transformation defined by the values on the basis S of $\mathcal{V}$ as follows. For any vector $(\{x,y\} \in S)$, the image is defined by
  $\mathcal{A}(\{x,y\}) = \{\tau(x), \tau(y)\}$ if x, y are not separated by Θ;
  $\mathcal{A}(\{x,y\}) = \sum_{i=0}^{p} \mathcal{A}(\{\theta_i, \theta_{i+1}\})$, where $\theta_0 = x$, $\theta_{p+1} = y$, and $\theta_i \in \Theta_{k_i} \in \Theta$, if $\{x,y\}$ has the separating set $\{k_1, \ldots k_p\} \neq \varnothing$.
- Let A be the matrix associated with the linear transformation $\mathcal{A}$ with respect to the basis S. As this matrix is non-negative, according to Perron–Frobenius Theorem, the spectral radius ρ of A is non-negative [16].

Theorem 1 (Gao). *Let P_d be a polynomial of degree d with a finite postcritical set, and let Θ be a critical marking for P_d. If ρ is the spectral radius of the matrix in Thurston's algorithm, then $h(H(P_d), P_d) = \log \rho$.*

A full proof of the above Theorem can be found in [9].

One of the advantages of studying the entropy in the critical portrait is the fact that each point of the postcritical set corresponds to an angle in the set $post(\Theta)$ in such a way that any arc of $H(P_d)$ between two vertices can be represented by some pair of angles, although possibly not in a unique way. Intuitively, one can think that the actions of P_d in those arcs induce a transformation in the space generated by the pair of angles in the set $post(\Theta)$ given by the matrix A of Thurston's algorithm.

2.2. Thurston's Algorithm in the Polynomial Family (1)

Let P_a be a polynomial in the family (1). The critical points of P_a are 0 and $-a$. The point 0 is the center of the fixed Fatou component B_a, and $-a$ is a free critical point. If $-a$ is the center of a Fatou component, then this component will be denoted by U_1.

We also define the following set of angles

$$\Theta_0 = \Theta(B_a),$$

$$\Theta_{-a} = \begin{cases} \Theta(U_1) & if \quad -a \text{ is the center of a Fatou component} \\ \Theta(c_1) & if \quad -a \in J_a, \end{cases}$$

where $\Theta(B_a)$, $\Theta(U_1)$ and $\Theta(c_1)$ are defined as in Section 2.1.

Assume that P_a has a finite postcritical set. The collection of angles $\Theta_{P_a} = \Theta_{-a}$ is called a restricted critical marking.

Let Θ_{P_a} be the restricted critical marking of P_a. We define the restricted critical set and the restricted postcritical set as

$$crit(\Theta_{P_a}) = \Theta_{-a} \text{ and } post(\Theta_{P_a}) = \bigcup_{n \geq 1} \tau^n crit(\Theta_{P_a}).$$

In the same way as we did before, we identify any point in $\partial \mathbb{D}$ with its argument in $\mathbb{T}$. The restricted critical marking of P_a viewed in $\mathbb{D}$ is denoted by Θ_a.

Lemma 1. *If Θ_a is the restricted critical portrait of the polynomial P_a, then $\mathbb{D} \setminus hull(\Theta_a)$ has 2 connected components with arcs in $\mathbb{T}$ of lengths $\frac{1}{d}$ and $\frac{d-1}{d}$, respectively.*

Proof. Since $deg(P|_{-a}) = 2$, Θ_a consists of two elements, the convex hull ($hull(\Theta_a)$) divides $\mathbb{D}$ in two regions. On the other hand, as the elements of Θ_a are obtained as inverse images of the same angle, the arc length between the elements of Θ_a is equal to $\frac{1}{d}$ (c.f. Proposition 2.31 in [13]). This completes the proof. $\square$

Definition 5. *Let Θ_a be a restricted critical portrait. Given any two angles $x, y \in \mathbb{T}$ (not necessarily different), we say that the chord $(\overline{xy})$ crosses the convex hull (hull(Θ_a) of Θ_a) if $x, y \notin \Theta_a$ and $\overline{xy} \cap hull(\Theta_a) \neq \emptyset$. Under these conditions, we say that x, y are separated by Θ_a.*

If P_a has a finite postcritical set, then the elements of the restricted critical portrait are rationals, and $post(\Theta_a)$ is a finite set. We define set S' as all pairs (not ordered) of $\{x, y\}$ with $x \neq y \in post(\Theta_a)$ as long as $card(post(\Theta_a)) \geq 2$. If $post(\Theta_a) = \{x\}$, then S' is the element $\{x, x\}$, and x is a fixed point of τ. Once we have defined set S', the adapted Thurston's algorithm that is used to approximate the entropy of P_a over its Hubbard tree is given by Algorithm 2.

Algorithm 2 The Adapted Thurston's Algorithm

- Define $\mathcal{V}'$ as the vector space over $\mathbb{R}$ which has the elements of S' as a basis.
- Define the linear transformation $\mathcal{A}' : \mathcal{V}' \to \mathcal{V}'$ by setting the values of A' on the basis S' of $\mathcal{V}'$ in the following way: For any vector $\{x, y\} \in S'$, the image is defined by
  $\mathcal{A}'(\{x, y\}) = \{\tau(x), \tau(y)\}$, if x, y are not separated by Θ; and
  $\mathcal{A}'(\{x, y\}) = \{\tau(x), \tau(\theta)\} + \{\tau(\theta), \tau(y)\}$, $\theta \in \Theta_a$ if x, y are separated by Θ.
- Let A' be the matrix associated with the linear transformation $\mathcal{A}'$ with respect to the basis S'. This matrix is non-negative, and, according to the Perron–Frobenius Theorem [16], the spectral radius ρ' of A' is non-negative.

Theorem 2. *Let P_a be a finite postcritical polynomial. If A denotes the matrix obtained by Algorithm 1, and A' is the matrix generated by Algorithm 2, then A and A' have the same spectral radius (ρ).*

Proof. In order to prove the Theorem, we have to consider two cases:

(1) If $-a \in B_a$, in this case, the core entropy is zero, and we show that in the restricted algorithm. The spectral radius of matrix A' is 1.

(2) If $-a \notin B_a$, we show that transformation $\mathcal{A}$ can be built without considering the line of separation of the critical point (0).

Let P_a be a polynomial of degree d with a finite postcritical set. In accordance with Böttcher's Theorem, a biholomorphism ϕ_a exists that conjugates P_a with the function z^d in a neighborhood of infinity. Since P_a is postcritically finite, its Julia set is locally connected; hence, ϕ_a can be extended continuously to J_a [21].

Moreover, the dynamics in B_a are conjugated to z^{d-1}, and the conjugation can be extended continuously to the boundary; hence, a fixed point p of P_a exists, with an internal angle of 0, that is, in ∂B_a. The Böttcher coordinate is chosen at infinity in such a way that the external angle of p is also 0.

According to the above and the construction of the critical portraits, the set of angles is $\Theta_0 = \left\{0, \frac{k_1}{d}, \dots \frac{k_{d-2}}{d}\right\}$, with $k_i \in \{1, 2, \dots, d-1\}$. Hence, the critical portrait of P_a is given by

$$\Theta = \left\{\left\{0, \frac{k_1}{d}, \dots \frac{k_{d-2}}{d}\right\}, \Theta_{-a}\right\} \text{ and } post(\Theta) = \bigcup_{n \geq 1} \tau^n(\Theta_{-a}) \bigcup \{0\},$$

where Θ_{-a} consists of two elements according to Lemma 1.

This shows that for a fixed d, the postcritical set varies only in the function of the critical point $(-a)$. On the other hand, the edges of the Hubbard tree are related to the pairs of angles in the critical portrait as follows: the interval of angles with extremes $\{\theta_1, \theta_2\}$ in the circumference represents a union of edges in the Hubbard tree, and the interval of angles in the circle given by the image of $\mathcal{A}(\{\theta_1, \theta_2\})$ is equivalent to the interval of angles that contains the image under P_a of the union of corresponding edges.

Remark 2. *If S denotes the basis of the vector space in Algorithm 1 and the pair $\{\theta_1, \theta_2\} \in S$ is separated with respect to the critical line Θ_{c_i}, then the corresponding edge or edges contain the critical point c_i.*

Case 1: Let P_a be a postcritically finite polynomial such that $-a \in B_a$. As P_a is conjugated to z^{d-1} in B_a, then the tree of a is star shaped with n edges. We can label the edges in the Hubbard tree such that the incidence matrix $\tilde{A} = (a_{i,j})$ is defined by $a_{i,i+1} = 1$ for $i = 1, \ldots, n-1$, $a_{n,j} = 1$ for some $j \in \{i, \ldots, n\}$ and zero otherwise.

The characteristical polynomial of the incidence matrix $\tilde{A}$ is

$$(-1)^n \lambda^{k-1} (\lambda^{n-(k-1)} - 1),$$

and its spectral radius is 1. Hence $h(P_a) = 0$. On the other hand, Theorem 1 says that the spectral radius of A obtained by Thurston's method is 1.

Due to the fact that the orbit of $-a$ is in B_a, the restricted critical portrait Θ_a consists of the external angles corresponding to the component B_a. Hence, pairs ($\{\theta_i, \theta_j\}$) separated with respect to the critical point $(-a)$ do not exist. Moreover, we can disregard the separation with respect to 0, as in the restricted algorithm. Thus, there is no pair that is separated. Consequently, all pairs $\{\theta_i, \theta_j\} \in S$ have only one image. Furthermore, all rows of matrix A' add up to 1; thus, the spectral radius is 1.

Case 2: If $-a \notin B_a$, we have the next claim.

Claim 1. If $[v_1, 0]$ and $[0, v_2]$ are the two edges of $H(a)$, and $\gamma = [v_1, 0] \cup [0, v_2]$, then $P_a(\gamma) = [P_a(v_1), P_a(v_2)]$.

Proof. If $-a$ is a periodic point, then there are no edges of the forms $[v_1, 0]$ and $[0, v_2]$ that have the same image. Hence, as 0 is a fixed point, $P_a(\gamma) = [P_a(v_1), P_a(v_2)]$.

On the other hand, if $-a$ is preperiodic with $-a \notin B_a$, then $-a$ eventually goes to a bifurcation point of ∂B_a; thus, in this case, there are no edges of the forms $[v_1, 0]$ and $[0, v_2]$ that go to the same image. $\square$

In set S, in order to obtain the image of a separated pair ($\{\theta_1, \theta_2\}$) we can discard the characteristic of being separated with respect to the critical line of Θ_0. Thus, for this family of polynomials, if the pair $\{\theta_1, \theta_2\}$ is separated with respect to Θ, its separation set consists only of one element—the one associated with Θ_{-a}.

Since the postcritical set of P_a only depends on the critical point $(-a)$, we can study them if we separate them into the following cases:

(1) If $-a$ is the center of a capture component, then the orbit of Θ_{-a} eventually contains the zero angle, which is a fixed angle. In this case, the postcritical set of Θ is $\bigcup_{n \geq 1} \tau^n(\Theta_{-a})$. Hence, $A = A'$.

(2) If $-a$ eventually goes to $p \in \partial B_a$, with a fixed p, then the orbit of Θ_{-a} contains the zero angle; thus, as above, $A = A'$.

(3) In any other case, the orbits of Θ_{-a} and Θ_0 are disjoint. Hence, the postcritical set is

$$Post(\Theta) = \{0, \theta_1, \ldots \theta_k\},$$

where $\theta_i = P_a^i(\theta)$ with $\theta \in \Theta_{-a}$.

If we write set S in such a way that the first k elements are of the form $\{0, \theta_i\}$, then S can be written as

$$S = \{\{0, \theta_1\}, \ldots \{0, \theta_k\}\} \bigcup \{\{\theta_i, \theta_j\} \, ; i \leq j\}.$$

Hence, the matrix associated with the transformation $\mathcal{A}$ is

$$A = \begin{pmatrix} B & X \\ N & C \end{pmatrix},$$

where B is the submatrix corresponding to the relations of the images of the pairs of the form $\{0, \theta_i\}$ with themselves, and C is the submatrix corresponding to the relations of the images of the pairs $\{\theta_i, \theta_j\}$ with themselves. The lower submatrix $\mathbf{N}$ represents the relations between the images of pairs $\{\theta_i, \theta_j\}$ with $\{0, \theta_i\}$.

Since we do not consider the separation with respect to the critical line Θ_0, and the orbit of Θ_{-a} does not contain 0, the image of a pair $\{\theta_i, \theta_j\}$ does not have a component of the form $\{0, \theta_l\}$. Hence, the matrix $\mathbf{N}$ is identically 0.

Since C is exactly the matrix A', to finish the proof of the theorem, it is enough to prove the following claim.

Claim 2. *The spectral radius of matrix B is 1.*

Proof. Notice that a pair $(\{0, \theta_i\})$ is fixed under $\mathcal{A}$ only when the angle θ_i is fixed. Due to the fact that the only fixed angle of τ is zero and $\theta_i \neq 0$, all the elements of the diagonal of B are zeros. On the other hand, if $\{0, \theta_i\}$ is not separated, then its image is $\{0, \theta_{i+1}\}$, and if it is separated, then its image is $\{0, \theta_1\} + \{\theta_1, \theta_i\}$. In the first case, this generates a 1 over the diagonal of B, and in the second case, it generates a 1 on the first column. Thus, B has the form

$$B = \begin{pmatrix} 0 & 1 & 0 & \ldots, & 0 & 0 \\ 0 & 0 & 1 & \ldots, & 0 & 0 \\ \vdots & \vdots & \ldots & 1 & \vdots & \vdots \\ 1 & 0 & 0 & \ldots & 0 & 0 \\ \vdots & 0 & 0 & \ldots & 0 & 0 \\ 1 & 0 & 0 & \ldots & 0 & 0 \end{pmatrix}$$

and its spectral radius is then 1. $\square$

$\square$

Example 1. *Taking $d = 3$ and $a = 1.07183814 + 0.1928507i$ in the polynomial family (1), we have a polynomial with the critical point $-a$ which is periodic with a period of 4. The Julia set is shown in Figure 1.*

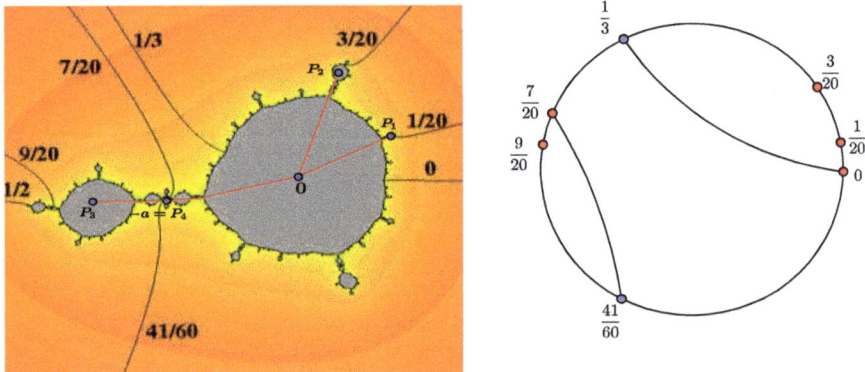

Figure 1. Julia set and critical portrait for $d = 3$ and $a = 1.07183814 + 0.1928507i$.

The critical portrait associated with P_a is $\Theta = \{\{0, \frac{1}{3}\}, \{\frac{7}{20}, \frac{41}{60}\}\}$ and $post(\Theta) = \{0, \frac{1}{20}, \frac{3}{20}, \frac{9}{20}, \frac{7}{20}\}$. It can be seen in Figure 1.

The basis S for the space $\mathcal{V}$ is

$$S = \left\{ \left\{0, \tfrac{1}{20}\right\}, \left\{0, \tfrac{3}{20}\right\}, \left\{0, \tfrac{9}{20}\right\}, \left\{0, \tfrac{7}{20}\right\}, \left\{\tfrac{1}{20}, \tfrac{3}{20}\right\}, \left\{\tfrac{1}{20}, \tfrac{9}{20}\right\}, \left\{\tfrac{1}{20}, \tfrac{7}{20}\right\}, \left\{\tfrac{3}{20}, \tfrac{9}{20}\right\}, \left\{\tfrac{3}{20}, \tfrac{7}{20}\right\}, \left\{\tfrac{9}{20}, \tfrac{7}{20}\right\} \right\}.$$

By applying the linear transformation $\mathcal{A}$ to the elements of the basis, we obtain

$$\left\{0, \tfrac{1}{20}\right\} \mapsto \left\{0, \tfrac{3}{20}\right\}$$
$$\left\{0, \tfrac{3}{20}\right\} \mapsto \left\{0, \tfrac{9}{20}\right\}$$
$$\left\{0, \tfrac{9}{20}\right\} \mapsto \left\{0, \tfrac{1}{20}\right\} + \left\{\tfrac{1}{20}, \tfrac{7}{20}\right\}$$
$$\left\{0, \tfrac{7}{20}\right\} \mapsto \left\{0, \tfrac{1}{20}\right\}$$
$$\left\{\tfrac{1}{20}, \tfrac{3}{20}\right\} \mapsto \left\{\tfrac{3}{20}, \tfrac{9}{20}\right\}$$

$$\left\{\tfrac{1}{20}, \tfrac{9}{20}\right\} \mapsto \left\{0, \tfrac{3}{20}\right\} + \left\{0, \tfrac{1}{20}\right\} + \left\{\tfrac{1}{20}, \tfrac{7}{20}\right\}$$
$$\left\{\tfrac{1}{20}, \tfrac{7}{20}\right\} \mapsto \left\{0, \tfrac{3}{20}\right\} + \left\{0, \tfrac{1}{20}\right\}$$
$$\left\{\tfrac{3}{20}, \tfrac{9}{20}\right\} \mapsto \left\{0, \tfrac{9}{20}\right\} + \left\{0, \tfrac{1}{20}\right\} + \left\{\tfrac{1}{20}, \tfrac{7}{20}\right\}$$
$$\left\{\tfrac{3}{20}, \tfrac{7}{20}\right\} \mapsto \left\{0, \tfrac{9}{20}\right\} + \left\{0, \tfrac{1}{20}\right\}$$
$$\left\{\tfrac{9}{20}, \tfrac{7}{20}\right\} \mapsto \left\{\tfrac{1}{20}, \tfrac{7}{20}\right\}.$$

Hence, the matrix associated with the linear transformation $\mathcal{A}$ is

$$A = \begin{pmatrix}
0 & 1 & 0 & 0 & 0 & 0 & 0 & 0 & 0 & 0 \\
0 & 0 & 1 & 0 & 0 & 0 & 0 & 0 & 0 & 0 \\
1 & 0 & 0 & 0 & 0 & 0 & 1 & 0 & 0 & 0 \\
1 & 0 & 0 & 0 & 0 & 0 & 0 & 0 & 0 & 0 \\
0 & 0 & 0 & 0 & 0 & 0 & 0 & 1 & 0 & 0 \\
1 & 1 & 0 & 0 & 0 & 0 & 1 & 0 & 0 & 0 \\
1 & 1 & 0 & 0 & 0 & 0 & 0 & 0 & 0 & 0 \\
1 & 0 & 1 & 0 & 0 & 0 & 1 & 0 & 0 & 0 \\
1 & 0 & 1 & 0 & 0 & 0 & 0 & 0 & 0 & 0 \\
0 & 0 & 0 & 0 & 0 & 0 & 1 & 0 & 0 & 0
\end{pmatrix},$$

and its spectral radius is 1.3953. In accordnce with Theorem 1, we conclude that the entropy of P_a, restricted to its Hubbard tree, is $\log 1.3953$.

On the other hand, the restricted critical portrait associated with P_a is $\Theta = \left\{\left\{\tfrac{7}{20}, \tfrac{41}{60}\right\}\right\}$, and post$(\Theta) = \left\{\tfrac{1}{20}, \tfrac{3}{20}, \tfrac{9}{20}, \tfrac{7}{20}\right\}$. It can be seen in Figure 2.

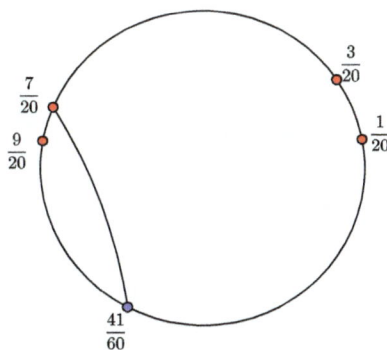

Figure 2. Restricted critical portrait for $d = 3$ and $a = 1.07183814 + 0.1928507i$.

The basis S of the space $\mathcal{V}$ is given by

$$S = \left\{ \left\{\tfrac{1}{20}, \tfrac{3}{20}\right\}, \left\{\tfrac{1}{20}, \tfrac{9}{20}\right\}, \left\{\tfrac{1}{20}, \tfrac{7}{20}\right\}, \left\{\tfrac{3}{20}, \tfrac{9}{20}\right\}, \left\{\tfrac{3}{20}, \tfrac{7}{20}\right\}, \left\{\tfrac{9}{20}, \tfrac{7}{20}\right\} \right\}.$$

The transformation $\mathcal{A}'$ on the basis S is

$$\left\{\tfrac{1}{20}, \tfrac{3}{20}\right\} \mapsto \left\{\tfrac{3}{20}, \tfrac{9}{20}\right\} \qquad\qquad \left\{\tfrac{3}{20}, \tfrac{9}{20}\right\} \mapsto \left\{\tfrac{1}{20}, \tfrac{9}{20}\right\} + \left\{\tfrac{1}{20}, \tfrac{7}{20}\right\}$$

$$\left\{\tfrac{1}{20}, \tfrac{9}{20}\right\} \mapsto \left\{\tfrac{1}{20}, \tfrac{3}{20}\right\} + \left\{\tfrac{1}{20}, \tfrac{7}{20}\right\} \qquad\qquad \left\{\tfrac{3}{20}, \tfrac{7}{20}\right\} \mapsto \left\{\tfrac{1}{20}, \tfrac{9}{20}\right\}$$

$$\left\{\left\{\tfrac{1}{20}, \tfrac{7}{20}\right\} \mapsto \left\{\tfrac{1}{20}, \tfrac{3}{20}\right\} \qquad\qquad \left\{\tfrac{9}{20}, \tfrac{7}{20}\right\} \mapsto \left\{\tfrac{1}{20}, \tfrac{7}{20}\right\}\right.$$

The associated matrix is

$$A = \begin{pmatrix} 0 & 0 & 0 & 1 & 0 & 0 \\ 1 & 0 & 1 & 0 & 0 & 0 \\ 1 & 0 & 0 & 0 & 0 & 0 \\ 0 & 1 & 1 & 0 & 0 & 0 \\ 0 & 1 & 0 & 0 & 0 & 0 \\ 0 & 0 & 1 & 0 & 0 & 0 \end{pmatrix},$$

with a spectral radius of 1.3953.

As the above example shows, the Thurston restricted algorithm allows us to reduce the dimensions of the matrix as well as the cardinality of the orbit of $-a$. Furthermore, the sum of the elements of any row of A' is, at most, 2, while the sum of the elements of a row in A can be greater than 2. Figure 3 shows the core entropy as a function of the external argument for $d = 3$.

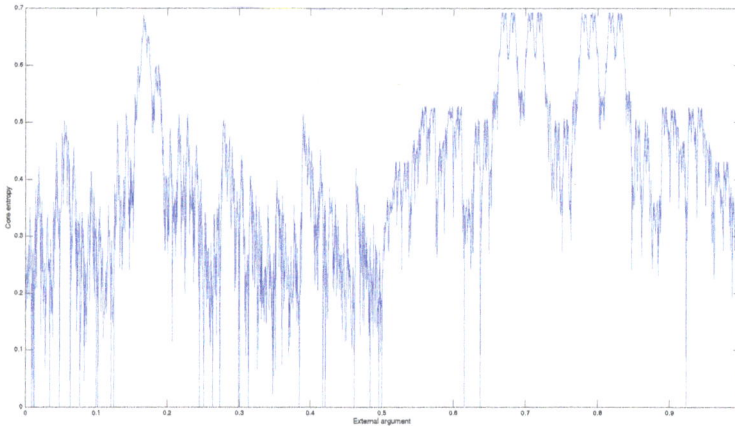

Figure 3. Core entropy for $d = 3$.

Author Contributions: Formal analysis, G.B. and D.G.

Funding: Consejo Nacional de Ciencia y Tecnología: 278289.

Acknowledgments: We would like to thank the referees for their valuable comments to improve this manuscript. The second author would like to thank CONACYT for the scholarship 278289.

References

1. Douady, A. *Topological Entropy of Unimodal Maps: Monotonicity for Quadratic Polynomials*; Springer: Berlin, Germany, 1993; pp. 65–87.
2. Milnor, J.; Thurston, W. *On Iterated Maps of the Interval, in Dynamical Systems*; Springer: Berlin, Germany, 1988, pp. 465–563.
3. Milnor, J.; Tresser, C. On entropy and monotonicity for real cubic maps, with an appendix by Adrien Douady and Pierrette Sentenac. *Commun. Math. Phys.* **2000**, *209*, 123–178. [CrossRef]

4. Radulescu, A. The connected isentropes conjecture in a space of quartic polynomials. *Discrete Contin. Dyn. Syst.* **2007**, *19*, 139–175. [CrossRef]
5. Bruin, H.; van Strien, S. Monotonicity of entropy for real multimodal maps. *J. Am. Math. Soc.* **2015**, *28*, 1–61. [CrossRef]
6. Li, T. A Monotonicity Conjecture for the Entropy of Hubbard Trees. Ph.D. Thesis, State University of New York at Stony Brook, Stony Brook, NY, USA, August 2007.
7. Poirier, A. Hubbard trees. *Fund. Math.* **2010**, *208*, 193–248. [CrossRef]
8. Thurston, W.P. Geometry and dynamics of iterated Rational Maps. In *Complex Dynamics*; Schleicher, D., Selinger, N., Eds.; AK Peters/CRC Press: Wellesley, MA, USA, 2009; pp. 3–137.
9. Gao, Y. On Thurston's core entropy algorithm. *arXiv* **2015**, arXiv:1511.06513v2.
10. Tiozzo, G. Entropy, dimension and combinatorial moduli for one-dimensional dynamical systems. Ph.D. Thesis, Harvard University, Ann Arbor, MI, USA, April 2013.
11. Tiozzo, G. Continuity of core entropy of quadratic polynomials. *Invent. Math.* **2016**, *203*, 891–921. [CrossRef]
12. Gao, Y.; Tiozzo, G. The core entropy for polynomials of higher degree. *arXiv* **2017**, arXiv:1703.08703.
13. Roesch, P. Hyperbolic components of polynomials with a fixed critical point of maximal order, (English, French summary). *Ann. Sci. École Norm. Super.* **2007**, *40*, 901–949. [CrossRef]
14. Milnor, J. *Periodic Orbits, External Rays and the Mandelbrot Set: An Expository Account*; Geometrie Complexe et Systemes Dynamiques, Astérisque; American Mathematical Society: Providence, RI, USA, 2000; Volume 261, pp. 277–333.
15. Milnor, J. Cubic polynomial maps with periodic critical orbit, Part I. In *Complex Dynamics: Families and Friends*; Schleicher, D., Peters, A.K., Ed.; CRC Press: Boca Raton, FL, USA, 2009; pp. 333–411.
16. Pole, D. *Linear Algebra, a Modern Introduction*; 2nd ed.; Thomson: Belmont, CA, USA, 2006.
17. Block, L.B.; Coppel, W.A. *Dynamics in One Dimension*; Lecture Notes in Mathematics; Springer: Berlin, Germany, 1992; p. 1513.
18. Carleson, L.; Gamelin, T.W. *Complex Dynamics*; Universitext: Tracts in Mathematics; Springer: New York, NY, USA, 1993.
19. Douady, A. Algorithms for computing angles in the Mandelbort set. In *Chaotic Dynamics and Fractals*; Academic Press: Cambridge, MA, USA, 1986; pp. 155–168.
20. Zakeri, S. Biaccessibility in quadratic Julia Sets. *Ergod. Theory Dyn. Syst.* **2000**, *20*, 1859–1883. [CrossRef]
21. Douady, A.; Hubbard, J.H. Étude dynamique des polynômes complexes. In *Publications Mathématiques d'Orsay*; Mathematical Publications of Orsay; Université de Paris-Sud, Département de Mathématiques: Orsay, France, 1984.

MDPI

Article

Strange Attractors Generated by Multiple-Valued Static Memory Cell with Polynomial Approximation of Resonant Tunneling Diodes

Jiri Petrzela [iD]

Department of Radio Electronics, FEEC, Brno University of Technology, 601 90 Brno, Czech Republic;
petrzelj@feec.vutbr.cz

Received: 20 August 2018; Accepted: 9 September 2018; Published: 12 September 2018

Abstract: This paper brings analysis of the multiple-valued memory system (MVMS) composed by a pair of the resonant tunneling diodes (RTD). Ampere-voltage characteristic (AVC) of both diodes is approximated in operational voltage range as common in practice: by polynomial scalar function. Mathematical model of MVMS represents autonomous deterministic dynamical system with three degrees of freedom and smooth vector field. Based on the very recent results achieved for piecewise-linear MVMS numerical values of the parameters are calculated such that funnel and double spiral chaotic attractor is observed. Existence of such types of strange attractors is proved both numerically by using concept of the largest Lyapunov exponents (LLE) and experimentally by computer-aided simulation of designed lumped circuit using only commercially available active elements.

Keywords: chaos; Lyapunov exponents; multiple-valued; static memory; strange attractors

---

1. Introduction

A general property of chaos is long-time unpredictability; i.e., random-like evolution of dynamical system even if the describing mathematical model does not contain stochastic functions or parameters. Because of its nature, chaotic behavior was often misinterpreted as noise. The first mention of this kind of the complex solution was in [1] where Lorenz noticed the extreme sensitivity of autonomous deterministic dynamics to tiny changes of the initial conditions. After this very milestone, chaos started to be reported in many distinct scientific fields as well as daily life situations. Chaotic motion has been observed in chemical reactions [2], classical mechanics [3], hydrodynamics [4], brain activity [5], models of biological populations [6], economy [7] and, of course, in many lumped circuits.

Two basic vector field mechanisms are required for evolution of chaos: stretching and folding. The first mechanism is responsible for exponential divergence of two neighboring state trajectories and second one bounds strange attractor within a finite state space volume. Pioneering work showing the presence of robust chaotic oscillation within dynamics of simple electronic circuit is [8]. So far, the so-called Chua´s oscillator was subject of laboratory demonstrations, deep numerical investigations and many research studies [9–11]. Several interesting strange attractors associated with different vector field local geometries have been localized within the dynamics of three-segment piecewise-linear Chua systems [12–14]. However, the inventors of Chua´s oscillator built it intentionally to construct a vector field capable of generating chaotic waveforms. Progress in computational power together with development of the parallel processing allows chaos localization in standard functional blocks of radiofrequency subsystems such as in harmonic oscillators [15,16], frequency filters [17,18], phase-locked loops [19], power [20] and dc-dc [21] converters, etc. From a practical point of view, chaos represents an unwanted operational regime that needs to be avoided. It can be recognized among regular behavior because of the specific features in the frequency domain: continuous and broadband frequency spectrum. However, an approach that is more sophisticated is to derive a set of describing differential equations and utilize the concept of LLE to find regions of chaotic solutions [22].

Searching for chaos in a mathematical model that describes simplified real electronic memory block is also topic of this paper. Three programs were utilized for the numerical analysis of MVMS: Matlab 2015 for the search-through-optimization algorithm including CUDA-based parallelization, Mathcad 15 for graphical visualization of results and Orcad Pspice 16 for circuit verification. The content of this paper is divided into four sections with the logical sequence: model description, numerical analysis, circuit realization and verification; both through simulation and measurement.

2. Dynamical Model of Fundamental MVMS

Basic mathematical model of MVMS [23,24] is given in Figure 1 and can be described by three first-order ordinary differential equations in the following form:

$$C_1 \frac{dv_1}{dt} = i - f_1(v_1) \quad C_2 \frac{dv_2}{dt} = i - f_2(v_2) \quad L \frac{di}{dt} = v_{bias} - v_1 - v_2 - R \cdot i \tag{1}$$

where the state vector is $\mathbf{x} = (v_1, v_2, i)^T$, C_1 and C_2 is parasitic capacitance of first and second RTD, L and R is summarized (RTDs are connected in series) lead inductance and resistance, respectively. Details about modeling of high frequency RTD including typical values of parasitic elements can be found in [25]. We can express both nonlinear functions (for $k = 1, 2$) as:

$$f_k(x) = a_k(x - d_k)^3 + b_k(x - d_k) + c_k \tag{2}$$

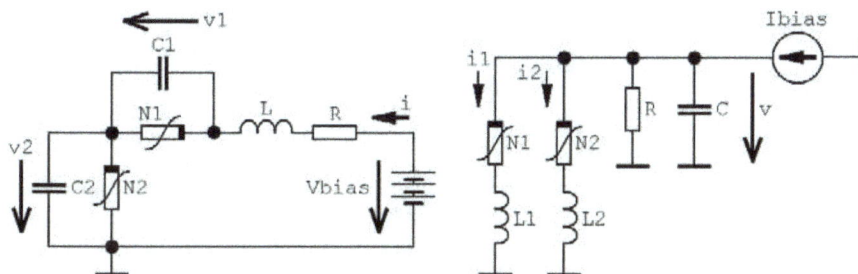

Figure 1. Basic network configurations of MVMS: original (**left** schematic), dual (**right** schematic).

Thus, AVC of each RTD is a cubic polynomial that should form an N-type curve with a negative segment. Fixed points x_e are all solutions of the problem $dx/dt = 0$. For further simplicity, let's assume that $R = 0\ \Omega$. We can determine global conditions and position for its existence within state space as each solution of system of the nonlinear algebraic equations, namely:

$$a_1(x_e - d_1)^3 + b_1(x_e - d_1) + c_1 = a_2(V_{bias} - x_e - d_2)^3 + b_2(V_{bias} - x_e - d_2) + c_2$$
$$y_e = V_{bias} - x_e \tag{3}$$
$$z_e = a_1(x_e - d_1)^3 + b_1(x_e - d_1) + c_1$$

Vector field geometry depends on the eigenvalues; i.e., roots of a characteristic polynomial. It can be calculated as $\det(s \cdot \mathbf{E} - \mathbf{J}) = 0$ where $\mathbf{E}$ is the unity matrix and $\mathbf{J}$ is the Jacobi matrix:

$$\mathbf{J} = \begin{pmatrix} -b_1 - 3a_1(x_e - d_1)^2 & 0 & 1 \\ 0 & -b_2 - 3a_2(y_e - d_2)^2 & 1 \\ -1 & -1 & 0 \end{pmatrix}. \tag{4}$$

Characteristic polynomial in symbolical form becomes:

$$s^3 + \{3a_2V_{bias}^2 - 6a_2V_{bias}(x_e + d_2) + b_1 + b_2 + 3a_1(x_e^2 + d_1^2) + 3a_2(x_e^2 + d_2^2) - 6x_e(a_1d_1 - a_2d_2)\}s^2 + \{[a_1a_2(9d_1^2 - 18d_1x_e + 9x_e^2) + 3b_1a_2]V_{bias}^2 + [a_1a_2(36d_1x_e^2 - 18d_1^2d_2 - 18d_1^2x_e + 36d_1d_2x_e - 18x_e^3 - 18d_2x_e^2) - 6b_1a_2x_e - 6b_1a_2d_2]V_{bias} + b_1b_2 + 3a_1d_1^2b_2 + 3b_1a_2d_2^2 + 3a_1b_2x_e^2 + 3b_1a_2x_e^2 - 6a_1d_1b_2x_e + 6b_1a_2d_2x_e + a_1a_2(9x_e^4 - 18d_1x_e^3 - 18d_2x_e^3 + 9d_1^2d_2^2 + 9d_1^2x_e^2 + 9d_2^2x_e^2 - 36d_1d_2x_e^2 - 18d_1d_2^2x_e + 18d_1^2d_2x_e) + 2\}s + b_1 + b_2 + 3V_{bias}^2a_2 + 3a_1d_1^2 + 3a_2d_2^2 + 3x_e^2(a_1 + a_2) - 6V_{bias}a_2(d_2 + a_2x_e) - 6x_e(a_1d_1 - a_2d_2) = 0,$$

(5)

Obviously, symbolical expressions for the individual eigenvalues are very complicated and cannot further contribute to the better understanding of a vector field configuration and chaos evolution; check well-known Cardan rules. In situation, where x_e and y_e coordinate of equilibrium point is close to offset voltages represented by d_1 and d_2, characteristic polynomial simplifies into the relation:

$$s^3 + (b_1 + b_2)s^2 + (b_1b_2 + 2)s + b_1 + b_2 = 0,$$

(6)

and the eigenvalues depend only on linear part of polynomial approximation of AVCs of RTDs.

Until very recently, analysis of MVMS was focused only on a high-frequency modeling of RTDs, influence of a pulse driving force on overall stability [26] and global dynamics [27] and specification of the boundary planes [28]. However, existence of chaos has not been uncovered and examined.

3. Numerical Results and Discussion

Numerical values of MVMS parameters can be obtained by the optimization technique described in [29]. In this case, the mathematical model of MVMS was considered piecewise-linear. Such kind of a vector field allows better understanding of chaos evolution, allows partial analytic solution and makes linear analysis generally more powerful. However, situation when AVCs of both RTDs are approximated by the polynomial functions is closer to reality. Thus, our problem stands as follows: couple of three-segment piecewise-linear functions needs to be transformed into the cubic (or higher-order if necessary) polynomial functions without losing robust chaotic solution having numerically close metric fractal dimension (Kaplan-Yorke is preferred over capacity because of rapid and precise calculation). For more details, readers should consult [30] where the inverse problem has been successfully solved. Finally, chaotic attractor like the so-called Rossler attractor [31] was localized.

Searching within smooth vector field (1) and by considering default normalized values $C_1 = 11$ F, $C_2 = 37$ F, $L = 100$ mH and $R = 0\ \Omega$ leads to the following optimal values of the cubic polynomials (2):

$$a_1 = 648.5, \quad b_1 = -23.1, \quad c_1 = -0.13, \quad d_1 = 0.3, \quad a_2 = 58.1, \quad b_2 = -21.9, \quad c_2 = -0.65, \quad d_2 = 0.5.\ (7)$$

Adopting these values and fourth-order Runge-Kutta numerical integration process we get the reference chaotic orbits provided in Figure 2. Accordingly to Equation (3) we have three fixed points: first located in position $x_e = (0.046, 0.704, -4.768)^T$ and characterized by eigenvalues $\lambda_1 = -101.241$, $\lambda_2 = 0.062$, $\lambda_3 = 13.915$, second equilibria in position $x_e = (0.292, 0.458, 0.043)^T$ having set of the eigenvalues $\lambda_1 = 0.09$, $\lambda_2 = 22.97$, $\lambda_3 = 21.72$ and $x_e = (0.55, 0.2, 4.299)^T$ with local behavior given by eigenvalues $\lambda_1 = -99$, $\lambda_2 = 0.13$, $\lambda_3 = 7.146$. This is an interesting geometric configuration of the vector field: a funnel-type strange attractor generated by saddle-node equilibria with the stability index 1, 0 and 0; saddle-focuses are completely missing. Developed optimization algorithm can be utilized to maximize unpredictability of a dynamical flow and increase system entropy. Starting with values (7) the following set of numerical values was obtained:

$$a_1 = 680, \quad b_1 = -23, \quad c_1 = -0.13, \quad d_1 = 0.3, \quad a_2 = 55, \quad b_2 = -25, \quad c_2 = -0.65, \quad d_2 = 0.5, \quad (8)$$

leading to a double-hook [12] like chaotic attractor visualized by means of Figure 3. The position of the first fixed point changes slightly to $x_e = (0.045, 0.705, -5.48)^T$ as well as the corresponding eigenvalues $\lambda_1 = -109.037$, $\lambda_2 = 0.049$, $\lambda_3 = 13.915$. The second equilibrium moves into position $x_e = (0.29, 0.46, 0.089)^T$ and possesses the set of eigenvalues $\lambda_1 = 0.084$, $\lambda_2 = 22.764$, $\lambda_3 = 24.817$. Finally, the last fixed point moves towards position $x_e = (0.554, 0.196, 5.31)^T$ where the local behavior will be given by the eigenvalues $\lambda_1 = -109.953$, $\lambda_2 = 0.085$, $\lambda_3 = 10.596$. The mentioned equilibria together with important state space sections are provided in Figure 4. Note that local geometries of a vector field are not affected since, in a closed neighborhood of the fixed points, dynamical movement is given by three eigenvectors as in the case of a funnel chaotic attractor. Also note that one direction of the flow (along the eigenvector associated with λ_1) is strongly attracting; this nature is evident from Figure 5. Interesting fragments of the bifurcation diagrams are depicted in Figure 6.

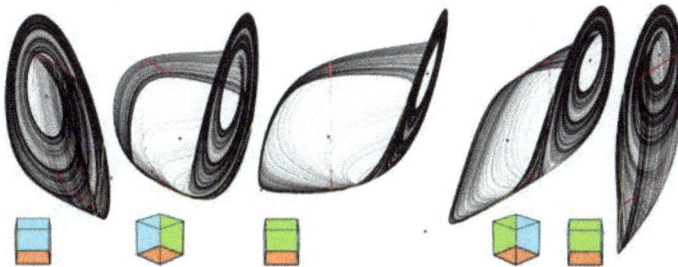

Figure 2. Three-dimensional perspective views on a typical chaotic attractor generated by polynomial MVMS for the initial conditions $\mathbf{x}_0 = (0.1, 0.3, 0)^T$; Poincaré sections in planes shifted by the offsets of polynomial functions: $x = c_1$ (red), $y = c_2$ (yellow); fixed point of the flow (blue dots), state space rotation (no deformations of axis system). Numerical integration with final time 10^4 and time step 0.01.

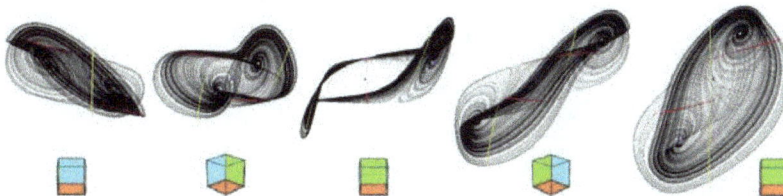

Figure 3. Three-dimensional perspective views on chaotic attractor with increased entropy generated by polynomial MVMS for the initial conditions $\mathbf{x}_0 = (0.1, 0.3, 0)^T$; Poincaré sections in planes shifted by offsets of polynomial functions: $x = c_1$ (red), $y = c_2$ (yellow); fixed point of the flow (blue dots), state space rotation (no deformations of axis system). Calculation with final time 10^4 and time step 0.01.

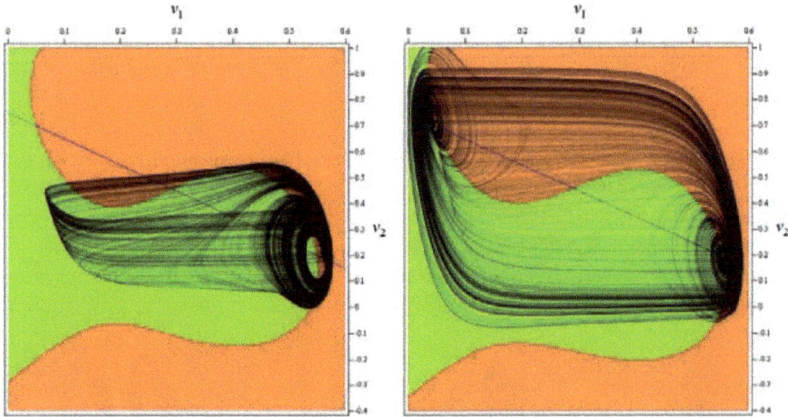

Figure 4. Important geometric structures located within state space: v_1–v_2 plane projection of function $f_1(v_1)$ (orange), function $f_2(v_2)$ (green) and $v_1 = V_{bias} - v_2$ (blue line), intersections of these planes are fixed points of dynamical flow. Discovered chaotic attractor put into the context of vector field.

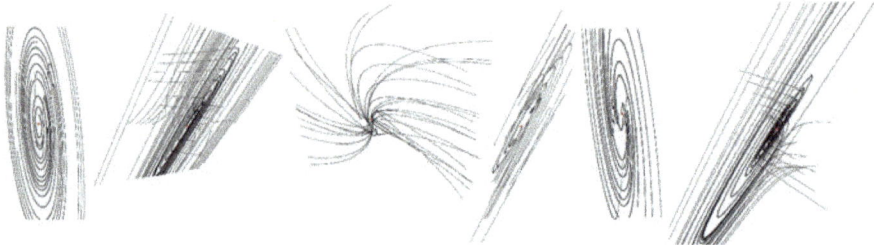

Figure 5. Graphical visualization of a dynamical behavior near the fixed points (red dots): equilibrium $x_e = 0.046$ (left two images), for fixed point located at $x_e = 0.292$ (middle two plots) and fixed point with $x_e = 0.55$ (right two state portraits).

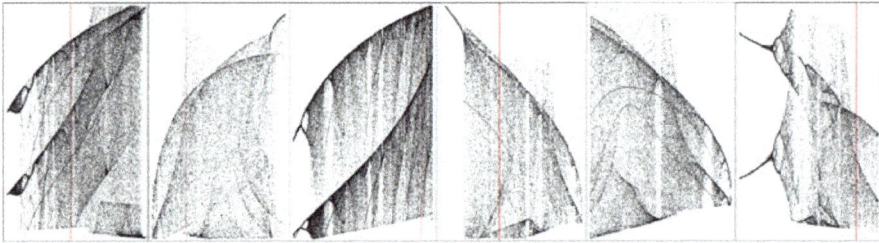

Figure 6. Gallery of one-dimensional bifurcation diagrams calculated with respect to the polynomial approximation of AVCs of RTDs; individual plots from left to right: horizontal axis represented by parameter range $a_1 \in (600, 700)$ calculated with step $\Delta = 0.1$; parameter range $b_1 \in (-30, 0)$ established with step $\Delta = 0.01$; parameter range $c_1 \in (-1.5, 0)$ together with step $\Delta = 0.001$; parameter $a_2 \in (30, 100)$ with step $\Delta = 0.1$; parameter $b_2 \in (-25, -20)$ together with step $\Delta = 0.01$ and parameter $c_2 \in (-4, 0)$ with step $\Delta = 0.01$.

Here, transient motion has been removed and plane $x = d_1$ has been used for individual slices. These are plotted against parameters a_k, b_k and c_k of polynomial approximations of AVC of k-th RTD. Parameters a_1 and a_2 can burn or bury chaotic attractor since these affect eigenvalues in "outer" parts of the active vector field (given by size of the state attractor) while geometry around "middle" fixed point remains almost unchanged.

Figure 7 shows numerical calculation of a gained energy with small time step with respect to the state space location; red regions mark large increment while dark blue stands for a small evolution.

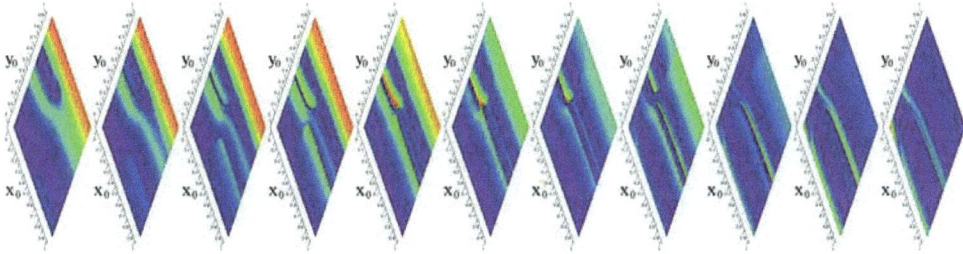

Figure 7. Rainbow-scaled contour plots of short-time evolution of MVMS energy for the state space slices forming unity cube (left to right): $z_0 = -5$, $z_0 = -4$, $z_0 = -3$, $z_0 = -2$, $z_0 = -1$, $z_0 = 0$, $z_0 = 1$, $z_0 = 2$, $z_0 = 3$, $z_0 = 4$, $z_0 = 5$.

As stated before, chaotic solution is sensitive to the changes of parameters a_k, b_k, c_k and d_k, where $k = 1, 2$. If we consider these values variable we create eight-dimensional hyperspace in which chaos alternates with periodic orbits or fixed-point solution. Two-dimensional subspaces hewed-out from such hyperspace are provided in Figure 8. Each graph is composed of $101 \times 101 = 10,201$ points, calculation routine deals with time interval $t \in (100, 10^4)$, random initial conditions inside basin of attraction and Gram-Smith orthogonalization [32]. Here, dark blue represents a trivial solution, light blue a limit cycle, green color stands for weak chaos and yellow marks strong chaotic behavior. Note that LLE for set of values (7) can be found in each visualized plot.

Maximal merit of LLE is 0.089 for a value set (7) and 0.103 for (8) and associated Kaplan–Yorke dimension [33] is 2.016 and 2.021 respectively. Specification of sufficiently large "chaotic" area is important also from practical viewpoint; as will be clarified later.

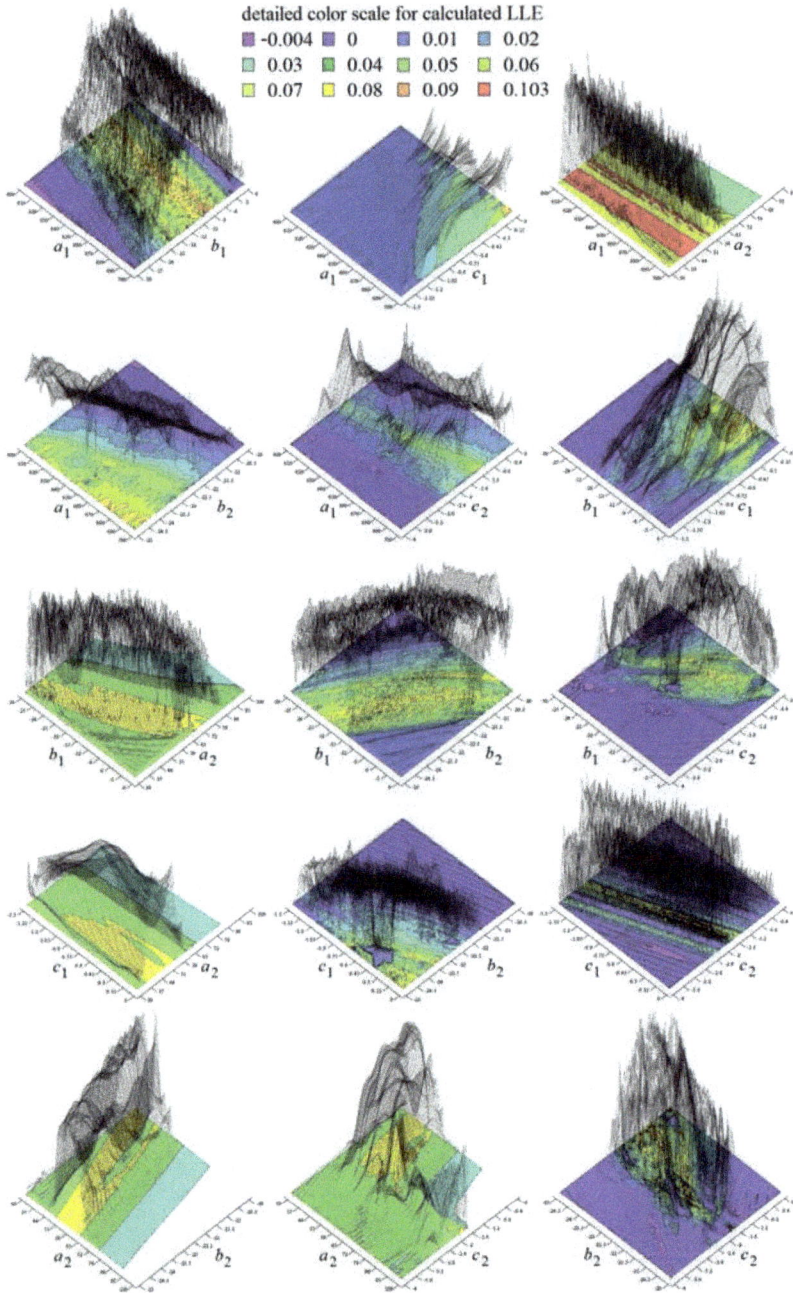

Figure 8. Gallery of the rainbow-scaled surface-contour plots of LLE as functions of two parameters; vertical range dedicated for LLE is -0.004 to 0.103, counting of plots from left to right and up to down: a_1–b_1, a_1–c_1, a_1–a_2, a_1–b_2, a_1–c_2, b_1–c_1, b_1–a_2, b_1–b_2, b_1–c_2, c_1–a_2, c_1–b_2, c_1–c_2, a_2–b_2, a_2–c_2, b_2–c_2.

4. Circuitry Realization of MVMS-Based Chaotic Oscillators

Design of analog equivalent circuit is common way how to prove existence of structurally stable strange attractors within the dynamics of a prescribed set of ordinary differential equations. Realization of such the so-called chaotic oscillators is a simple and straightforward task that we can solve by using several approaches [34–38], both using discrete components and in integrated form. A favorite method that allows us to utilize commercially available active elements follows the concept of analog computers. Thus, only three building blocks are required for circuit construction: inverting summing integrator, summing amplifier and, in the case of a polynomial nonlinearity, four-segment analog multiplier. Fully analog circuit implementation is shown in Figure 9.

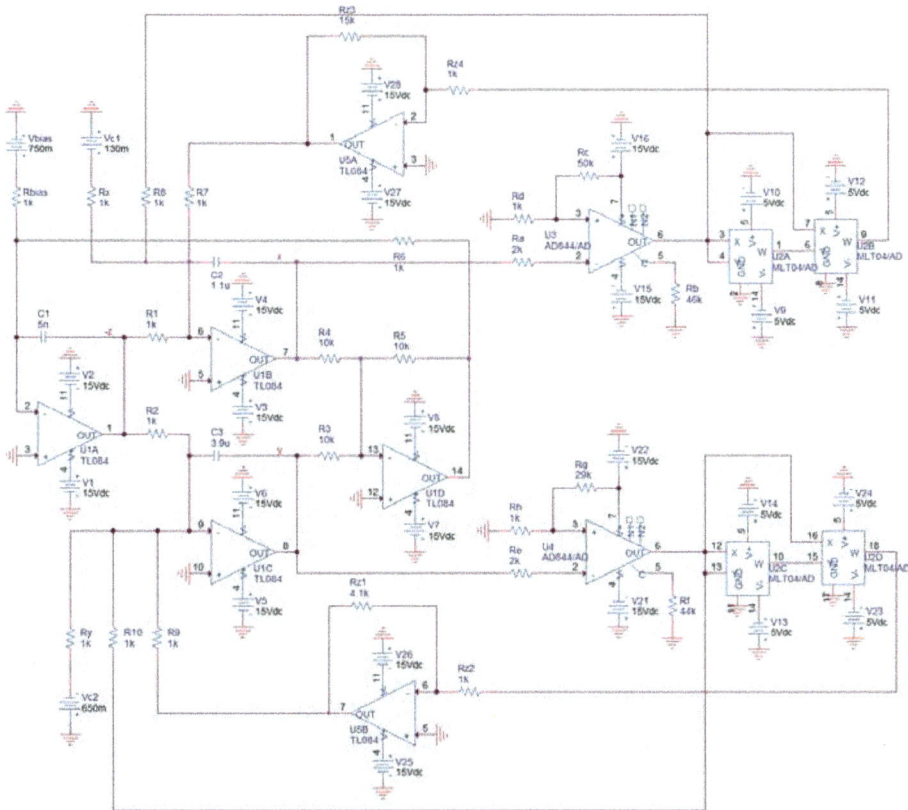

Figure 9. Chaotic oscillator designed to audio band based on integrator block schematic associated with mathematical model of MVMS, numerical values of passive network components are included.

Note that this circuit synthesis requires many active devices: two TL084, two AD844 and a single four channel four quadrant analog multiplier MLT04. Supply voltage is symmetrical ±15 V but, for MLT04, this voltage is lowered to ±5 V. Majority of the analog realizations of the chaotic systems with a polynomial nonlinearity utilize AD633, i.e., single channel multiplier. It is possible also in our case but with the cost of eight active devices. Dynamical range for correct operation of MLT04 is only ±2 V. However, prescribed strange attractor is smaller in v_1–v_2 dimension. Advantage of this circuit is that individual MVMS parameters can be adjusted independently using potentiometers. Theoretically, using different decomposition of the polynomial functions, total number of the active elements can be lowered to four. Proposed chaotic oscillator is uniquely described by the following set of the differential equations:

$$\frac{dv_1}{dt} = -\frac{1}{C_2}\left[-\frac{v_3}{R_1} - \frac{R_b}{R_a}\left(v_1 - \frac{R_d}{R_d+R_c}V_{cc}\right)\left\{\frac{1}{R_8} - \frac{R_{23}}{R_{24}}\left(\frac{R_b}{R_a}\right)^3 K^2\left(v_1 - \frac{R_d}{R_d+R_c}V_{cc}\right)\frac{1}{R_7}\right\} - \frac{V_{c1}}{R_x}\right]$$

$$\frac{dv_2}{dt} = -\frac{1}{C_3}\left[-\frac{v_3}{R_2} - \frac{R_f}{R_e}\left(v_2 - \frac{R_h}{R_h+R_g}V_{cc}\right)\left\{\frac{1}{R_{10}} - \frac{R_{21}}{R_{22}}\left(\frac{R_f}{R_e}\right)^3 K^2\left(v_2 - \frac{R_h}{R_h+R_g}V_{cc}\right)\frac{1}{R_9}\right\} - \frac{V_{c2}}{R_y}\right] \qquad (9)$$

$$\frac{dv_3}{dt} = \frac{1}{C_1}\left[-\frac{R_5}{R_6}\left(\frac{v_1}{R_4} + \frac{v_2}{R_5}\right) + \frac{V_{bias}}{R_{bias}}\right]$$

where V_{C1}, V_{C2} and V_{bias} are independent dc voltage sources and $K = 0.4$ is internally-trimmed scaling constant of the analog multiplier cells MLT04. Fundamental time constant of this chaotic oscillator is chosen to be $\tau = R \cdot Cv = 10^3 \cdot 10^{-7} = 100$ μs.

Of course, it is also possible to build both analog networks provided in Figure 1 directly. Instead of nonlinear two-ports we must construct a couple of resistors with polynomial AVC; systematic design towards these network elements can be found in [39,40]. Circuitry realization of original MVMS with state vector $\mathbf{x} = (v_1, v_2, i)^T$ is demonstrated by means of Figure 10; i.e., the state variables are voltages across grounded capacitors and current flowing through the inductor. Note that both polynomial function (2) need to be rewritten into the form $i = f(v) = a \cdot v^3 + b \cdot v^2 + c \cdot v + d$. Thus, a new set of the dimensionless ordinary differential equations to be implemented as lumped analog electronic circuit as follows:

$$\frac{dx}{dt} = -z - 648 \cdot x^3 - 583 \cdot x^2 - 152 \cdot x - 10.7 \qquad \frac{dy}{dt} = z - 58 \cdot y^3 + 87 \cdot y^2 - 20 \cdot y - 3.3$$

$$\frac{dz}{dt} = x - y + 0.75 \qquad (10)$$

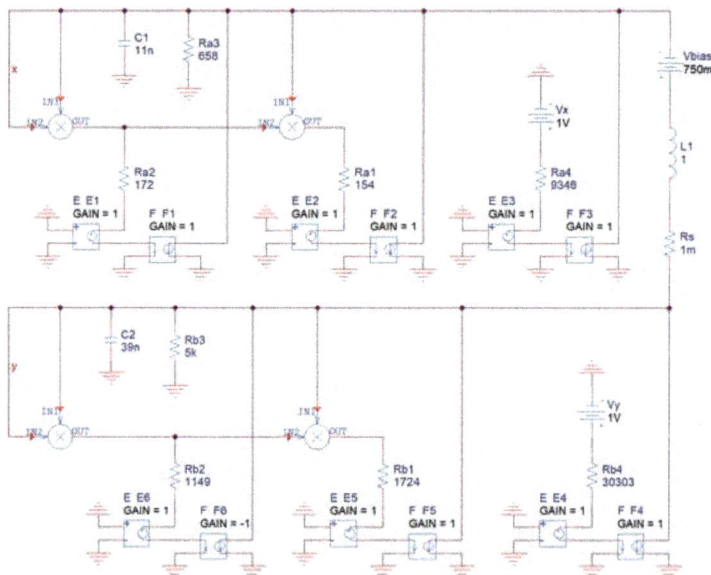

Figure 10. Chaotic system obtained directly from fundamental MVMS with ideal multipliers and ideal second-generation current-conveyors, numerical values of the circuit components are included.

Now assume that impedance and frequency norm are 10^5 and 10^4, respectively. Such values lead to the nominal inductance 1H. This simplified concept of chaotic oscillators is given in Figure 10 and described by following set of the ordinary differential equations:

$$C_1 \frac{dv_1}{dt} + i_L + \frac{v_1^3}{R_{a1}} + \frac{v_1^2}{R_{a2}} + \frac{v_1}{R_{a3}} + \frac{V_X}{R_{a4}} = 0$$
$$C_2 \frac{dv_2}{dt} + \frac{v_2^3}{R_{b1}} + \frac{v_2}{R_{b3}} + \frac{V_Y}{R_{b4}} = i_L + \frac{v_2^2}{R_{b2}} \qquad (11)$$
$$L \frac{di_L}{dt} + R_S i_L + v_2 = v_1 + V_{bias}$$

where a small value R_S can still model lead to resistances of both RTDs that are parts of MVMS in Figure 1.

Note that this kind of realization utilizes second generation current conveyors implemented by using ideal voltage-controlled voltage-source E and current-controlled current-source F. A positive variant of this active three-port element is commercially available as the AD844 while a negative variant is EL2082 (only one negative device is required). In practice, the inductor should be substituted by the synthetic equivalent; i.e., active floating gyrator (Antoniou's sub-circuit) with a capacitive load, check Figure 11. In this case, the number of the active elements raises to eight: a single TL084, six AD844s and a single MLT04. Dynamical behavior is uniquely determined by the following mathematical model:

$$\frac{dv_1}{dt} = \frac{1}{C_1} \left[-i - \frac{K^2 v_1^3}{R_{a1}+R_{in}} - \frac{K v_1^2}{R_{a2}+R_{in}} - \frac{v_1}{R_{a3}} - \frac{V_X}{R_{a4}+R_{in}} \right]$$
$$\frac{dv_2}{dt} = \frac{1}{C_2} \left[-i - \frac{K^2 v_2^3}{R_{b1}+R_{in}} - \frac{K v_2^2}{R_{b2}+R_{in}} - \frac{v_2}{R_{b3}} - \frac{V_y}{R_{b4}+R_{in}} \right] \qquad (12)$$
$$\frac{di}{dt} = \frac{R_{g2}}{R_{g1} R_{g3} R_{g4} C_{g1}} [v_1 - v_2 + V_{bias}]$$

where R_{in} represents input resistance of current input terminal of AD844. Its typical value is 50 Ω and, due to the high values of the polynomial coefficients and in the case of small impedance norm chosen, it cannot be generally neglected. On the other hand, we must avoid output current saturation of each AD844 and consider frequency limitations of each active device. Thus, choice of both normalization factors is always a compromise. Thanks to the symmetry inside floating Antoniou's structure $R_{g5} = R_{g3}$, $R_{g7} = R_{g1}$, $R_{g6} = R_{g2}$ and $C_{g2} = C_{g1}$. Behavior of this chaotic oscillator is extremely sensitive to the working resistors connected in the nonlinear two-terminal devices. Thus, calculated values were specified by Orcad Pspice optimizer where fitness functions (several should be defined to create tolerance channel) are absolute difference between polynomials in (11) and actual input resistance of designed circuit. Corresponding dc sweep analysis were estimated for input voltages from 0 V to 2 V with step 10 mV.

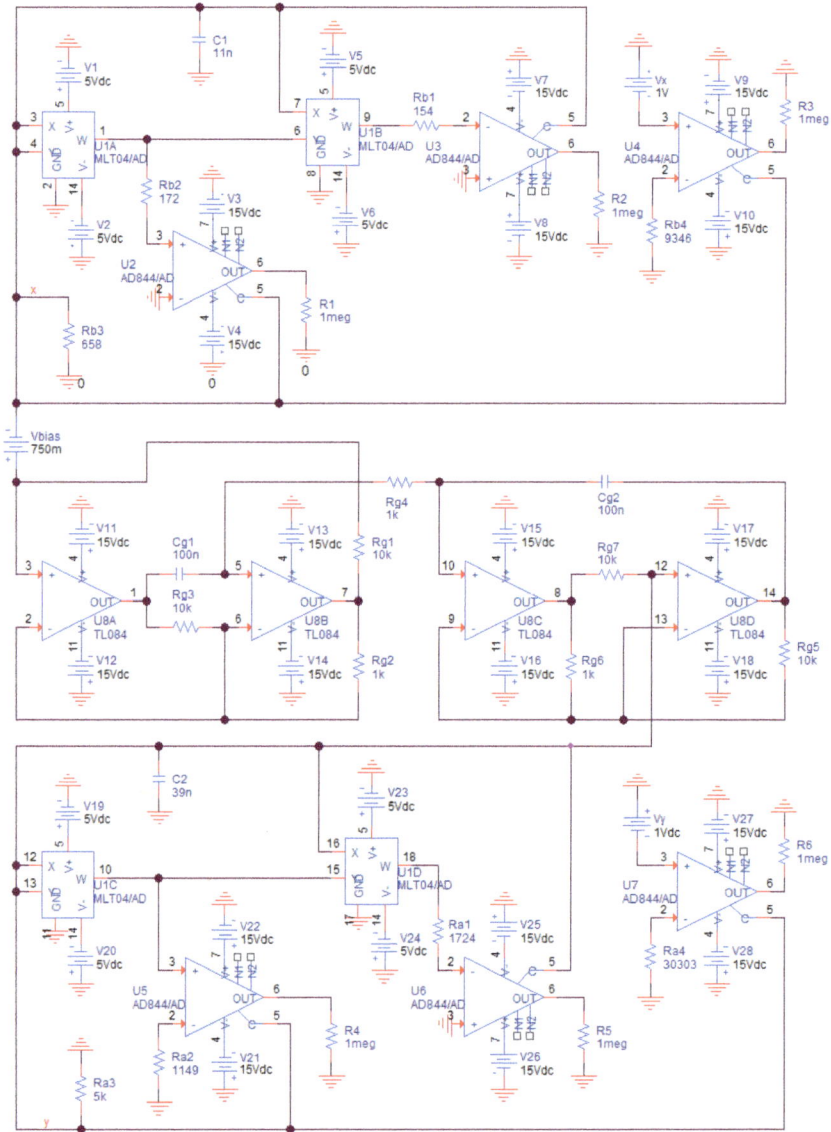

Figure 11. Chaotic system obtained directly from fundamental MVMS network with floating synthetic inductor, numerical values of the passive circuit components are included, ready for verification.

Last circuitry implementation is provided in Figure 12; namely dual network to the original MVMS. Impedance norm is chosen to be 10^3 and frequency normalization factor is 10^5. To get the reasonable values of the resistors further impedance rescaling is possible. Set of describing differential equations can be expressed as follows:

$$\frac{di_{L1}}{dt} = \frac{1}{L_1}\left[-R_{s1}i_{L1} + r_{T2}\left\{\frac{(r_{T1}i_{L1})^3}{R_1} + \frac{r_{T1}i_{L1}}{R_3} - \frac{V_1}{R_4}\right\} - r_{T3}\frac{(r_{T1}i_{L1})^2}{R_2}\right]$$

$$\frac{di_{L2}}{dt} = \frac{1}{L_2}\left[-R_{s2}i_{L2} + r_{T5}\left\{\frac{(r_{T4}i_{L2})^3}{R_5} + \frac{r_{T4}i_{L2}}{R_7} - \frac{V_2}{R_8}\right\} - r_{T6}\frac{(r_{T4}i_{L2})^2}{R_6}\right] \quad (13)$$

$$\frac{dv_Z}{dt} = \frac{1}{C_X}\left(-\frac{v_Z}{R_p} - i_{L1} - i_{L2} + I_1\right)$$

where r_{Tk} is trans-resistance of k-h ideal current-controlled voltage-source and I_1 is independent dc current source.

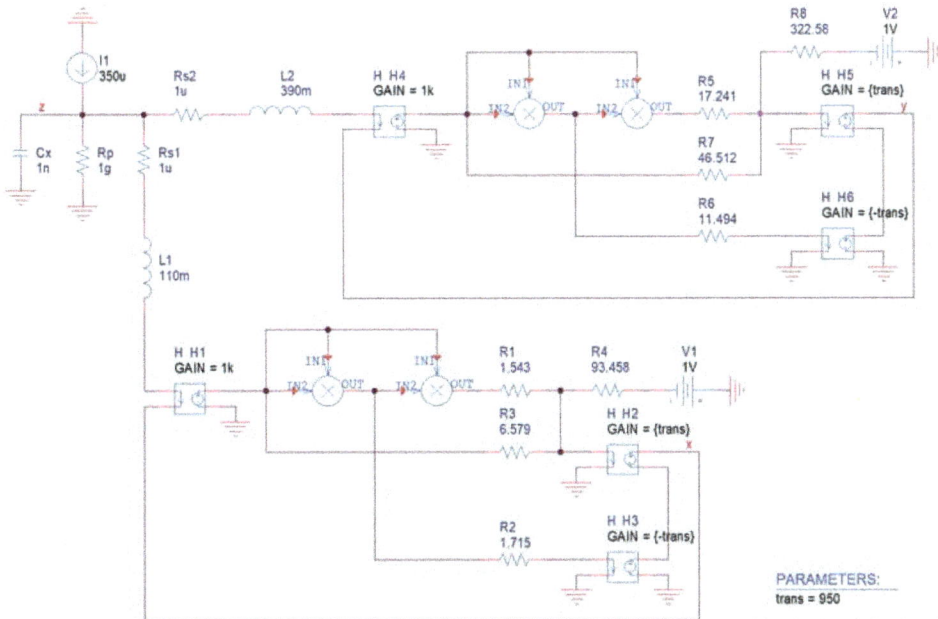

Figure 12. Hypothetic analog circuit realization dual to original MVMS network with ideal controlled sources: trans-resistance of output current-to-voltage conversion is considered as global parameter.

Polynomial nonlinearity is implemented by ideal multiplication using block MULT. Current flowing through both inductors can be sensed via small resistor R_{S1} and R_{S2} respectively. However, these resistors represent error terms that are inserted into describing differential equations; similarly as a parasitic shunt resistor R_p. Such parasitic properties change global dynamics, can boost system dimensionality (not in this situation) and desired chaotic behavior can eventually disappear.

Unfortunately, active elements where output voltage is controlled by input current are not off-the-shelf components. However, trans-resistance amplifier can be constructed using single standard operational amplifier and feedback resistor. To do this, input is fed directly to − terminal, node + is connected to ground, resistor between − and OUT. Node OUT also represents output of a designed transimpedance amplifier. Equivalently, the AD844 can also do the trick. Input can be connected to − terminal, ground to + terminal, resistor between C and ground and output to OUT. Considering the latter case, number of the active devices becomes seven: one MLT04 and six AD844.

Colored plots provided in Figure 8 demonstrate that region of chaos around discovered values (7) is wide enough to provide a structurally stable strange attractor; both funnel and double-scroll type. The same analysis was performed for set (8); this strange attractor should be also observable.

5. Circuit Simulation, Experimental Verification and Comparison

As mentioned before, AVC of both nonlinear resistors in Figure 11 should be as precise as possible to reach desired state space attractor. To fulfil this requirement, build-in Orcad Pspice optimizer can be adopted; see Figure 13 where optimal AVC of one nonlinear two-terminal device is reached.

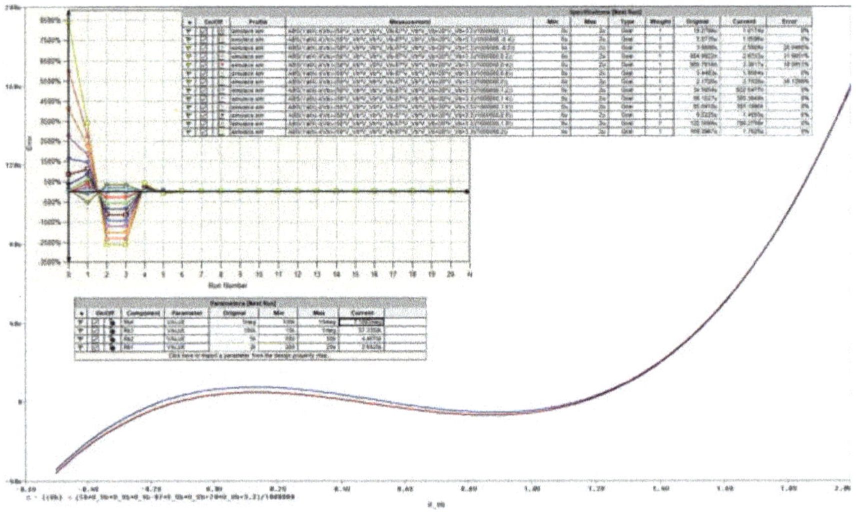

Figure 13. Orcad Pspice optimization toolbox: definitions of the objective functions and requested maximal errors (upper right field), error graph (upper left), new values of resistors (middle table).

Of course, the operational regime of this nonlinear resistor needs to be limited; in our case to input voltages into range starting with −0.5 V and ending with 2 V. Note that up to thirteen fitness functions were supposed to cover predefined operational range. Also note that optimization was stopped while some objective functions still have nonzero error (expressed in terms of the differential percentages). Important is the matter of similarity between simulated and numerically integrated strange attractor.

Implementation of the chaotic oscillator based on the integrator block schematic is more robust; i.e., the desired strange attractor is less vulnerable to the passive component matching: fabrication series (E12) and component tolerances (1%) were considered for design. We are also experiencing superior parasitic properties of the voltage-mode active elements: both integrated circuits TL084 and MLT04 have very high input and very low output resistances. Values of resistors that form external network connected to AD844 were chosen such that parasitic input and output impedances can be neglected. This is the main reason why this kind of MVMS realization was picked and forwarded into practical experiments and undergoes laboratory measurement.

Generally, parasitic properties of the active elements play important role in a design process of analog chaotic oscillators and, of course, should be minimized. Besides input and output impedances also roll-off effects of transfer constants need to be analyzed. Several publications have been devoted to reveal and study problems associated with the parasitic properties of the specific active devices and how these affect global dynamics; for example [41].

In simulation profile of transient response, final time was set to 1 s and maximal time step 10 μs. This setup is kept for each simulation mentioned in this section. Circuit simulation associated with Figure 9 is provided in Figure 14. To transform the funnel into a double-scroll chaotic attractor we should change value of the resistors R_{z3}, R_{z1} and R_{10}. Simulation results associated with direct realization of original MVMS provided in Figure 10 is demonstrated by means of Figure 15. Computer-aided analysis of chaotic oscillator with idealized controlled sources is showed in Figure 16.

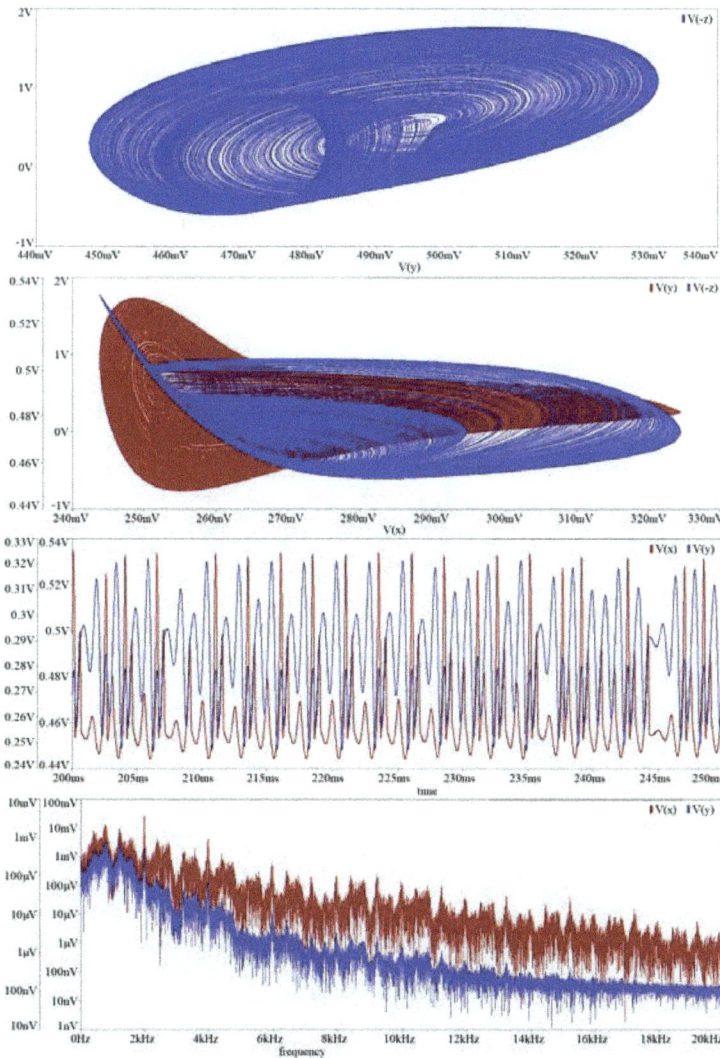

Figure 14. Orcad Pspice circuit simulation associated with chaotic circuit given in Figure 9: selected plane projection v_3–v_2 (blue, upper graph), v_2–v_1 (red) and v_3–v_1 (blue) of the chaotic attractor, generated chaotic signal v_1 (red) and v_2 (blue) in the time domain, chaotic waveform v_1 (red) and v_2 (blue) in the frequency domain. Note that significant frequency components are concentrated in audio range.

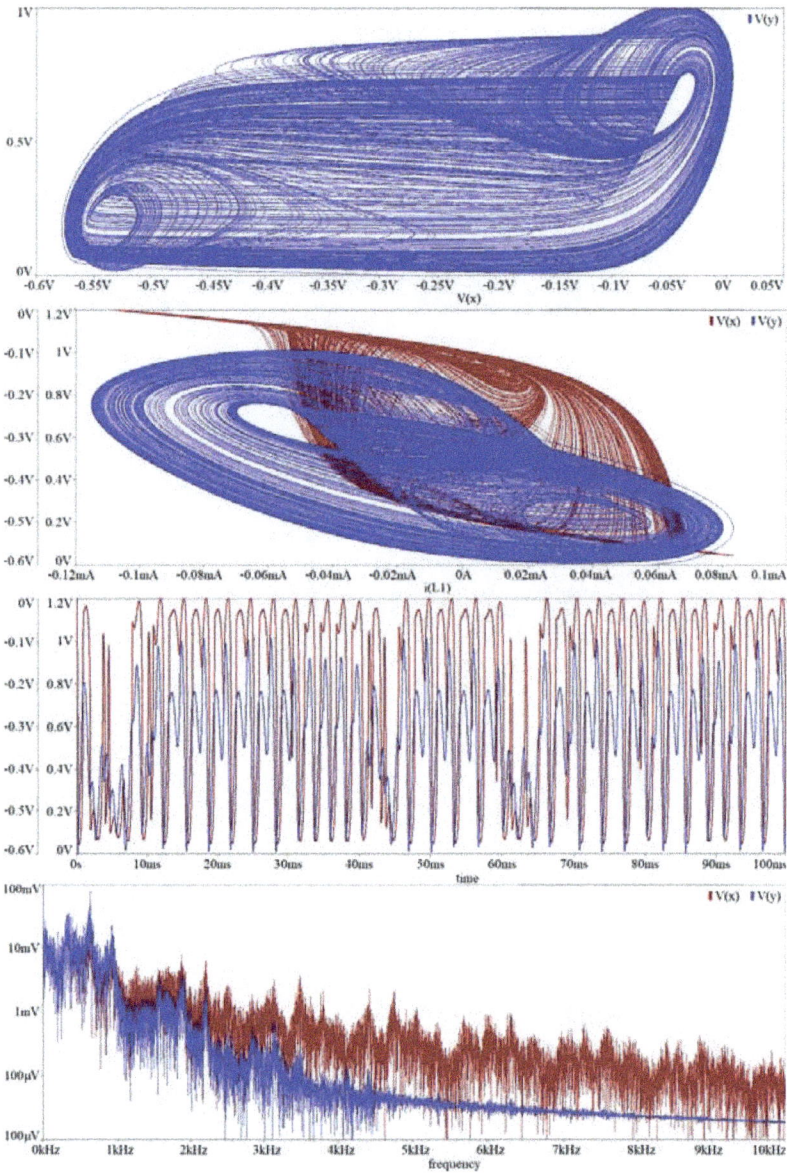

Figure 15. Orcad Pspice circuit simulation associated with network given in Figure 10: selected plane projection v_1–v_2 (blue, upper plot), v_1–i_L (red) and v_2–i_L (blue) of chaotic attractor, generated signal v_1 (red) and v_2 (blue) in time domain, chaotic waveform v_1 (red) and v_2 (blue) in frequency domain.

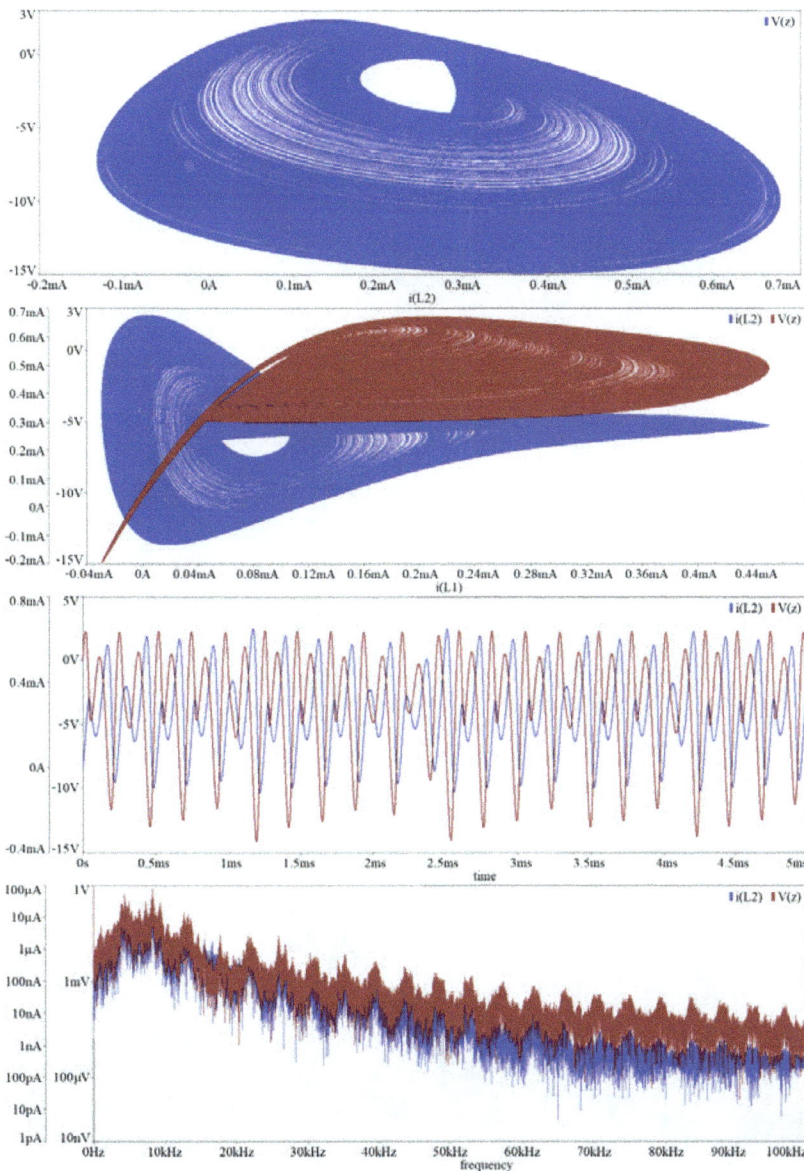

Figure 16. Orcad Pspice circuit simulation associated with network given in Figure 10: plane projection v_Z–i_{L2} (blue, upper plot), i_{L2}–i_{L1} (red) and v_Z–i_{L1} (blue) of the observed strange attractor, generated chaotic signal i_{L2} (red) and v_Z (blue) in the time domain, chaotic waveform i_{L2} (red) and v_Z (blue) in the frequency domain. Note that generated chaos is not affected by saturation levels of the active devices.

Existence of observable strange attractors and numerically expected routing-to-chaos scenarios within dynamics of the fundamental MVMS has been proved experimentally; by its construction on the bread-board (see Figure 17) and consequent laboratory measurement using the analog oscilloscope. Due to its simpler realization and increased robustness, only circuitry illustrated by means of Figure 9 was decided for a real measurement. Selected chaotic waveforms in a time domain are provided by means of Figure 18. Independent voltage source V_{bias} = 750 mV in series with R_{bias} = 1 kΩ can be replaced by the positive supply voltage +5 V and the fixed series resistance R_{bias} = 6.7 kΩ. Analogically, combinations V_{c1} = −130 mV, R_x = 1 kΩ and V_{c2} = −650 mV, R_y = 1 kΩ can be replaced by the negative supply voltage −5 V, R_x = 38.5 kΩ and the same voltage −5 V, R_y = 7.7 kΩ respectively.

Figure 17. Practical realization of the integrator-based chaotic MVMS using bread-board.

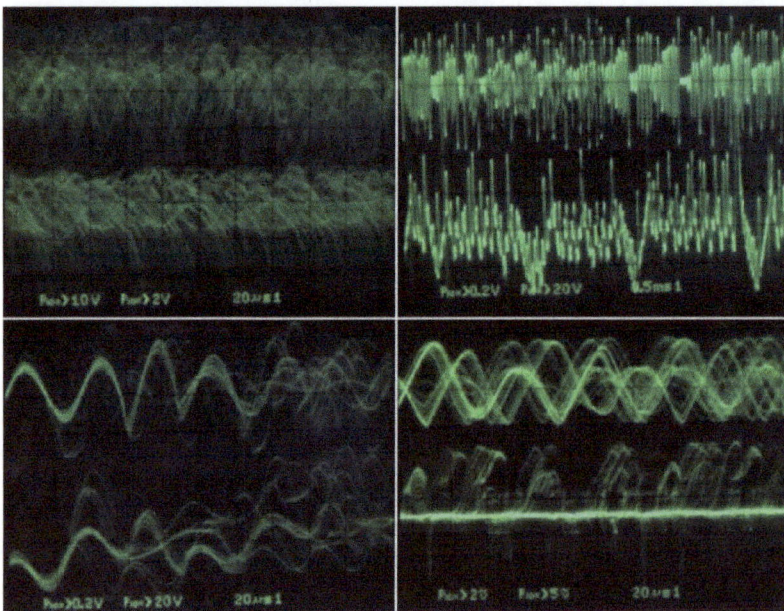

Figure 18. Selected chaotic waveforms in time domain generated by integrator-based MVMS.

Thus, there is no need to introduce three additional independent dc voltage sources into oscillator. However, to trace route-to-chaos scenarios, MVMS parameters c_1 and c_2 represented by voltages V_{c1} and V_{c2} have been considered as the variables. In practice, two voltage dividers based on potentiometers followed by two voltage buffers (remaining part of fifth integrated circuit TL084) did the trick. Since extreme values of V_{c1}, V_{c2} voltages can be managed observed Monge projections given in Figure 19 (plane v_3 vs. v_2) originate both inside and outside prescribed bifurcation scheme pictured in Figure 6 (third and sixth plot). Of course, very small steps of the hand-swept parameters cannot be captured by the oscilloscope. Different plane projections, namely v_3 vs. v_1, are shown in Figure 20. The provided screenshots are centered on captured state attractors not the origin of the state coordinates. Of course, the mentioned plane projections are not in full one-to-one correspondence with the theoretical results given in Figure 14, simply because transfer functions of the nonlinear two-ports cannot be defined precisely using imperfect discrete resistors; it does not get better shape even if the working resistors are replaced by the standard hand potentiometers. Finally, the sequence of periodic and chaotic windows with continuous change of the individual coefficients of the polynomials has been confirmed; roughly proving the correctness of bifurcation diagrams in Figure 6. During laboratory experiments, unexpected and a very interestingly-shaped strange attractors were identified (see Figure 21). Unfortunately, numerical values of the mathematical model parameters were not found, and reference state trajectories cannot be created. Initial conditions need not to be imposed on the outputs of the inverting integrators (capacitors need not to be pre-charged) since neighborhood of zero belongs into the basin of attraction for both strange attractors.

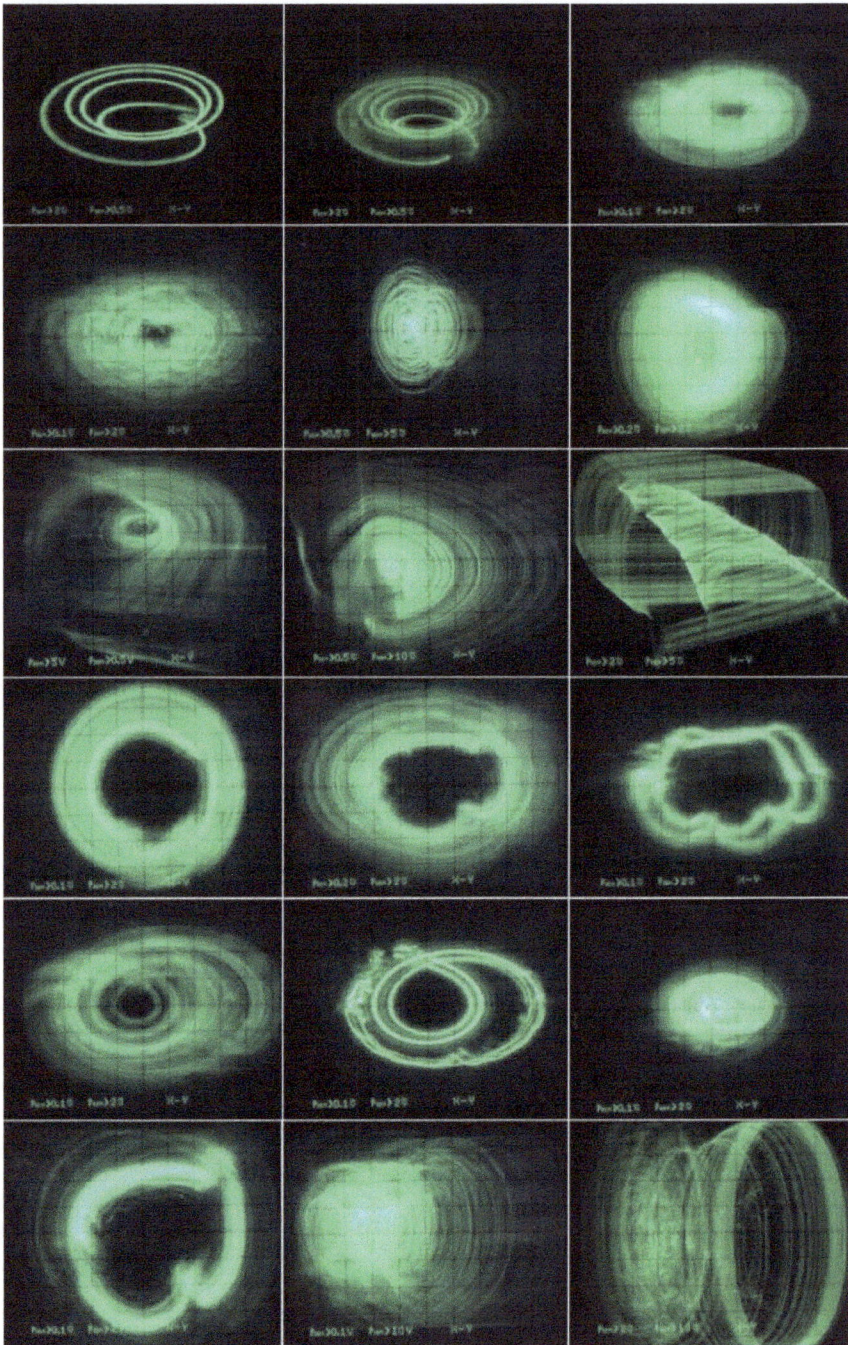

Figure 19. Selected v_3–v_2 plane projections measured of integrator-based MVMS realization, see text.

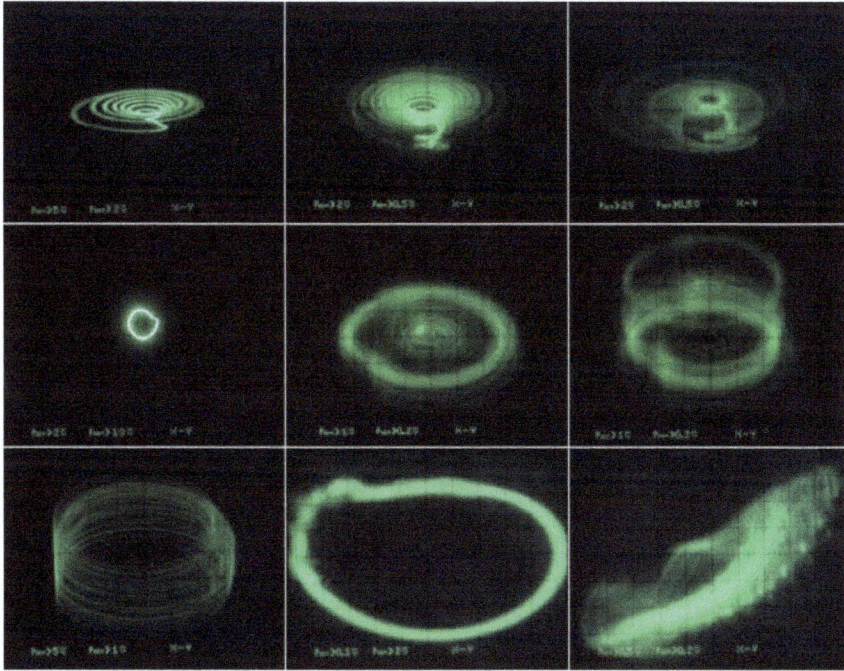

Figure 20. Few selected v_3–v_1 plane projections measured within dynamics of the integrator-based implementation MVMS, voltage sources V_{c1} and V_{c2} are swept, i.e., system parameters c_1 and c_2 are considered as variable parameters, change of the voltages starts within bifurcation diagrams provided by means of Figure 6 but finally goes far beyond it.

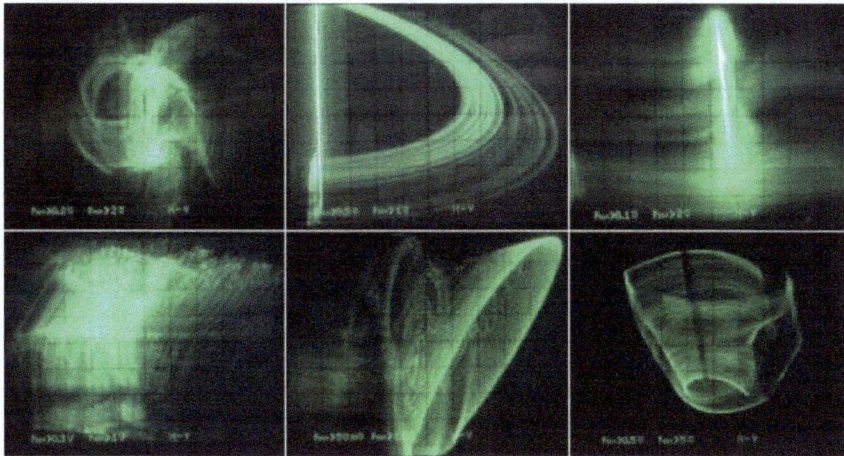

Figure 21. Gallery of interesting (still robust and real-time observable) strange attractors generated by the integrator-based realization of MVMS, different plane projections, see text for details.

6. Conclusions

Existence of robust chaotic attractors in the smooth vector field of MVMS have been demonstrated in this paper. This work represents a significant extension of the discoveries discussed in [29]. Both referred chaotic attractors are self-excited attractors; hidden attractors were not sought after and, in the case of analyzed MVMS dynamics, remain a mystery.

The proposed mathematical model: (1) together with the nonlinear functions (2) and parameters (7) or (8) can be also considered as a new chaotic dynamical system. This is still an up-to-date problem for many engineers and the topic of many recent scientific papers. However, algebraically much simpler dynamical systems that exhibit chaotic attractors exist; reading [42–47] is recommended.

Three different circuit realizations are presented. Each was verified by circuit simulation and the most robust implementation undergoes experimental measurement. Two designed oscillators utilize off-the-shelf active elements and can serve for various demonstrations.

The fundamental motivation of this work is to show that structurally stable strange attractors can be observed in smooth dynamical system naturally considered as nonlinear but with non-chaotic limit sets. However, designed autonomous chaotic oscillators can serve as the core circuits in many practical applications such as cryptography [48], spread-spectrum modulation techniques [49], useful signal masking [50,51], random number generators [52,53], etc.

This work also leaves several places for future research. For example, following interesting questions needs to be answered: Are there some hidden attractors? Are multi-scroll chaotic attractors observable if many RTD will be connected appropriately? Can MVMS generate chaos if values of the internal parameters are close to microelectronic memory cell fabricated in common technology?

Funding: Research described in this paper was financed by the National Sustainability Program under Grant LO1401. For the research, infrastructure of the SIX Center was used.

Conflicts of Interest: The author declares no conflict of interest.

References

1. Lorenz, E.N. Deterministic nonperiodic flow. *J. Atmos. Sci.* **1963**, *20*, 130–141. [CrossRef]
2. Rossler, O.E. Chemical turbulence: Chaos in a simple reaction-diffusion system. *J. Phys. Sci.* **1976**, *31*, 1168–1172. [CrossRef]
3. Shaw, S.W.; Rand, R.H. The transition to chaos in a simple mechanical system. *Int. J. Non-Linear Mech.* **1989**, *24*, 41–56. [CrossRef]
4. Hsieh, D.Y. Hydrodynamics instability, chaos and phase transition. *Nonlinear Anal. Theory Methods Appl.* **1997**, *30*, 5327–5334. [CrossRef]
5. Babloyantz, A.; Salazar, J.M.; Nicolis, C. Evidence of chaotic dynamics of brain activity during the sleep cycle. *Phys. Lett. A* **1985**, *111*, 152–156. [CrossRef]
6. May, R.M.; Wishart, D.M.G.; Bray, J.; Smith, R.L. Chaos and the dynamics of biological populations. *Proc. R. Soc. Lond. A* **1987**, *413*, 27–44. [CrossRef]
7. Day, R.H.; Pavlov, O.V. Computing economic chaos. *Comput. Econ.* **2004**, *23*, 289–301. [CrossRef]
8. Matsumoto, T. A chaotic attractor from Chua's circuit. *IEEE Trans. Circuits Syst.* **1984**, *31*, 1055–1058. [CrossRef]
9. Guzan, M. Variations of boundary surface in Chua's circuit. *Radioengineering* **2015**, *24*, 814–823. [CrossRef]
10. Pivka, L.; Spany, V. Boundary surfaces and basin bifurcations in Chua's circuit. *J. Circuits Syst. Comput.* **1993**, *3*, 441–470. [CrossRef]
11. Spany, V.; Galajda, P.; Guzan, M.; Pivka, L.; Olejar, M. Chua's singularities: Great miracle in circuit theory. *Int. J. Bifurc. Chaos* **2010**, *20*, 2993–3006. [CrossRef]
12. Bartissol, P.; Chua, L.O. The double hook. *IEEE Trans. Circuits Syst.* **1988**, *35*, 1512–1522. [CrossRef]
13. Parker, T.; Chua, L.O. The dual double scroll equation. *IEEE Trans. Circuits Syst.* **1987**, *31*, 1059–1073. [CrossRef]
14. Yang, T.; Chua, L.O. Piecewise-linear chaotic systems with a single equilibrium point. *Int. J. Bifurc. Chaos* **2000**, *10*, 2015–2060. [CrossRef]
15. Kennedy, M.P. Chaos in the Colpitts oscillator. *IEEE Trans. Circuits Syst.* **1994**, *41*, 771–774. [CrossRef]
16. Kilic, R.; Yildrim, F. A survey of Wien bridge-based chaotic oscillators: Design and experimental issues. *Chaos Solitons Fractals* **2008**, *38*, 1394–1410. [CrossRef]

17. Petrzela, J. On the existence of chaos in the electronically adjustable structures of state variable filters. *Int. J. Circuit Theory Appl.* **2016**, *11*, 605–653. [CrossRef]
18. Petrzela, J. Chaotic behavior of state variable filters with saturation-type integrators. *Electron. Lett.* **2015**, *51*, 1159–1160. [CrossRef]
19. Endo, T.; Chua, L.O. Chaos from phase-locked loops. *IEEE Trans. Circuits Syst.* **1988**, *35*, 987–1003. [CrossRef]
20. Hamill, D.C.; Jeffries, D.J. Subharmonics and chaos in a controlled switched-mode power converter. *IEEE Trans. Circuits Syst.* **1988**, *35*, 1059–1061. [CrossRef]
21. Zhou, X.; Li, J.; Youjie, M. Chaos phenomena in dc-dc converter and chaos control. *Procedia Eng.* **2012**, *29*, 470–473. [CrossRef]
22. Gotthans, T.; Petrzela, J. Experimental study of the sampled labyrinth chaos. *Radioengineering* **2011**, *20*, 873–879.
23. Smith, K.C. The prospects for multivalued logic: A technology and applications view. *IEEE Trans. Comput.* **1981**, *30*, 619–634. [CrossRef]
24. Buttler, J.T. Multiple-valued logic. *IEEE Potentials* **1995**, *14*, 11–14. [CrossRef]
25. Liou, W.R.; Roblin, P. High frequency simulation of resonant tunneling diodes. *IEEE Trans. Electron. Devices* **1994**, *41*, 1098–1111. [CrossRef]
26. Galajda, P.; Guzan, M.; Spany, V. The state space mystery with negative load in multiple-valued logic. *Radioengineering* **2008**, *17*, 19–24.
27. Guzan, M. Analysis of 6(4)-valued memory. *Elektron. Elektrotech.* **2014**, *20*, 89–92. [CrossRef]
28. Spany, V.; Pivka, L. Boundary surfaces in sequential circuits. *Int. J. Circuit Theory Appl.* **1990**, *18*, 349–360. [CrossRef]
29. Petrzela, J. Multi-valued static memory with resonant tunneling diodes as natural source of chaos. *Nonlinear Dyn.* **2018**, *93*, 1–21. [CrossRef]
30. Petrzela, J. Optimal piecewise-linear approximation of quadratic chaotic dynamics. *Radioengineering* **2012**, *21*, 20–28.
31. Rossler, O.E. An equation for continuous chaos. *Phys. Lett.* **1976**, *57A*, 397–398. [CrossRef]
32. Sprott, J.C. *Chaos and Time Series Analysis*; Oxford University Press: Oxford, UK, 2003; pp. 115–122. ISBN 978-0198508397.
33. Frederickson, P.; Kaplan, J.L.; Yorke, E.D.; Yorke, J.A. The Liapunov dimension of strange attractors. *J. Differ. Equ.* **1983**, *49*, 185–207. [CrossRef]
34. Trejo-Guerra, R.; Tlelo-Cuautle, E.; Jimenez-Fuentes, J.M.; Munoz-Pacheco, J.M.; Sanchez-Lopez, C. Multiscroll floating gate-based integrated chaotic oscillator. *Int. J. Circuit Theory Appl.* **2011**, *41*, 831–843. [CrossRef]
35. Itoh, M. Synthesis of electronic circuits for simulating nonlinear dynamics. *Int. J. Bifurc. Chaos* **2001**, *11*, 605–653. [CrossRef]
36. Trejo-Guerra, R.; Tlelo-Cuautle, E.; Carbajal-Gomez, V.H.; Rodriguez-Gomez, G. A survey on the integrated design of chaotic oscillators. *Appl. Math. Comput.* **2013**, *219*, 5113–5122. [CrossRef]
37. Petrzela, J.; Hrubos, Z.; Gotthans, T. Modeling deterministic chaos using electronic circuits. *Radioengineering* **2011**, *20*, 438–444.
38. Trejo-Guerra, R.; Tlelo-Cuautle, E.; Jimenez-Fuentes, J.M.; Sanchez-Lopez, C.; Munoz-Pacheco, J.M.; Espinosa-Flores-Verdad, G.; Rocha-Perez, J.M. Integrated circuit generating 3- and 5-scroll attractors. *Commun. Nonlinear Sci. Numer. Simul.* **2012**, *17*, 4328–4335. [CrossRef]
39. Zhong, G.-Q.; Ko, K.-T.; Man, K.-F.; Tang, K.-S. A systematic procedure for synthesizing two-terminal devices with polynomial non-linearity. *Int. J. Circuit Theory Appl.* **2001**, *29*, 241–249. [CrossRef]
40. Petrzela, J.; Pospisil, V. Nonlinear resistor with polynominal AV characteristics and its application in chaotic oscillator. *Radioengineering* **2004**, *13*, 20–25.
41. Munoz-Pacheco, J.M.; Tlelo-Cuautle, E.; Toxqui-Toxqui, I.; Sanchez-Lopez, C.; Trejo-Guerra, R. Frequency limitations in generating multi-scroll chaotic attractors using CFOAs. *Int. J. Electron.* **2014**, *101*, 1559–1569. [CrossRef]
42. Sprott, J.C. Simple chaotic flows and circuits. *Am. J. Phys.* **2000**, *68*, 758–763. [CrossRef]
43. Sprott, J.C. Simplest dissipative chaotic flow. *Phys. Lett. A* **1997**, *228*, 271–274. [CrossRef]
44. Sprott, J.C. Some simple chaotic flows. *Phys. Rev. E* **1994**, *50*, 647–650. [CrossRef]
45. Sprott, J.C. Some simple chaotic jerk functions. *Am. J. Phys.* **1997**, *65*, 537–543. [CrossRef]

46. Eichhorn, R.; Linz, S.J.; Hanggi, P. Simple polynomial classes of chaotic jerky dynamics. *Chaos Solitons Fractals* **2002**, *13*, 1–15. [CrossRef]
47. Gottlieb, H.P.W.; Sprott, J.C. Simplest driven conservative chaotic oscillator. *Phys. Lett. A* **2001**, *291*, 385–388. [CrossRef]
48. Xu, G.; Shekofteh, Y.; Akgul, A.; Li, C.; Panahi, S. A new chaotic system with a self-excited attractor: Entropy measurement, signal encryption, and parameter estimation. *Entropy* **2018**, *20*, 86. [CrossRef]
49. Itoh, M. Spread spectrum communication via chaos. *Int. J. Bifurc. Chaos* **1999**, *9*, 155–213. [CrossRef]
50. Morgul, O.; Feki, M. A chaotic masking scheme by using synchronized chaotic systems. *Phys. Lett. A* **1999**, *251*, 169–176. [CrossRef]
51. Pan, J.; Ding, Q.; Du, B. A new improved scheme of chaotic masking secure communication based on Lorenz system. *Int. J. Bifurc. Chaos* **2012**, *22*, 1250125. [CrossRef]
52. Drutarovsky, M.; Galajda, P. Chaos-based true random number generator embedded in a mixed-signal reconfigurable hardware. *J. Electr. Eng.* **2006**, *57*, 218–225.
53. Drutarovsky, M.; Galajda, P. A robust chaos-based true random number generator embedded in reconfigurable switched-capacitor hardware. *Radioengineering* **2007**, *16*, 120–127.

![entropy]

Article

The Co-existence of Different Synchronization Types in Fractional-order Discrete-time Chaotic Systems with Non–identical Dimensions and Orders

Samir Bendoukha [1] [ID], Adel Ouannas [2], Xiong Wang [3], Amina-Aicha Khennaoui [4], Viet-Thanh Pham [5,*], Giuseppe Grassi [6] and Van Van Huynh [5]

[1]   Electrical Engineering Department, College of Engineering at Yanbu, Taibah University, Medina 42353, Saudi Arabia; sbendoukha@taibahu.edu.sa
[2]   Department of Mathematics and Computer Science, University of Larbi Tebessi, Tebessa 12002, Algeria; Ouannas@mail.univ-tebessa.dz
[3]   Institute for Advanced Study, Shenzhen University, Shenzhen 518060, China; wangxiong8686@szu.edu.cn
[4]   Department of Mathematics and Computer Sciences, University of Larbi Ben M'hidi, Oum El Bouaghi 04000, Algeria; kamina_aicha@yahoo.fr
[5]   Modeling Evolutionary Algorithms Simulation and Artificial Intelligence, Faculty of Electrical & Electronics Engineering, Ton Duc Thang University, Ho Chi Minh City, Vietnam; huynhvanvan@tdt.edu.vn
[6]   Dipartimento Ingegneria Innovazione, Universita del Salento, 73100 Lecce, Italy; giuseppe.grassi@unisalento.it
*   Correspondence: phamvietthanh@tdt.edu.vn

Received: 25 August 2018; Accepted: 13 September 2018; Published: 14 September 2018

Abstract: This paper is concerned with the co-existence of different synchronization types for fractional-order discrete-time chaotic systems with different dimensions. In particular, we show that through appropriate nonlinear control, projective synchronization (PS), full state hybrid projective synchronization (FSHPS), and generalized synchronization (GS) can be achieved simultaneously. A second nonlinear control scheme is developed whereby inverse full state hybrid projective synchronization (IFSHPS) and inverse generalized synchronization (IGS) are shown to co-exist. Numerical examples are presented to confirm the findings.

Keywords: fractional discrete chaos; entropy; projective synchronization; full state hybrid projective synchronization; generalized synchronization; inverse full state hybrid projective synchronization; inverse generalized synchronization

1. Introduction

Discrete-time chaotic systems have been the center of attention in the fields of control [1,2] and secure communications in the last few years [3–6]. This attention can be attributed to two main characteristics. First, the chaotic nature of the dynamical systems, which seems random-like but is in fact completely determined and can be predicted once the initial conditions are known. For instance, this allows for the generation of pseudo–random sequences in secret or private-key encryption. The second interesting property is their discrete nature, which allows for simple implementation and reduced computational complexity. Among the well known discrete-time chaotic systems proposed throughout the years are the Hénon map [7], the Lozi system [8], the generalized Hénon map [9] and the Baier–Klein system [10].

In recent years, researchers have picked an interest in fractional discrete-time chaotic systems. These involve fractional calculus, where the differences in the system's dynamics are fractional. Numerous studies have been dedicated to establishing a framework for fractional discrete calculus such as [11–14]. A good summary of the subject is given in [15].

In general, chaotic systems became of interest in science and engineering in the early 1990s after synchronization was demonstrated. The earliest studies include [16–19]. Since then, various types of synchronization have been proposed in the literature including projective synchronization (PS) [20], generalized synchronization (GS) [21], full state hybrid projective synchronization (FSHPS) [22], and many more. Some modification have also been made to these synchronization types leading, for instance, to inverse generalized synchronization (IGS) [23] and inverse FSHPS (IFSHPS) [24]. With the emergence of fractional chaotic maps such as the fractional Hénon map [25] and the fractional generalized Hénon map [26], the synchronization of such maps became of interest. Very few studies can be found on the subject including [27–32].

Naturally, curiosity grew as to the possibility of multiple synchronization types being achieved simultaneously for the states of the response system. This phenomenon is commonly referred to as the co-existence of synchronization types. Many studies can be found in the literature proposing linear and nonlinear control laws that give rise to the co-existence phenomenon for continuous-time integer-order systems [33], continuous-time fractional systems [34–38], and discrete-time integer-order systems [39–41]. However, to the best of the authors' knowledge, no such studies have been made for fractional-order discrete-time systems. This has motivated us to examine the phenomenon and develop suitable control laws for various types co-existing.

The next section of this paper describes the model for the drive and response systems and defines the necessary notation and synchronization types. Section 3 presents the control law that guarantees the co-existence of PS, FSHPS, and GS as the control laws that establish the co-existence of IFSHPS and IGS. Section 4 presents numerical examples that confirm the validity of the findings. Finally, Section 6 summarizes the work carried out in this paper.

2. System Model

In order to establish the co-existence of different synchronization types in fractional order discrete-time chaotic systems, we consider the generic n-dimensional drive and response pair of the form

$$\begin{cases} {}^C\Delta_a^v x_i\left(t\right) = F_i(X\left(t+\alpha-1\right)), \\ {}^C\Delta_a^v y_i\left(t\right) = G_i(Y\left(t+\beta-1\right)) + u_i, \end{cases} \quad t \in \mathbb{N}_{a+1-v} \tag{1}$$

where $X\left(t\right) = \left(x_1\left(t\right),\dots,x_n\left(t\right)\right)^T$, $Y\left(t\right) = \left(y_1\left(t\right),\dots,y_n\left(t\right)\right)^T$ represent the states of the drive and response systems, respectively, F_i, G_i are functions from $\mathbb{R}^n$ to $\mathbb{R}$ for $1 \le i \le n$, and $u_i, 1 \le i \le n$, denote control parameters to be identified by means of the synchronization strategy.

The notation ${}^C\Delta_a^v X\left(t\right)$ denotes the v–Caputo type delta difference of a function $X\left(t\right) : \mathbb{N}_a \to \mathbb{R}$ with $\mathbb{N}_a = \{a, a+1, a+2, \dots\}$ [12], which is of the form

$$ {}^C\Delta_a^v X\left(t\right) = \Delta_a^{-(n-v)} \Delta^n X\left(t\right) = \frac{1}{\Gamma\left(n-v\right)} \sum_{s=a}^{t-(n-v)} \left(t-\sigma\left(s\right)\right)^{(n-v-1)} \Delta^n X\left(s\right), \tag{2}$$

for $v \notin \mathbb{N}$ is the fractional order, $t \in \mathbb{N}_{a+n-v}$, and $n = [v] + 1$. In (2), the v–th fractional sum of $\Delta_s^n X\left(t\right)$ is defined similar to [11] as

$$\Delta_a^{-v} X\left(t\right) = \frac{1}{\Gamma\left(v\right)} \sum_{s=a}^{t-v} \left(t-\sigma\left(s\right)\right)^{(v-1)} X\left(s\right), \tag{3}$$

with $v > 0$, $\sigma(s) = s+1$. The term $t^{(v)}$ denotes the falling function defined in terms of the Gamma function Γ as

$$t^{(v)} = \frac{\Gamma\left(t+1\right)}{\Gamma\left(t+1-v\right)}. \tag{4}$$

Let us, now, define the types of synchronization with which we are interested in our study. The idea is to show that multiple types of synchronization may exist simultaneously for a pair of fractional-order discrete-time chaotic systems.

Definition 1. *If there exists a controller $U = (u_i)_{1 \leq i \leq n}$ and either constants $\gamma \in \mathbb{R}^*$, a matrix Φ, a map $\phi : \mathbb{R}^n \longrightarrow \mathbb{R}^n$, a matrix Θ, or a map $\varphi : \mathbb{R}^n \longrightarrow \mathbb{R}^n$ such that*

$$\lim_{t \to +\infty} \|Y(t) - \gamma X(t)\| = 0 \quad \implies \quad \text{Pair (1) is projective synchronized (PS).}$$

$$\lim_{t \to +\infty} \|Y(t) - \Phi X(t)\| = 0 \quad \implies \quad \text{Pair (1) is full state hybrid projective synchronized (FSHPS).}$$

$$\lim_{t \to +\infty} \|Y(t) - \phi(Y(t))\| = 0 \quad \implies \quad \text{Pair (1) is generalized synchronized (GS).}$$

$$\lim_{t \to +\infty} \|X(t) - \Theta Y(t)\| = 0 \quad \implies \quad \text{Pair (1) is inverse full state hybrid projective synchronized (IFSHPS).}$$

$$\lim_{t \to +\infty} \|X(t) - \varphi(Y(t))\| = 0 \quad \implies \quad \text{Pair (1) is inverse generalized synchronized (IGS).}$$

Note that in Definition 1 above, γ is a constant used to scale the master state vector. Matrices Φ and Θ represent linear transformation of the master and slave state vectors, respectively, and are usually referred to as scaling matrices. The terms ϕ and φ denote some arbitrary maps from $\mathbb{R}^n$ towards $\mathbb{R}^n$. In general, these are nonlinear maps that represent scaling functions. We are now ready to present the main findings of our study.

3. Results

3.1. Co-existence of PS, FSHPS and GS

Let us consider the 2-dimensional drive system and a 3-dimensional response system given, respectively, by

$$^C\Delta_a^v x_i(t) = f_i(X(t + v - 1)), \quad i = 1, 2, \tag{5}$$

and

$$^C\Delta_a^v y_i(t) = \sum_{j=1}^{3} b_{ij} y_j(t + v - 1) + g_i(Y(t + v - 1)) + u_i, \quad i = 1, 2, 3, \tag{6}$$

where $t \in \mathbb{N}_{a+1-v}$, $0 < v \leq 1$, $f_i : \mathbb{R}^2 \longrightarrow \mathbb{R}$, $1 \leq i \leq 2$, $(b_{ij}) \in \mathbb{R}^{3 \times 3}$ is the linear part of the drive system, $g_i : \mathbb{R}^3 \longrightarrow \mathbb{R}$, $1 \leq i \leq 3$, are nonlinear functions, and u_i, $i = 1, 2, 3$, are controllers to be designed. Based on Definition 1, we may define the co-existence of PS, FSHPS and GS for the coupled systems (5) and (6) as follows.

Definition 2. *It is said that PS, FSHPS and GS co-exist in the synchronization of the drive system (5) and the response systems (6) if there exist a controller $U = (u_i)_{1 \leq i \leq 3}$, a constant $\gamma \in \mathbb{R}^*$, a constant matrix $\Phi = (\Phi_{ij})_{1 \times 2}$, and nonlinear map $\phi : \mathbb{R}^2 \longrightarrow \mathbb{R}$ such that the synchronization errors*

$$\begin{cases} e_1(t) = y_1(t) - \gamma x_1(t), \\ e_2(t) = y_2(t) - \Phi \times (x_1(t), x_2(t))^T, \\ e_3(t) = y_3(t) - \phi(x_1(t), x_2(t)), \end{cases} \tag{7}$$

all satisfy the asymptotic rule

$$\lim_{t \to +\infty} \|e_i(t)\| = 0 \quad \text{for } i = 1, 2, 3. \tag{8}$$

Remark 1. *From the error system (7), it is obvious that states y_1 and x_1 are projective synchronized, y_2 is full state hybrid projective synchronized with x_1 and x_2, and y_3 is generalized synchronized with x_1 and x_2.*

We also need to state the following theorems, which are necessary for the proofs to come.

Theorem 1 ([42]). *The zero equilibrium of the linear fractional-order discrete-time system*

$$^{C}\Delta_a^v e(t) = \mathbf{D}e(t+v-1),\tag{9}$$

where $e(t) = (e_1(t),...,e_n(t))^T$, $0 < v \le 1$, $\mathbf{D} \in \mathbb{R}^{n \times n}$ *and* $\forall t \in \mathbb{N}_{a+1-v}$, *is asymptotically stable if*

$$\lambda \in \left\{ z \in \mathbb{C}: \ |z| < \left(2\cos\frac{|\arg z| - \pi}{2 - v} \right)^v \ and \ |\arg z| > \frac{v\pi}{2} \right\},\tag{10}$$

for all the eigenvalues λ *of* $\mathbf{D}$.

Next, we propose control laws that achieve the co-existence rule (7). Let us define the matrix $B = (b_{ij})_{3\times3}$.

Theorem 2. *PS, FSHPS and GS co-exist for the pair (5)–(6) subject to*

$$\begin{cases} u_1 = \sum_{j=1}^{3}(c_{1j} - b_{1j})\,e_j(t) - \sum_{j=1}^{3}b_{1j}y_j(t) - g_1(Y(t+v-1)) + \gamma f_1(X(t+v-1)),\\ u_2 = \sum_{j=1}^{3}(c_{2j} - b_{2j})\,e_j(t) - \sum_{j=1}^{3}b_{2j}y_j(t) - g_2(Y(t+v-1)) + \Phi_1 f_1(X(t+v-1))\\ \quad + \Phi_1 f_2(X(t+v-1)),\\ u_3 = \sum_{j=1}^{3}(c_{3j} - b_{3j})\,e_j(t) - \sum_{j=1}^{3}b_{3j}y_j(t) - g_3(Y(t+v-1)) + {}^{C}\Delta^{\beta}\phi(x_1(t),x_2(t)), \end{cases}\tag{11}$$

where $C = (c_{ij})_{3\times3}$ *is a constant matrix chosen such that all the eigenvalues* λ_i *of* $B - C$ *satisfy*

$$-2^v < \lambda_i < 0, \quad i = 1,2,3.\tag{12}$$

Proof. The difference equations corresponding to the error system (7) are given by

$$\begin{cases} {}^{C}\Delta_a^v e_1(t) = {}^{C}\Delta_a^v y_1(t) - \gamma\, {}^{C}\Delta_a^v x_1(t),\\ {}^{C}\Delta_a^v e_2(t) = {}^{C}\Delta_a^v y_2(t) - \Phi\, {}^{C}\Delta_a^v(x_1(t),x_2(t))^T,\\ {}^{C}\Delta_a^v e_3(t) = {}^{C}\Delta_a^v y_3(t) - {}^{C}\Delta_a^v \phi(x_1(t),x_2(t)). \end{cases}\tag{13}$$

Substituting the system nonlinearities yields

$$\begin{cases} {}^{C}\Delta_a^v e_1(t) = \sum_{j=1}^{3}b_{1j}y_j(t+v-1) + g_1(Y(t+v-1)) + u_1 - \gamma f_1(X(t+v-1)),\\ {}^{C}\Delta_a^v e_2(t) = \sum_{j=1}^{3}b_{2j}y_j(t+v-1) + g_2(Y(t+v-1)) + u_2 - \Phi_1 f_1(X(t+v-1))\\ \quad - \Phi_1 f_2(X(t+v-1)),\\ {}^{C}\Delta_a^v e_3(t) = \sum_{j=1}^{3}b_{3j}y_j(t+v-1) + g_3(Y(t+v-1)) + u_3 - {}^{C}\Delta_a^v \phi(x_1(t),x_2(t)). \end{cases}\tag{14}$$

Substituting the proposed control law (11) in (14) yields

$$\begin{cases} {}^{C}\Delta_a^v e_1(t) = \sum_{j=1}^{3}(b_{1j} - c_{1j})\,e_j(t+v-1),\\ {}^{C}\Delta_a^v e_2(t) = \sum_{j=1}^{3}(b_{2j} - c_{2j})\,e_j(t+v-1),\\ {}^{C}\Delta_a^v e_3(t) = \sum_{j=1}^{3}(b_{3j} - c_{3j})\,e_j(t+v-1). \end{cases}\tag{15}$$

In order to show that the zero solution of (16) is globally asymptotically stable, we use the linearization method as described in Theorem 1. The error system (15) can be written in the compact form

$$^{C}\Delta_a^v e(t) = (B - C)e(t+v-1).\tag{16}$$

where $e(t) = (e_1(t), e_2(t), e_3(t))^T$. According to condition (12), it is easy to see that all the eigenvalues of the matrix $B - C$ satisfy $|\arg \lambda_i| = \pi > \frac{v\pi}{2}$ and $|\lambda_i| < \left(2\cos\frac{|\arg \lambda_i| - \pi}{2-v}\right)^v$, for $i = 1,2,3$. It, then, follows immediately from Theorem 1 that the zero solution of (16) is globally asymptotically stable and consequently, systems (5) and (6) are synchronized in 3–dimensions according to Definition 2. $\square$

3.2. Co-existence of IFSHPS and IGS

We, now, would like to achieve similar results for the inverse synchronization types listed in Definition 1. Consider the drive and response pair of the form

$$\begin{cases} {}^C\Delta_a^v x_i\left(t\right) = \sum_{j=1}^2 a_{ij} x_j\left(t+v-1\right) + f_i\left(X\left(t+v-1\right)\right), & i=1,2, \\ {}^C\Delta_a^v y_i\left(t\right) = g_i\left(Y\left(t+v-1\right)\right) + u_i, & i=1,2,3, \end{cases} \tag{17}$$

where $t \in \mathbb{N}_{a+1-v}$, $A = \left(a_{ij}\right) \in \mathbb{R}^{2\times 2}$ and $f_i : \mathbb{R}^2 \to \mathbb{R}$, $1 \le i \le 2$, are nonlinear functions, and $g_i : \mathbb{R}^3 \to \mathbb{R}$, $1 \le i \le 3$. Based on Definition 1, we can state what is meant by the co-existence of IFSHPS and IGS for (17) as summarized in the following definition.

Definition 3. *IFSHPS and IGS are said to co-exist in the synchronization of the pair (17) if there exist controllers u_i, $i=1,2,3$, a constant matrix $\Theta = \left(\Theta_{ij}\right)_{1\times 3}$, and a map $\varphi : \mathbb{R}^3 \longrightarrow \mathbb{R}$ such that the synchronization errors*

$$\begin{cases} e_1\left(t\right) = x_1\left(t\right) - \Theta \times \left(y_1\left(t\right), y_2\left(t\right), y_3\left(t\right)\right)^T, \\ e_2\left(t\right) = x_2\left(t\right) - \varphi\left(y_1\left(t\right), y_2\left(t\right), y_3\left(t\right)\right), \end{cases} \tag{18}$$

all satisfy the asymptotic rule

$$\lim_{t\to+\infty} e_i\left(t\right) = 0 \ \text{for } i=1,2. \tag{19}$$

Remark 2. *From the error system (18), it is apparent that x_1 is inverse full state hybrid projective synchronized with $y_1\left(t\right)$, $y_2\left(t\right)$ and $y_3\left(t\right)$, and that $x_2(t)$ is inverse generalized synchronized with $y_1\left(t\right)$, $y_2\left(t\right)$ and $y_3\left(t\right)$.*

Suppose that the function φ can be factorized in the form

$$\varphi\left(y_1\left(t\right), y_2\left(t\right), y_3\left(t\right)\right) = \sum_{j=1}^3 \theta_j y_j\left(t\right) + \psi\left(y_1\left(t\right), y_2\left(t\right), y_3\left(t\right)\right), \tag{20}$$

where θ_j, $j=1,2,3$, are real numbers and $\psi : \mathbb{R}^3 \to \mathbb{R}$ is a nonlinear function. The error dynamics (18) yield the difference equations

$$\begin{cases} {}^C\Delta_a^v e_1\left(t\right) = {}^C\Delta_a^v x_1\left(t\right) - \Theta_1 \, {}^C\Delta_a^v y_1\left(t\right) - \Theta_2 \, {}^C\Delta_a^v y_2\left(t\right) \\ \qquad\qquad - \Theta_3 \, {}^C\Delta_a^v y_3\left(t\right), \\ {}^C\Delta_a^v e_2\left(t\right) = {}^C\Delta_a^v x_2\left(t\right) - \theta_1 \, {}^C\Delta_a^v y_1\left(t\right) - \theta_2 \, {}^C\Delta_a^v y_2\left(t\right) \\ \qquad\qquad - \theta_3 \, {}^C\Delta_a^v y_3\left(t\right) - {}^C\Delta_a^v \psi\left(y_1\left(t\right), y_2\left(t\right), y_3\left(t\right)\right). \end{cases} \tag{21}$$

To simplify the equations, we can define

$$R_1 = \sum_{j=1}^2 a_{1j} x_j\left(t\right) + f_1\left(X\left(t\right)\right) - \sum_{j=1}^3 \Theta_j g_i\left(Y\left(t\right)\right), \tag{22}$$

and

$$R_2 = \sum_{j=1}^2 a_{2j} x_j\left(t+v-1\right) + f_2\left(X\left(t\right)\right) - \sum_{j=1}^3 \theta_j g_i\left(Y\left(t\right)\right) - {}^C\Delta_a^v \psi\left(y_1\left(t\right), y_2\left(t\right), y_3\left(t\right)\right). \tag{23}$$

Using (22) and (23), (21) may be written in the reduced form

$$\begin{cases} {}^C\Delta_a^v e_1\left(t\right) = R_1 - \sum_{j=1}^3 \Theta_j u_j, \\ {}^C\Delta_a^v e_2\left(t\right) = R_2 - \sum_{j=1}^3 \theta_j u_j, \end{cases} \tag{24}$$

or more compactly as

$$^C\Delta_a^v e(t) = R - M \times (u_1, u_2)^T - (\Theta_3 u_3, \theta_3 u_3)^T, \tag{25}$$

where $R = (R_1, R_2)^T$ and

$$M = \begin{pmatrix} \Theta_1 & \Theta_2 \\ \theta_1 & \theta_2 \end{pmatrix}. \tag{26}$$

To establish the co-existence of IFSHPS and IGS, we assume that M is invertible and denote its inverse by M^{-1}. The control law is, then, given by

$$(u_1, u_2)^T = M^{-1} \times [(L - A) e(t) + R] \text{ and } u_3 = 0, \tag{27}$$

where $L \in \mathbb{R}^{2\times 2}$ is a control matrix to be determined. Substituting (27) into Equation (25), we get

$$^C\Delta_a^v e(t) = (A - L) e(t + v - 1). \tag{28}$$

The following result follows in a similar manner to Theorem 2. The proof has been omitted as it can be inferred directly from that of Theorem 2.

Theorem 3. *If the control matrix L is chosen such that all the eigenvalues of $A - L$ such that $-2^v < \lambda_i < 0$, $i = 1, 2$, then IFSHPS and IGS co-exist for (17) as described in (18) subject to control law (27).*

4. Numerical Examples

We will now put the theoretical results presented in Section 3 to the test. We consider the 2D fractional Hénon map proposed in [25] as the drive system and the 2D fractional-order generalized Hénon map [26] as the response system. The pair is described as

$$\begin{cases} ^C\Delta_a^v x_1(t) = x_2(t + v - 1) - x_1(t + v - 1) + 1 - a_1 x_1^2(t + v - 1), \\ ^C\Delta_a^v x_2(t) = b_1 x_1(t + v - 1) - x_2(t + v - 1), \end{cases} \tag{29}$$

and

$$\begin{cases} ^C\Delta_a^v y_1(t) = -y_1(t + v - 1) - b_2 y_3(t + v - 1) + u_1(t + v - 1), \\ ^C\Delta_a^v y_2(t) = b_2 y_3(t + v - 1) + y_1(t + v - 1) - y_2(t + v - 1) + u_2(t + v - 1), \\ ^C\Delta_a^v y_3(t) = 1 + y_2(t + v - 1) - a_2 y_3^2(t + v - 1) - y_3(t + v - 1) + u_3(t + v - 1). \end{cases} \tag{30}$$

The linear and nonlinear parts of the drive system (29) and the response system (30) are given by, respectively,

$$A = \begin{pmatrix} -1 & 1 \\ b_1 & -1 \end{pmatrix}, f = \begin{pmatrix} -a_1 x_1^2(t) + 1 \\ 0 \end{pmatrix},$$

and

$$B = \begin{pmatrix} -1 & 0 & -b_2 \\ 1 & -1 & b_2 \\ 0 & 1 & -1 \end{pmatrix}, g = \begin{pmatrix} 0 \\ 0 \\ 1 - a_2 y_3^2(t) \end{pmatrix}.$$

These two systems were proposed in the literature and shown to exhibit chaotic behaviors. For instance, when $(a_1, b_1) = (1.4, 0.3)$, $(a_2, b_2) = (0.99, 0.2)$, $a = 0$ and $v = 0.984$. Figures 1 and 2 show the chaotic trajectories of the drive system (29) and response system (29), respectively.

Previous research in information theory has established that entropy quantifies the rate of transfer or generation of information in a particular system. In general, Kolmogorov–Sinai (KS) entropy is applied to measure dynamical systems. A direct time–series approximation of the KS entropy was proposed in [43] named ER entropy, which indicates the level of chaos in a particular system.

Because calculating the exact ER entropy experimentally is difficult, an approximate entropy (ApEn) measure was introduced in [44,45]. Approximate entropy has been used to investigate chaotic systems recently [46,47].

In our work, the approximate entropy values of the drive and response systems have been calculated by using the reported scheme in [44,45]. As a brief summary of the approximation scheme, consider N data samples generated by our fractional map $x(1)$, $x(2)$, $\ldots$, $x(N)$. The data is arranged in a sequence of vectors with an embedding dimension m of the form

$$X(i) = [x(i), x(i+1), ..., x(i+m-1)] \text{ with } 1 \leq i \leq N-m+1. \tag{31}$$

The distance between two distinct vectors $X(i)$ and $X(j)$ is denoted by $d(X(i), X(j))$. We also define a threshold for our entropy calculation similar to [44,45] as

$$r = 0.2\text{std}(x), \tag{32}$$

with std(x) being the standard deviation of x. We, then, iterate over the regresser vectors and calculate the number of vectors K that yield a distance $d(X(i), X(j)) \leq r$. The approximate entropy is, then, given by

$$\text{ApEn} = \phi^m(r) - \phi^{m+1}(r), \tag{33}$$

where

$$\phi^m(r) = \frac{1}{N-m-1} \sum_{i=1}^{N-m+1} \log\left(\frac{K_i}{N-m+1}\right). \tag{34}$$

The approximate entropy of the 2D fractional-order Hénon map is ApEn = 0.4159. The approximate entropy of the 2D fractional-order generalized Hénon map is ApEn = 0.0114. The results agree with trajectories illustrated in Figures 1 and 2.

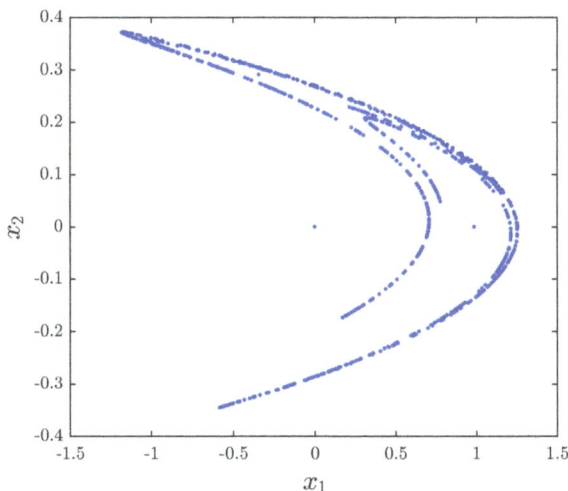

Figure 1. Phase space plot for the fractional Hénon map with $(a_1, b_1) = (1.4, 0.3)$, $v = 0.984$, and $(x_1(0), x_2(0)) = (0,0)$.

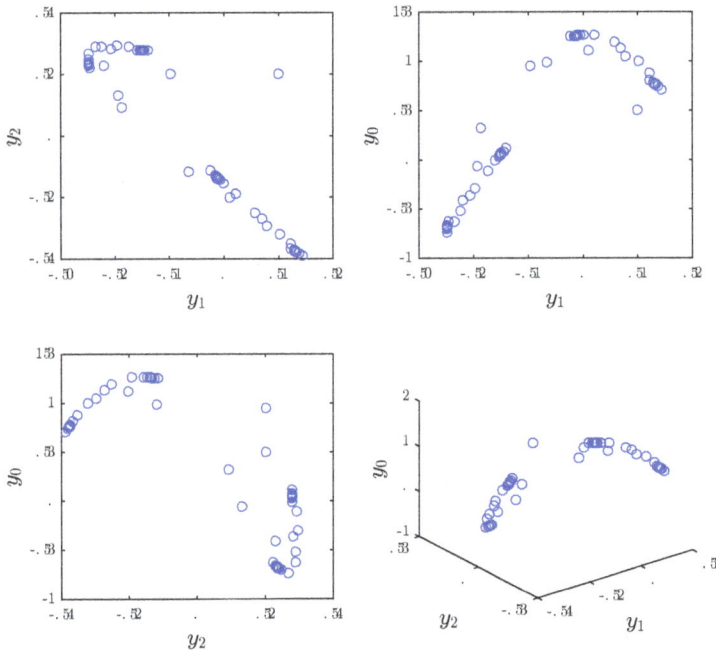

Figure 2. Phase portraits for the fractional generalized Hénon map with $(a_2, b_2) = (0.99, 0.2)$, $v = 0.984$, and $(y_1(0), y_2(0), y_3(0)) = (0.1, 0.2, 0.5)$.

Example 1. *The error system for the PS-FSHPS-GS synchronization scheme was described in Definition 2. We let*

$$\gamma = 3, \ \Phi = (1,3) \ and \ \phi(x_1(t), x_2(t)) = (x_1(t)x_2(t)). \tag{35}$$

Theorem 2 requires the selection of a control matrix C such that all the eigenvalues of B − C satisfy condition (12). For instance, the control matrix C can be chosen as

$$C = \begin{pmatrix} 0 & 0 & 0 \\ 1 & 0 & 0 \\ 0 & 1 & 0 \end{pmatrix}. \tag{36}$$

Simply, we can show that all eigenvalues of B − C are: $\lambda_1 = \lambda_2 = \lambda_3 = -1$ and therefor condition of Theorem 2 is satisfied. We can use the matrix C to construct the following controllers

$$\begin{cases} u_1(t) = & -e_1(t) - b_2 e_3(t) + y_1(t) + b_2 y_3(t) + 3x_2(t) \\ & -3x_1(t) + 3 - 3a_1 x_1^2(t), \\ u_2(t) = & -e_2(t) + b_2 e_3(t) - b_2 y_3(t) - y_1(t) + y_2(t) - 2x_2(t) \\ & + (3b_1 - 1) x_1(t) + 1 - a_1 x_1^2(t) \\ u_3(t) = & -e_3(t) - 1 - y_2(t) + a_2 y_3^2(t) + y_3(t) \\ & + {}^{C}\Delta_a^v x_1(t) x_2(t). \end{cases} \tag{37}$$

These controllers leads to the simplified error system

$$\begin{cases} {}^C\Delta_a^v e_1(t) = -e_1(t+v-1) - b_2 e_3(t+v-1), \\ {}^C\Delta_a^v e_2(t) = -e_2(t+v-1) + b_2 e_3(t+v-1), \\ {}^C\Delta_a^v e_3(t) = -e_3(t+v-1). \end{cases} \tag{38}$$

Figure 3 shows the errors as functions of time for parameter sets $(a_1, b_1) = (1.4, 0.3)$ and $(a_2, b_2) = (0.99, 0.2)$, starting point $a = 0$, fractional order $v = 0.984$, and initial errors $(e_1(0), e_2(0), e_3(0)) = (0.1, 0.2, 0.5)$. Clearly, the errors converge towards the zero solution implying that the three slave states are PS–FSHPS–GS synchronized.

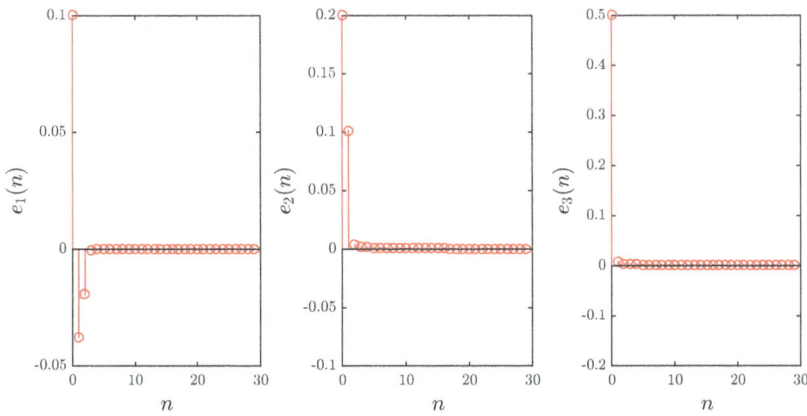

Figure 3. The evolution of errors over time for Example 1.

Example 2. *The second case is concerned with the co-existence of IFSHPS and IGS in 2D. The error system is defined according to Definition 3 where*

$$\Theta = (1,0,3) \text{ and } \varphi(y_1(t), y_2(t), y_3(t)) = y_1(t) + y_2(t) + y_3^2(t). \tag{39}$$

Following the approach of Theorem 3, we start with a factorization of φ as

$$\varphi(y_1(t), y_2(t), y_3(t)) = \sum_{j=1}^{3} \theta_j y_j(t) + \psi(y_1(t), y_2(t), y_3(t)). \tag{40}$$

It can be easily shown that

$$(\theta_1, \theta_2, \theta_3) = (1,2,0) \text{ and } \psi(y_1(t), y_2(t), y_3(t)) = y_3^2(t), \tag{41}$$

are sufficient. The proposed synchronization scheme rearranges Θ and $(\theta_1, \theta_2, \theta_3)$ into the matrix

$$M = \begin{pmatrix} 1 & 0 \\ 1 & 2 \end{pmatrix}, \tag{42}$$

which is invertible with inverse

$$M^{-1} = \begin{pmatrix} 1 & 0 \\ -\frac{1}{2} & \frac{1}{2} \end{pmatrix}. \tag{43}$$

Theorem 3 requires the choice of a matrix L. This may be achieved with

$$L = \begin{pmatrix} 1 & \frac{13}{4} \\ b_1 - 1 & -2 \end{pmatrix}. \tag{44}$$

The controllers can, thus, be constructed according to (27) based on R_1 and R_2 defined in (22) and (23), respectively. We end up with

$$\begin{cases} u_1(t) = -2e_1 - \frac{9}{4}e_2 + x_2(t) - x_1(t) - a_1 x_1^2(t) + y_1(t) \\ \qquad + (b_2 + 3) y_3(t) - 2 - 3y_2(t) + 3a_2 y_3^2(t), \\ u_2(t) = \frac{3}{2}e_1 + \frac{13}{8}e_2 - y_1(t) + \frac{5}{2}y_2(t) - \left(\frac{3}{2} + b_2\right) y_3(t) \\ \qquad -x_2(t) + \frac{(b_1+1)}{2}x_1(t) + \frac{1}{2}a_1 x_1^2(t) \\ \qquad -\frac{3a_2}{2}y_3^2(t) - \frac{1}{2} {}^C \Delta^\beta y_3^2(t) + 1, \\ u_3(t) = 0. \end{cases} \tag{45}$$

and

$$\begin{cases} {}^C \Delta_a^v e_1(t) = -2e_1(t+v-1) - \frac{9}{4}e_2(t+v-1), \\ {}^C \Delta_a^v e_2(t) = e_1(t+v-1) + e_2(t+v-1). \end{cases} \tag{46}$$

Figure 4 depicts the stabilized states subject to parameter sets $(a_1, b_1) = (1.4, 0.3)$ and $(a_2, b_2) = (0.99, 0.2)$, starting point $a = 0$, fractional order $v = 0.984$, and initial errors $(e_1(0), e_2(0)) = (-1.6, -0.325)$. It is easy to from Figure 4 that the errors converge towards zero in sufficient time proving that the controllers (45) in fact achieve IFSHPS–IGS synchronization for the pair (29).

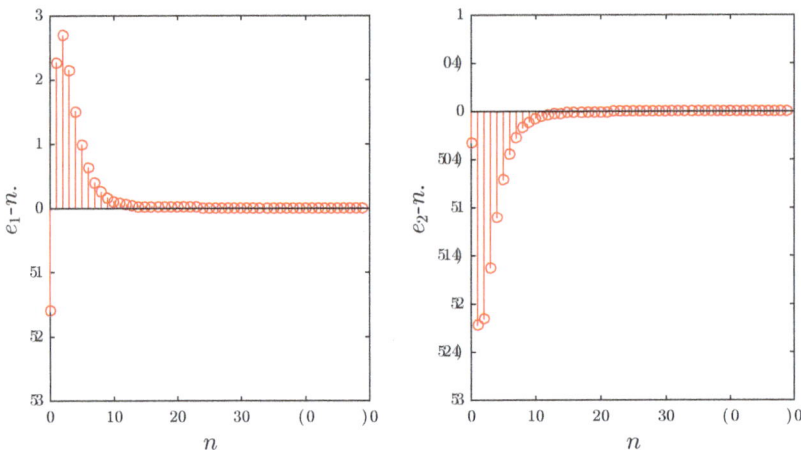

Figure 4. The evolution of errors over time for Example 2.

5. Discussion

In this paper, we have presented novel results concerning the co-existence of multiple synchronization types in Caputo-type fractional chaotic maps. To the best of our knowledge, the topic of co-existence has not been considered before for this type of system, which motivated this research. The synchronization types considered are rather general, which allows for multiple applications, especially in the fields of secure communications and data encryption. In fact, as we mentioned before, very few studies can be found in the literature concerning the synchronization of fractional chaotic maps, which makes this work all the more interesting.

Perhaps the most interesting studies related to the subject are [27–32]. In [27], the authors merely consider a pair of identical fractional logistic maps and propose a simple direct synchronization controller. In [28], an identical synchronization scheme is proposed based on the results of [48,49]. The authors of [29], again, consider the synchronization of identical fractional Hénon maps. The same can be said regarding [32]. As for [31], the authors propose a simple linear feedback controller suitable for a variety of maps. However, it is only shown to achieve complete synchronization, which is the most basic form of synchronization. In [30], the fractional difference operator used is different from the one used here and thus comparison is difficult.

Generally speaking, it is difficult to compare our results to those reported in the above mentioned studies as the scope of our work is much wider. In addition, we are mainly concerned with co-existence, which has not been considered before for this type of systems.

6. Concluding Remarks

In this work, we have shown that different types of synchronization can co-exist for fractional-order discrete-time chaotic systems. We assumed a two dimensional drive system and a three dimensional response system. The main results of the study were two fold. First, we presented a nonlinear control scheme whereby PS, FSHPS, and GS are achieved simultaneously for the three states of the response system. The stability of the zero solution, and consequently the convergence of the synchronization error, was established by means of the stability theory of linear fractional-order discrete-time systems. The second main result concerns the co-existence of IFSHPS and IGS for the same drive-response pair. The three response states are simultaneously IFSHPS synchronized with the first drive state and IGS synchronized with the second drive state. Numerical results have confirmed the findings of the study. Simulations were carried out on Matlab to ensure that the errors converge to zero subject to the proposed control laws.

Author Contributions: Conceptualization, S.B. and A.O.; Formal analysis, X.W. and G.G.; Investigation, S.B.; Methodology, A.O. and V.-T.P.; Software, X.W., A.-A.K. and G.G.; Validation, V.V.H.; Writing – original draft, A.-A.K. and V.-T.P.; Writing – review & editing, V.V.H.

Funding: The author X.W. was supported by the National Natural Science Foundation of China (No. 61601306) and Shenzhen Overseas High Level Talent Peacock Project Fund (No. 20150215145C).

Conflicts of Interest: The authors declare no conflict of interest.

References

1. Kocarev, L.; Szczepanski, J.; Amigo, J.M.; Tomovski, I. Discrete Chaos–I: Theory. *IEEE Trans. Circuits Syst. I Reg. Pap.* **2006**, *53*, 1300–1309. [CrossRef]
2. Li, C.; Song, Y.; Wang, F.; Liang, Z.; Zhu, B. Chaotic path planner of autonomous mobile robots based on the standard map for surveillance missions. *Math. Prob. Eng.* **2015**, *2015*, 263964. [CrossRef]
3. Papadimitriou, S.; Bezerianosa, A.; Bountisb, T.; Pavlides, G. Secure communication protocols with discrete nonlinear chaotic maps. *J. Syst. Archit.* **2001**, *47*, 61–72. [CrossRef]
4. Kwok, H.S.; Tang, W.K.S.; Man, K.F. Online secure chatting system using discrete chaotic map. *Int. J. Bifurc. Chaos* **2004**, *14*, 285. [CrossRef]
5. Banerjee, S.; Kurth, J. Chaos and cryptography: A new dimension in secure communications. *Eur. Phys. J. Spec. Top.* **2014**, *223*, 1441–1445. [CrossRef]
6. Fataf, N.A.A.; Mukherjee, S.; Said, M.R.M.; Rauf, U.F.A.; Hina, A.D.; Banerjee, S. Synchronization between two discrete chaotic systems for secure communications. In Proceedings of the 2016 IEEE Sixth International Conference on Communications and Electronics (ICCE), Ha Long, Vietnam, 27–29 July 2016; pp. 477–481.
7. Hénon, M. A two-dimensional mapping with a strange attractor. *Commun. Math. Phys.* **1976**, *50*, 69–77.
8. Lozi, R. Un atracteur étrange du type attracteur de Hénon. *J. Physique* **1978**, *39*, 9–10.
9. Hitzl, D.L.; Zele, F. An exploration of the Hénon quadratic map. *Phys. D Nonlinear Phenom.* **1985**, *14*, 305–26. [CrossRef]
10. Baier, G.; Sahle, S. Design of hyperchaotic flows. *Phys. Rev. E* **1995**, *51*, 2712–2714. [CrossRef]

11. Atici, F.M.; Eloe, P.W. Discrete fractional calculus with the nabla operator. *Electron. J. Qual. Theory Differ. Equ.* **2009**, *4*, 1–12. [CrossRef]

12. Abdeljawad, T. On Riemann and Caputo fractional differences. *Comput. Math. Appl.* **2011**, *62*, 1602–1611. [CrossRef]

13. Abdeljawad, T.; Baleanu, D.; Jarad, F.; Agarwal, R.P. Fractional sums and differences with binomial coefficients. *Discret. Dyn. Nat. Soc.* **2013**, *2013*, 104173. [CrossRef]

14. Baleanu, D.; Wu, G.; Bai, Y.; Chen, F. Stability analysis of Caputo–Like discrete fractional systems. *Commun. Nonlinear Sci. Numer. Simul.* **2017**, *48*, 520–530. [CrossRef]

15. Goodrich, C.; Peterson, A.C. *Discrete Fractional Calculus*; Springer: Berlin, Germany, 2015.

16. Yamada, T.; Fujisaca, H. Stability theory of synchronized motion in coupled-oscillator. Systems. II: The mapping approach. *Prog. Theor. Phys.* **1983**, *70*, 1240–1248. [CrossRef]

17. Yamada, T.; Fujisaca, H. Stability theory of synchronized motion in coupled-oscillator Systems. III: Mapping model for continuous system. *Prog. Theor. Phys.* **1984**, *72*, 885–894. [CrossRef]

18. Afraimovich, V.S.; Verochev, N.N.; Robinovich, M.I. Stochastic synchronization of oscillations in dissipative systems. *RadioPhys. Quantum Electron.* **1984**, *29*, 795–803. [CrossRef]

19. Pecora, L.M.; Carrol, T.L. Synchronization in chaotic systems. *Phys. Rev. A* **1990**, *64*, 821–824.

20. Grassi, G.; Miller, D.A. Projective synchronization via a linear observer: Application to time-delay, continuous-time and discrete-time systems. *Int. J. Bifurc. Chaos* **2007**, *17*, 1337–1344. [CrossRef]

21. Ouannas, A.; Odibat, Z. Generalized synchronization of different dimensional chaotic dynamical systems in discrete-time. *Nonlinear Dyn.* **2015**, *81*, 765–771. [CrossRef]

22. Ouannas, A. On full state hybrid projective synchronization of general discrete chaotic systems. *J. Nonlinear Dyn.* **2014**, *2014*, 983293. [CrossRef]

23. Ouannas, A. On inverse generalized synchronization of continuous chaotic dynamical systems. *Int. J. Appl. Comp. Math.* **2016**, *2*, 1–11. [CrossRef]

24. Ouannas, A.; Grassi, G. Inverse full state hybrid projective synchronization for chaotic maps with different dimensions. *Chin. Phys. B* **2016**, *25*, 090503. [CrossRef]

25. Hu, T. Discrete Chaos in Fractional Hénon Map. *Appl. Math.* **2014**, *5*, 2243–2248. [CrossRef]

26. Shukla, M.K.; Sharma, B.B. Investigation of chaos in fractional order generalized hyperchaotic Hénon map. *Int. J. Elec. Comm.* **2017**, *78*, 265–273. [CrossRef]

27. Wu, G.; Baleanu, D. Chaos synchronization of the discrete fractional logistic map. *Signal Process.* **2014**, *102*, 96–99. [CrossRef]

28. Wu, G.; Baleanu, D.; Xie, H.; Chen, F. Chaos synchronization of fractional chaotic maps based on the stability condition. *Physica A* **2016**, *460*, 374–383. [CrossRef]

29. Liu, Y. Chaotic synchronization between linearly coupled discrete fractional Hénon maps. *Indian J. Phys.* **2016**, *90*, 313–317. [CrossRef]

30. Megherbi, O.; Hamiche, H.; Djennoune, S.; Bettayeb, M. A new contribution for the impulsive synchronization of fractional-order discrete-time chaotic systems. *Nonlinear Dyn.* **2017**, *90*, 1519–1533. [CrossRef]

31. Xin, B.; Liu, L.; Hou, G.; Ma, Y. Chaos synchronization of nonlinear fractional discrete dynamical systems via linear control. *Entropy* **2017**, *19*, 351. [CrossRef]

32. Khennaoui, A.A.; Ouannas, A.; Bendoukha, S.; Wang, X.; Pham, V.T. On chaos in the fractional-order discrete-time unified system and its control synchronization. *Entropy* **2018**, *20*, 530. [CrossRef]

33. Ouannas, A.; Azar, A.T.; Sundarapandian, A.V. New hybrid synchronization schemes based on coexistence of various types of synchronization between master-slave hyperchaotic systems. *Int. J. Comp. Apps. Tech.* **2017**, *55*, 112–120. [CrossRef]

34. Ouannas, A.; Abdelmalek, S.; Bendoukha, S. Coexistence of some chaos synchronization types in fractional-order differential equations. *Electron. J. Differ. Equ.* **2017**, *2017*, 1–15.

35. Ouannas, A.; Odibat, Z.; Hayat, T. Fractional analysis of co-existence of some types of chaos synchronization. *Chaos Solitons Fractals* **2017**, *105*, 215–223. [CrossRef]

36. Ouannas, A.; Wang, X.; Pham, V.T.; Grassi, G.; Ziar, T. Coexistence of identical synchronization, antiphase synchronization and inverse full state hybrid projective synchronization in different dimensional fractional-order chaotic systems. *Adv. Differ. Equ.* **2018**, *2018*, 35. [CrossRef]

37. Ouannas, A.; Wang, X.; Pham, V.T.; Ziar, T. Dynamic analysis of complex synchronization scheme between integer-order and fractional-order chaotic systems with different dimensions. *Complexity* **2017**, *2017*, 1–12. [CrossRef]
38. Ouannas, A.; Zehrour, O.; Laadjal, Z. Nonlinear methods to control synchronization between fractional-order and integer-order chaotic systems. *Nonlinear Stud.* **2018**, *25*, 1–13.
39. Ouannas, A.; Azar, A.T.; Abu-Saris, R. A new type of hybrid synchronization between arbitrary hyperchaotic maps. *Int. J. Mach. Learn. Cyb.* **2017**, *8*, 1–8. [CrossRef]
40. Ouannas, A. Co-existence of various types of synchronization between hyperchaotic maps. *Nonlinear Dyn. Syst. Theory* **2016**, *16*, 312–321.
41. Ouannas, A.; Grassi, G. A new approach to study co-existence of some synchronization types between chaotic maps with different dimensions. *Nonlinear Dyn.* **2016**, *86*, 1319–1328. [CrossRef]
42. Cermak, J.; Gyori, I.; Nechvatal, L. On explicit stability condition for a linear fractional difference system. *Fract. Calc. Appl. Anal.* **2015**, *18*, 651–672. [CrossRef]
43. Eckmann, J.P.; Ruelle, D. Ergodic theory of chaos and strange attractors. *Rev. Mod. Phys.* **1985**, *57*, 617. [CrossRef]
44. Pincus, S.M. Approximate entropy as a measure of system complexity. *Proc. Natl. Acad. Sci. USA* **1991**, *88*, 2297–2301. [CrossRef] [PubMed]
45. Pincus, S. Approximate entropy (ApEn) as a complexity measure. *Chaos Interdiscipl. J. Nonlinear Sci.* **1995**, *5*, 110–117. [CrossRef] [PubMed]
46. Xu, G.H.; Shekofteh, Y.; Akgal, A.; Li, C.B.; Panahi, S.A. New chaotic system with a self-excited attractor: Entropy measurement, signal encryption, and parameter estimation. *Entropy* **2018**, *20*, 86. [CrossRef]
47. Wang, C.; Ding, Q. A new two-dimensional map with hidden attractors. *Entropy* **2018**, *20*, 322. [CrossRef]
48. Abu-Saris, R.; Al-Mdallal, Q. On the asymptotic stability of linear system of fractional order difference equations. *Fract. Calc. Appl. Anal.* **2013**, *16*, 613–629. [CrossRef]
49. Mozyrska, D.; Wyrwas, M. The Z–transform method and Delta type fractional difference operators. *Discrete Dyn. Nat. Soc.* **2013**, *2013*, 852734. [CrossRef]

Article

The Complexity and Entropy Analysis for Service Game Model Based on Different Expectations and Optimal Pricing

Yimin Huang [1], Xingli Chen [2], Qiuxiang Li [3],* and Xiaogang Ma [4],*

[1] The College of Management & Mconomics, North China University of Water Resources and Electric Power, Zhengzhou 450046, China; huangyimin@ncwu.edu.cn
[2] Business School, Henan University, Kaifeng 475004, China; 104754170920@vip.henu.edu.cn
[3] The Institute of Management Science and Engineering, Henan University, Kaifeng 475004, China
[4] Management School, Wuhan Textile University, Wuhan 430073, China
* Correspondence: lqxkycg@henu.edu.cn (Q.L.); xgma@wtu.edu.cn (X.M.)

Received: 3 October 2018; Accepted: 5 November 2018; Published: 8 November 2018

Abstract: The internet has provided a new means for manufacturers to reach consumers. On the background of the widespread multichannel sales in China, based on a literature review of the service game and multichannel supply chain, this paper builds a multichannel dynamic service game model where the retailer operates an offline channel and the manufacturer operates an online channel and offers customers the option to buy online and pick up from the retailer's store (BOPS). The manufacturer and the retailer take maximizing the channel profits as their business objectives and make channel service game under optimal pricing. We carry on theoretical analysis of the model and perform numerical simulations from the perspective of entropy theory, game theory, and chaotic dynamics. The results show that the stability of the system will weaken with the increase in service elasticity coefficient and that it is unaffected by the feedback parameter adjustment of the retailer. The BOPS channel strengthens the cooperation between the manufacturer and the retailer and moderates the conflict between the online and the offline channels. The system will go into chaotic state and cause the system's entropy to increase when the manufacturer adjusts his/her service decision quickly. In a chaotic state, the system is sensitive to initial conditions and service input is difficult to predict; the manufacturer and retailer need more additional information to make the system clear or use the method of feedback control to delay or eliminate the occurrence of chaos.

Keywords: multichannel supply chain; service game; chaos; entropy; BOPS

1. Introduction

In recent years, some Chinese retailers have profited enormously from the development of online marketing channels, however these profits have come at the expense of traditional retailers who once dominated the market. These changes in power of the channels have also created channel conflicts [1]. Scholars have studied pricing strategies in dual-channel supply chains from several perspectives. Huang et al. [2] considered the effects of production disruption and demand disruption on the pricing and production decisions in a dual-channel supply chain. Li et al. [3] studied the effects of green costs on the retail prices and green degrees of a competitive dual-channel green supply chain. Kouvelis et al. [4] discussed supply chain contracting in environments with volatile input prices and frictions, and presented a game-theoretic study of a bilateral monopoly supply chain with stochastic demand, stochastic input costs, production lead times, and working capital constraints. Li et al. [5] studied the effects of risk attitude of retailer and uncertain demand on the pricing strategy and coordination. Li et al. [6] studied the pricing strategy of a dual-channel supply chain considering a perishable product and risk preference. Radhi and Zhang [7] discussed the effects of customer preference and customer return rate on the pricing in a dual-channel supply chain. Li et al. [8] analyzed the effects of pricing strategy of the retailer on the manufacturer's choice to open the direct selling channel. Ji et al. [9] developed four case models to study the optimal pricing and return policies considering false failure returns, and analyzed the influences of buy-back contract on optimal pricing and return policies. Xie et al. [10] built the revenue-sharing contract and cost-sharing contract to address the problem of forward channel conflicts and introduced the Stackelberg game to investigate the contract coordination mechanism. In contrast to this stream of work, this paper focuses on the service game of a multichannel supply chain where the manufacturer offers customers the option to buy online and pick up from the retailer's store (BOPS), which is a new retail environment today.

Channel service is an important factor affecting customers' channel choices and is broadly surveyed in the literature [11–15]. Considering service factors, Ma and Guo [16] studied the complex dynamics of a bivariate game model, in which the recursive least-squares (RLS) estimation is introduced to substitute naive estimation. Ali et al. [17] examined the effects of potential market demand disruptions on the prices and service levels of competitive retailers, and showed that the price and investment decision of service level are significant influenced by demand disruptions. Li and Li [18] found the entire supply chain could not be coordinated with a constant wholesale price when the retailer provided a value-added service and had concern for fairness. Zhou et al. [19] considered a dual-channel supply chain—where the retailer provided customers with certain pre-sales services and where the manufacturer free-rides the retailer's pre-sales services by sharing the retailer's sales effort cost—and investigated the influence of free riding on the pricing and service strategies of the two members. Zhou and Zhao [20] analyzed how the manufacturer used wholesale prices and slotting allowances to practice his signaling strategy with asymmetric information considering retailer's value-added services, respectively. Chen et al. [21] studied a retail service supply chain with an online-to-offline (O2O) mixed channel under different power structures. Kong et al. [22] studied the pricing and service level of CLSC under centralized and decentralized decision-making, respectively, and analyzed the effects of system's parameters on the system's performance. The above papers studied the impacts of service on channel pricing strategy; no paper studies the channel service decision under optimal pricing.

There are channel conflicts between online and offline channels, such as inconsistent goals, business scopes, and consumer purchasing behavior. The new retail model enables for online embracing offline to achieve channel integration; many enterprises practice this new retail model according to their realities, such as Jing Dong, Tmall, and Uniqlo. Many scholars have also studied channel conflicts and cooperation contracts of O2O channel from different perspectives. Cai et al. [23] used the price discount contracts and pricing schemes to coordinate the dual-channel supply chain, and found price discount contracts and consistent pricing scheme can reduce channel conflict. Tao and Li [24] developed an O2O channel model and analyzed the influence of service level, the free-riding coefficient, and a bonus strategy on pricing policies and channel performance. Zhao et al. [25] investigated pricing problem of a dual-channel supply chain considering complementary products and different market power structures, and discussed the effects of important parameters on the pricing strategies.

Channel integration management has received a lot of attention in marketing; the topic was broadly surveyed in literature [26–28]. Jin et al. [29] studied BOPS theoretical model in which a physical retailer adopting BOPS used a recommended service area to fulfill orders from both online and offline customers; the size of the BOPS service area is determined by the ratio of unit inventory cost to BOPS customers. Liu and Zhou [30] found whether corporations adopt the BOPS model or not depended on the size of BOPS-consumer and consumer's service sensitivity degree. Assuming a supply chain is comprised of a wholesaler and two retailers, Moon et al. [31] showed the process of collapsing the supply chain through interaction between subsystems by developing a system dynamics simulation model. Yan et al. [32] introduced the WeChat channel into multichannel supply chain system, and found that the WeChat channel could allow retailers to obtain increased profits and uncertainty for manufacturers. Matsui [33] investigated a multichannel supply chain model, where a manufacturer produces and sells products to retailers, and analyzed the optimal timing and level of wholesale and retail prices considering observable delay game.

The above literatures studied the impacts of service on channel conflicts and channel integration management. However, it is difficult for decision-makers to get all of the information in the market, so decision-makers have limited rational behavior. Because of their different expectations, few literatures have studied the service game of multichannel supply chain under the optimal price.

The supply chain system will be in an unstable state because of the manager's behavior and customer's behavior. Some scholars analyzed the complexity of supply chain based on entropy theory in literatures [34,35]. Kriheli and Levner [36] analyzed the complexity between the supply chain components under uncertainty environment using the information entropy. Levner and Ptuskin [37] presented the entropy-based optimization model for reducing the supply chain model size and assessing the economic loss. Lou et al. [38] analyzed the bullwhip effect in a supply chain with a sales game and consumer returns via the theory of entropy and complexity. Han et al. [39] built a duopoly game model with double delays in the hydropower market and analyzed the effect of time delay parameters on system entropy and stability.

The above researches are mainly focused on the pricing and service decisions of a dual-channel supply chain from the perspective of static operation. This paper will build a multichannel dynamic service game model and analyze its dynamic evolution characteristics using dynamics theory, game theory, and entropy theory.

The main contributions of this paper are as follows:

(1) This paper broadens and enriches the research of the multichannel service supply chain and proposes a new perspective for multichannel research and decision references for multichannel enterprises, because decision-makers hope to draw up service strategies for the multichannel supply chain to solve the practical troubles of firms;

(2) This paper studies the dynamic service strategy under optimal pricing which further widens the research scope of the multichannel supply chain;

(3) This paper uses the entropy theory and dynamics theory to study the complexity and characteristics of the multichannel service supply chain and reveals that decision variables and parameters have great impact on the stability of the multichannel service supply chain.

The rest of this paper is organized as follows. The model description and model construction are given in Section 2. Section 3 analyzes the stability of the system. Section 4 analyzes the complexity entropy and dynamic characteristics of the system. The feedback control model is designed to make the system return to the stable state in Section 5. Finally, Section 6 presents the conclusions of this paper.

2. Model Description and Model Construction

2.1. Model Description and Assumptions

In this paper, a service game model is developed in which a retailer operates an offline channel and a manufacturer operates an online channel and offers customers the option to buy online and pick up at the retailer's store (BOPS), as shown in Figure 1. The manufacturer and retailer take maximizing the channel profits as business objectives and make channel service game under optimal pricing, where p_i and s_i, $i = 1, 2, 3$ represent the sale prices of the product and channel service levels in three channels and w is the wholesale price that the manufacturer offers to the retailer. The manufacturer and retailer allocate the cost and profit of BOPS channel in a certain proportion. The retailer does not participate in service decisions of BOPS channel. Therefore, it is reasonable for the manufacturer to make the service decision of BOPS channel based on channel profit.

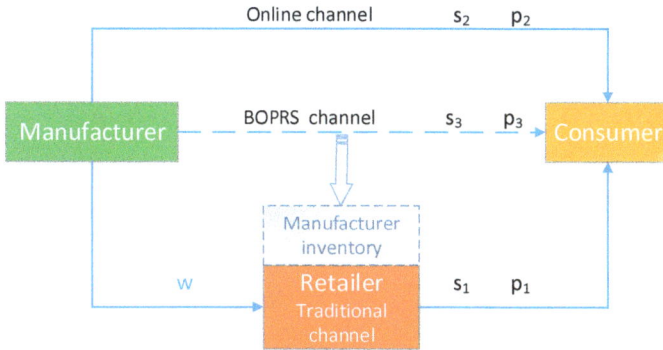

Figure 1. Multichannel service supply chain.

The following assumptions are used to facilitate our model in this paper:

(1)   The manufacturer and retailer aim at maximizing channel profits, the manufacturer takes s_2 and s_3 as decision variables and the retailer takes s_1 as a decision variable under the optimal price decision.
(2)   Channel service does not affect the demands of other channels. The production costs and sales costs of products are zero.
(3)   The inventories of the manufacturer and retailer are large enough to meet customer needs.

2.2. Model Construction

According to the actual market competition and extending the demand functions in Yao et al. [40] and Dan et al. [41], we assume that the primary demand functions in this paper are decided by p_i and s_i as follows

$$\begin{cases} D_1 = a_1 - bp_1 + kp_2 + kp_3 + \gamma_1 s_1 \\ D_2 = a_2 - bp_2 + kp_1 + kp_3 + \gamma_2 s_2 \\ D_3 = a_3 - bp_3 + kp_2 + kp_1 + \gamma_3 s_3 \end{cases} \tag{1}$$

where a_i, $i = 1, 2, 3$ are the base demands of products for traditional channel, direct channel and BOPS channel. $b\ (b > 0)$ represents the price sensitive coefficient of the product; $k\ (k > 0)$ is the cross-price sensitivity coefficient which reflects the substitution degrees of the products; $\gamma_i(\gamma_i > 0)$, $i = 1, 2, 3$ represent service sensitivity coefficients and satisfies $b > k$ and $b > \gamma_i$.

According to past literature [17], the service costs in the three channels satisfy: $c(s_i) = \frac{\eta_i s_i^2}{2}$, $i = 1, 2, 3$. $\eta_i > 0$ and $i = 1, 2, 3$ are unit service costs of each channel.

The decision-making process of the manufacturer and retailer is as follows (1) the manufacturer and retailer first make price decision simultaneously based on channel profits maximization and (2) then make service decision under the optimal price decision.

The channel profits of the manufacturer and retailer are represented as follows

$$\begin{cases} \pi_1 = (p_1 - w)D_1 - \frac{\eta_1 s_1^2}{2} \\ \pi_2 = p_2 D_2 - \frac{\eta_2 s_2^2}{2} \\ \pi_3 = p_3 D_3 - \frac{\eta_3 s_3^2}{2} \end{cases} \tag{2}$$

where π_1 is the retailer's profit from the traditional channel and π_2 and π_3 are the manufacturer' profits from online channel and BOPS channel. w is a constant which represents the wholesale price that the manufacturer provides for the retailer.

Supposing s_1, s_2, and s_3 are known, make a first-order partial derivative of π_i for p_i, the channel marginal profits of the manufacturer and retailer are as follows

$$\begin{cases} \frac{\partial \pi_1}{\partial p_1} = a_1 - 2bp_1 + kp_2 + kp_3 + r_1s_1 + bw \\ \frac{\partial \pi_2}{\partial p_2} = a_2 - 2bp_2 + kp_1 + kp_3 + r_2s_2 \\ \frac{\partial \pi_3}{\partial p_3} = a_3 - 2bp_3 + kp_1 + kp_2 + r_3s_3 \end{cases} \tag{3}$$

By solving $\frac{\partial \pi_i}{\partial p_i} = 0$, the optimal prices of the manufacturer and retailer are obtained:

$$\begin{cases} p_1^* = \frac{2a_1b - a_1k + a_2k + a_3k + 2b^2w - bkw + 2b\gamma_1s_1 - k\gamma_1s_1 + k\gamma_2s_2 + k\gamma_3s_3}{2(b-k)(2b+k)} \\ p_2^* = \frac{2a_2b + a_1k - a_2k + a_3k + bkw + 2b\gamma_2s_2 + k\gamma_1s_1 - k\gamma_2s_2 + k\gamma_3s_3}{2(b-k)(2b+k)} \\ p_3^* = \frac{2a_3b + a_1k + a_2k - a_3k + bkw + 2b\gamma_3s_3 + k\gamma_1s_1 + k\gamma_2s_2 - k\gamma_3s_3}{2(b-k)(2b+k)} \end{cases} \tag{4}$$

Substituting p_1^*, p_2^*, and p_3^* into the Equation (2), and making a first-order partial derivatives of π_i for s_i. By solving the equations $\frac{\partial \pi_i}{\partial s_i} = 0$, the optimal channel service levels of the manufacturer and the retailer are obtained as follows:

$$\begin{cases} s_1^* = \frac{A_0 + \left(2b^2k\gamma_1\gamma_2 - bk^2\gamma_1\gamma_2\right)s_2 + \left(2b^2k\gamma_1\gamma_3 - bk^2\gamma_1\gamma_3\right)s_3}{8\eta_1b^4 8k\eta_1b^3 - 4b^3\gamma_1^2 - 6\eta_1b^2k^2 + 4kb^2\gamma_1^2 + 4\eta_1bk^3 - bk^2\gamma_1^2 + 2\eta_1k^4} \\ s_2^* = \frac{B_0 + \left(2b^2k\gamma_1\gamma_2 - bk^2\gamma_1\gamma_2\right)s_1 + \left(2b^2k\gamma_2\gamma_3 - bk^2\gamma_2\gamma_3\right)s_3}{8\eta_1b^4 8k\eta_1b^3 - 4b^3\gamma_1^2 - 6\eta_1b^2k^2 + 4kb^2\gamma_1^2 + 4\eta_1bk^3 - bk^2\gamma_1^2 + 2\eta_1k^4} \\ s_3^* = \frac{C_0 + \left(2b^2k\gamma_1\gamma_3 - bk^2\gamma_1\gamma_3\right)s_1 + \left(2b^2k\gamma_2\gamma_3 - bk^2\gamma_2\gamma_3\right)s_2}{8\eta_1b^4 8k\eta_1b^3 - 4b^3\gamma_1^2 - 6\eta_1b^2k^2 + 4kb^2\gamma_1^2 + 4\eta_1bk^3 - bk^2\gamma_1^2 + 2\eta_1k^4} \end{cases} \tag{5}$$

where

$$A_0 = 4a_1\gamma_1b^3 - 4\gamma_1wb^4 + a_1b\gamma_1k^2 + 2k\gamma_1b^2(a_3 + a_2 - 2a_1) + b\gamma_1k^2(a_3 - a_2)$$
$$+ bk\gamma_1w(4b^2 + 3bk - 2k^2)$$

$$B_0 = 4a_2\gamma_2b^3 - a_1\gamma_2bk^2 + 2a_1k\gamma_2b^2 + a_2b\gamma_2k^2 - 4a_2k\gamma_2b^2 - a_3b\gamma_2k^2 + 2a_3k\gamma_2b^2$$
$$+ 2k\gamma_2wb^3 - b^2k^2\gamma_2w$$

$$C_0 = 4a_3b^3\gamma_3 - a_1bk^2\gamma_3 + 2a_1b^2k\gamma_3 - a_2bk^2\gamma_3 + 2a_2b^2k\gamma_3 + a_3bk^2\gamma_3 - 4a_3b^2k\gamma_3$$
$$+ 2b^3k\gamma_3w - b^2k^2\gamma_3w$$

The expressions of the optimal service levels are very intricate; the relationship between variables and parameters cannot see intuitively from the expression functions. Next, we will structure a dynamic game model to research the dynamic characteristics of the multichannel supply chain system.

The service decisions of the manufacturer and the retailer are not completely rational because they cannot get all the necessary information in the market, so the manufacturer and the retailer have incomplete rational behavior when they make decisions. The retailer adopts an adaptive expectation in the decision-making process as follows

$$s_1(t+1) = s_1(t) + \beta[s_1(t) - s_1^*(t)], \qquad 0 < \beta < 1 \tag{6}$$

where β is the service feedback parameter.

The manufacturer makes service decision based on bounded rationality expectation for direct channel and static expectation for BOPS channel:

$$s_2(t+1) = s_2(t) + \xi s_2(t)\frac{\partial \pi_2(t)}{\partial s_2} \tag{7}$$

$$s_3(t+1) = s_3^*(t) \tag{8}$$

where ζ is the service adjustment parameter which reflects the manufacturer's learning behavior and positive managerial behavior. When the marginal profit in period t exceeds zero, the manufacturer will increase the service level in period $t + 1$; contrarily, the manufacturer will decrease the service level in period $t + 1$. Namely, the service level of period $t + 1$ will be adjusted according to marginal profit of period t.

The three-dimensional dynamic service game system considering BOPS channel is as follows

$$
\begin{cases}
s_1(t+1) = s_1(t) + \beta \left[\dfrac{A_0 + \left(2b^2 k\gamma_1\gamma_2 - bk^2\gamma_1\gamma_2\right)s_2(t) + \left(2b^2 k\gamma_1\gamma_3 - bk^2\gamma_1\gamma_3\right)s_3(t)}{8\eta_1 b^4 - 8\eta_1 b^3 k - 4b^3\gamma_1^2 - 6\eta_1 b^2 k^2 + 4b^2 k\gamma_1^2 + 4\eta_1 bk^3 - bk^2\gamma_1^2 + 2\eta_1 k^4} - s_1(t) \right] \\[2mm]
s_2(t+1) = s_2(t) + \zeta s_2(t) \left[\dfrac{B_1 + B_2 s_1(t) + B_3 s_2(t) + \left(4b^2 k\gamma_2\gamma_3 - 2bk^2\gamma_2\gamma_3\right)s_3(t)}{(2b+k)^2 (2b-2k)^2} - \eta_2 s_2(t) \right] \\[2mm]
s_3(t+1) = \dfrac{C_0 + \left(2b^2 k\gamma_1\gamma_3 - bk^2\gamma_1\gamma_3\right)s_1(t) + \left(2b^2 k\gamma_2\gamma_3 - bk^2\gamma_2\gamma_3\right)s_2(t)}{8\eta_3 b^4 - 8\eta_3 b^3 k - 4b^3\gamma_3^2 - 6\eta_3 b^2 k^2 + 4b^2 k\gamma_3^2 + 4\eta_3 bk^3 - bk^2\gamma_3^2 + 2\eta_3 k^4}
\end{cases} \tag{9}
$$

where

$$
\begin{aligned}
B_1 &= 8a_2\gamma_2 b^2(b-k) + 2bk\gamma_2(a_1 k - wbk + 2wb^2 + 4k\gamma_2 b^2(a_1 + a_3) - 2b\gamma_2 k^2(a_1 + a_3), \\
B_2 &= 2bk\gamma_1\gamma_2(2b-k), \\
B_3 &= 2bk^2\gamma_2^2 + 8b^2\gamma_2^2(b-k).
\end{aligned}
$$

3. The Stability of System (9)

In this section, we will study the stable characteristics of system (9). Because of the particularity of the model, the Nash equilibrium solutions of system (9) are very complicated, and we cannot judge directly the interaction between variables and parameters. Here, we will study the stability of system (9) through numerical simulation [42], according to the current state and reality of the multichannel supply chain enterprises, the parameter values are as follows, $a_1 = 4$, $a_2 = 3$, $a_3 = 2$, $b = 1$, $\eta_1 = 0.7$, $\eta_2 = 0.75$, $\eta_3 = 0.7$, $\gamma_1 = 0.55$, $\gamma_2 = 0.4$, $\gamma_3 = 0.5$, $k = 0.6$, $w = 2$.

When $s_1(t+1) = s_1(t)$, $s_2(t+1) = s_2(t)$, $s_3(t+1) = s_3(t)$, the eight equilibrium solutions of system (9) are,

$$
\begin{aligned}
E_1 &= (9.5227,\ 6.3236,\ 8.7891), \\
E_2 &= (0,\ 4.6685,\ 6.3699), \\
E_3 &= (7.9027,\ 0,\ 7.3865), \\
E_4 &= (7.0220,\ 4.8424,\ 0), \\
E_5 &= (6.0508,\ 0,\ 0), \\
E_6 &= (0,\ 3.859,\ 0), \\
E_7 &= (0,\ 3.8509,\ 0), \\
E_8 &= (0,\ 0,\ 5.6042).
\end{aligned}
$$

Obviously, E_2, E_3, E_4, E_5, E_6, E_7, and E_8 are boundary equilibrium points which do not meet our expectations, because the decision variables obviously are not allowed to be zero in economics for decision makers, E_2–E_8 are unstable and E_1 is the only Nash equilibrium point. Because it is not significant to study the unstable equilibrium points we only consider the stability of the Nash equilibrium point in the following.

The Jacobian matrix of system (9) at E_1 is as follows

$$
J(E_1) = \begin{vmatrix}
1 - \beta & 0.2621\beta & 0.3277\beta \\
0.9611\zeta & 1 - 10.4062\zeta & 0.8737\zeta \\
3.3863 & 1.8452 & 0
\end{vmatrix}
$$

The characteristic polynomial of $J(E_1)$ takes the following form:

$$
f(\lambda) = \lambda^3 + A\lambda^2 + B\lambda + C \tag{10}
$$

where

$$A = \beta + 3.820\xi - 2,$$
$$B = 3.6164\beta\xi - 1.0560\beta - 3.9052\xi + 1,$$
$$C = -0.3410\beta\xi + 0.0565\beta + 0.0805\xi.$$

In order to guarantee E_1 is locally stable, A, B, and C must meet the following conditions.

$$\begin{cases} f(1) = A + B + C + 1 > 0 \\ -f(-1) = -A + B - C + 1 > 0 \\ |C| < 1 \\ |BC - AB| < |C^2 - A^2| \end{cases} \tag{11}$$

By solving condition (11), the stability domain of system (9) can be obtained. Due to these limitations being so complex, solving the inequality of Equation (11) is very difficult. If E_1 satisfies the inequality of Equation (11), we may judge that system (9) is locally stable. We will prove the stable region of system (9) through numerical simulation.

According to the inequality Equation (11), Figure 2 gives the stable region and unstable region of system (9) when $\gamma_1 = 0.25$ and $\gamma_1 = 0.55$.

We can see that the stable region of system (9) becomes smaller with increasing γ_1, the change of β has no effect on the stability of system (9) no matter what value the γ_1 takes. Namely, the greater the service elasticity coefficient is, the smaller the stable region is; the increase of the service elasticity coefficient will weaken the market competition; thus, the choice of irrational decision-making mode has a great influence on the stability of system (9). When ξ and β take values in the stable range, the system (9) will stabilize at Nash equilibrium point after a finite period game. When ξ and β take values in the unstable range, the system (9) will be unstable and enter into either a bifurcation state or chaotic state, the uncertainty of system (9) increases at this time and more information is needed to maintain the stability of the system (9).

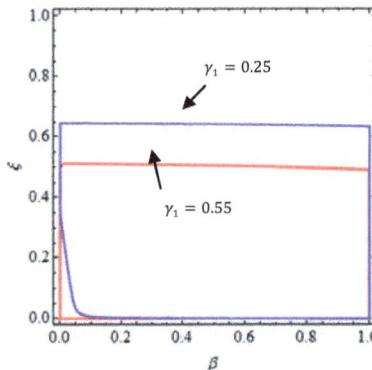

Figure 2. The stable regions of system (9) with $\gamma_1 = 0.25$ and $\gamma_1 = 0.55$.

4. The Entropy Complexity Analysis of System (9)

In order to better understand the dynamic characteristics of system (9), in this section, numerical simulation is used to explore the entropy complexity and dynamic behavior of system (9) using the bifurcation diagram, system's entropy, and the largest Lyapunov exponents (LLE), etc.

We know that entropy can measure the chaotic degree of the system; the system entropy is small when the system is in stable state and the system entropy is large when the system is in chaotic state. The equation of entropy used in this paper is as follows

$$S(p_1, \ p_2, \ \cdots, p_n) = -\sum_{i=1}^{n} p_i \log_2 p_i$$

Inevitably there will be many uncertain factors in the complex and changeable market. In this paper, through simulation analysis, we can clearly see the effect of parameter changing on the entropy of the system of the dual-channel supply chain, and then quantify the stability of the supply chain system using entropy, which lays the foundation for further effectively controlling the complexity of the whole supply chain.

4.1. The Entropy Complexity Analysis of System (9) with the Change of ξ

In this part, we also suppose the parameters values as above, the entropy and dynamic behavior of system (9) are described with the change of ξ when $\beta = 0.3$. Figure 3 shows the service level and the entropy of system (9) as ξ changes with $\gamma_1 = 0.55$. We can see that, for the multichannel supply chain system in this paper, the service level in traditional selling channel is the highest and the one of direct channel is the lowest, which accords with the operation of the real market. System (9) is stable when $\xi < 0.49$, and the bifurcation and chaos in system (9) occur through period-doubling bifurcation when ξ increases. System (9) has low entropy when it is in a stable state, and has high entropy when it is in a chaotic state. High entropy implies the system is more unstable; there exist many uncertainties in the complex and changeable market, and we require more information to keep system (9) in a stable state.

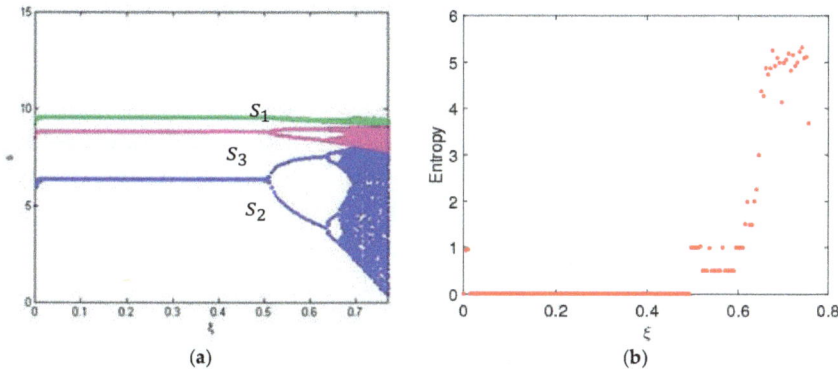

Figure 3. The evolution process of system (9) with change of ξ when $\gamma_1 = 0.55$. (a) The bifurcation diagram and (b) the entropy diagram.

Figure 4 shows the service levels and the entropy of system (5) as ξ changes with $\gamma_1 = 0.7$. We find that when consumers are more sensitive to service levels, system (9) will stabilize at a higher Nash equilibrium point, and the uncertainty of system (9) will appear earlier with high entropy, which is agreement with Figure 2. In the chaotic state, market competition is complex and unpredictable.

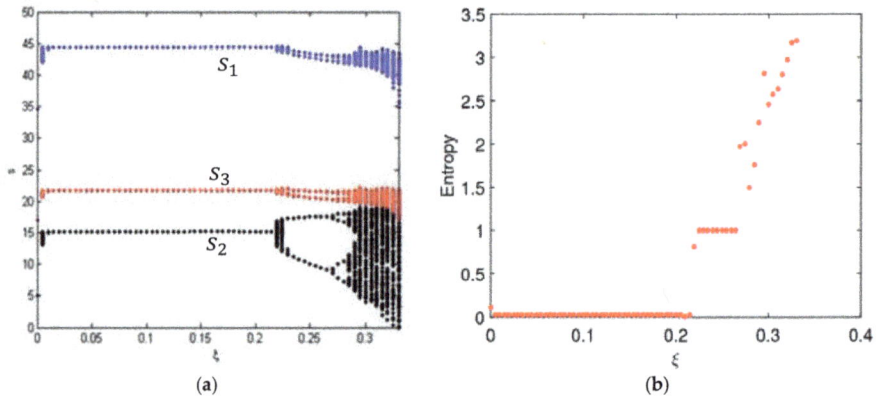

Figure 4. The evolution process of system (9) with change of ξ with $\gamma_1 = 0.7$. (**a**) The bifurcation diagram and (**b**) the entropy diagram.

In short, when consumers are more sensitive to channel services, the manufacturer and retailer will provide higher levels of channel services, and the uncertainty of system (9) will appear earlier with high entropy. In other words, the greater the service elasticity coefficient is, the easier system (9) goes into a chaotic state and the smaller the system's stable region becomes.

4.2. The Entropy Complexity Analysis of System (9) with Feedback Parameter (β)

In this section, we analyze the effects of feedback parameter (β) on the stability of system (9). Figure 5 is the bifurcation diagrams of system (9) with β changing. From Figure 5a, when $\xi = 0.3$, the system (9) gradually returns to the Nash equilibrium from the initial value no matter how β changes, system's entropy is zero and the market is in a stable state at this time. From Figure 5b, the system (9) is in the stable state with $0 < \beta \leq 0.08$ and makes 2-period bifurcation with $0.08 < \beta \leq 1$. So it can be seen that the feedback parameter (β) has little effect on system stability and system entropy.

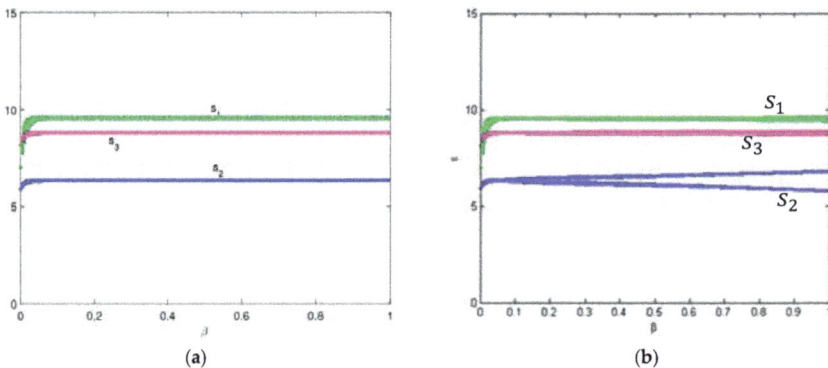

Figure 5. The bifurcation diagrams of the system (9) with change of β. (**a**) $\xi = 0.3$ and (**b**) $\xi = 0.5$.

When the system is in a stable state, the system's attractor is stable at fixed point; when the system goes into a chaotic state, the system's attractor will occupy a larger space and the structure of the chaotic attractor will be more complicated. When $\beta = 0.3$ and $\xi = 0.48$, system (9) is in stable state according to the bifurcation diagram shown in Figure 3, the chaotic attractor in this condition is shown in Figure 6a; when $\beta = 0.3$ and $\xi = 0.58$, the system (9) is in 2-period bifurcation state, the chaotic attractor is shown in Figure 6b. When $\beta = 0.3$ and $\zeta = 0.74$, the system (9) is in the chaotic state and the chaotic attractor is shown in Figure 6c.

From Figures 2–6, it can be seen that for the multichannel supply chain, the change of ζ can affect the period of the market entering chaos state even if the initial value of service variable is fixed. The more quickly the manufacturer adjusts channel service, the more easily the market falls into chaos. Therefore, in this competitive multichannel supply chain, the manufacturer and retailer should make their decisions with an overall consideration about the market situation and competitor's response rather than adjust their service level quickly and blindly.

According to information theory, when the market is in an orderly competitive state, the probability of optimal service level under optimal pricing will be large and system entropy will be low; when the market is in a disorderly competitive state, the service decisions under optimal pricing are out of order and the system entropy will be high.

Another obvious feature of the chaotic system is sensitive to the initial values. In other words, if there is a slight change in the initial values of the system's parameters, the results of system evolution will change greatly with time. When $\beta = 0.3$ and $\zeta = 0.48$, system (9) stays stable according to the above analysis. The initial values of the service level (s_2) are taken 5 and 5.001, after multiple iterations the differences between the two sets of numerical solutions are shown in Figure 7. We can find that, at the beginning of iterations, there is a little difference, but after approximately 40 iterations, the difference gradually reduces to zero.

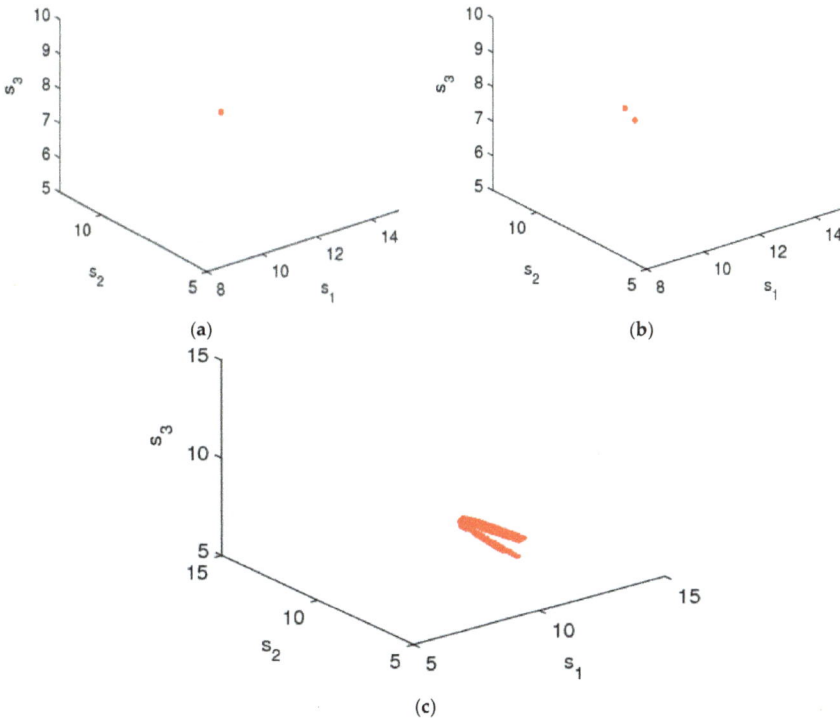

Figure 6. Chaotic attractor of system (9), (a) $\beta = 0.3$, $\zeta = 0.48$; (b) $\beta = 0.3$, $\zeta = 0.58$; and (c) $\beta = 0.3$, $\zeta = 0.74$.

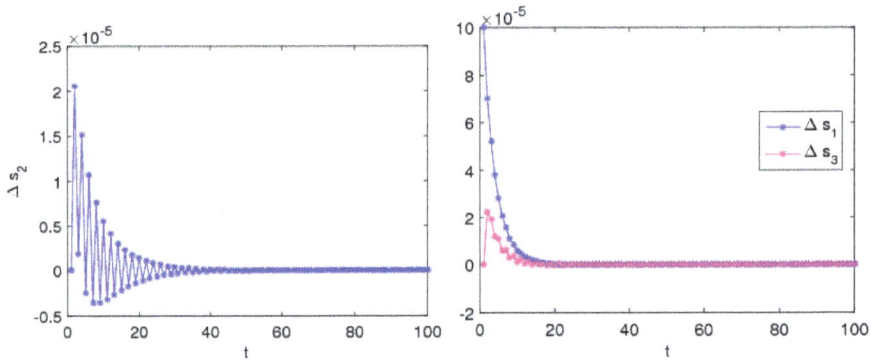

Figure 7. Initial value sensitivity of service level in stable system.

When $\beta = 0.3$ and $\xi = 0.72$, system (9) stays in a chaotic state according to the above analysis. The initial values of the service level (s_2) are taken 5 and 5.001, after multiple iterations, the differences between the two sets of numerical solutions are shown in Figure 8. We can see that, during the initial iterations, the values of channel services are no difference, but after approximately 20 iterations, the differences in channel services increase greatly.

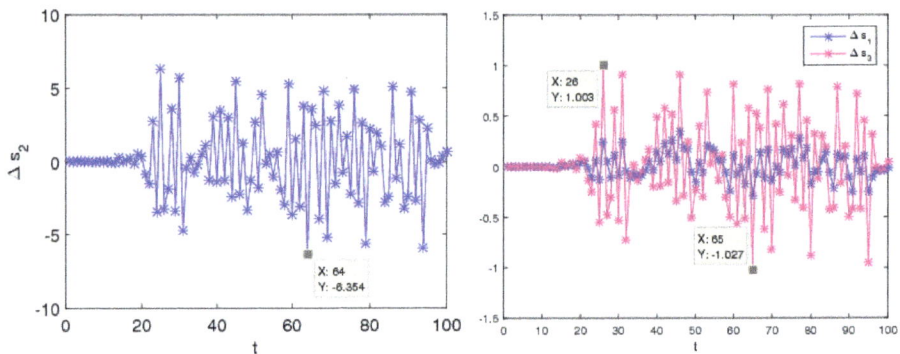

Figure 8. Initial value sensitivity of service level in a chaotic system.

Thus, it can see that the system is very sensitive to the initial value when the system is in chaos; small differences in initial values can cause a huge deviation after multiple iterations, which give us reassurance that decision-makers should choose the initial values of their decision variables more prudently.

4.3. The Influence of Parameter Changes on System's Profit

From the analysis above, we can find that when ξ is oversized, the uncertainty of system (9) will increase obviously, which can cause the market be complex and increase the system's entropy to a very large value making it difficult for decision-makers to make service decisions. So, we suspect that the profits of the system's participators will also be influenced. Figures 9 and 10 show the profit bifurcation diagrams of system (9) with respect to the change of ξ, respectively. As we have predicted, the profits of the system (9) stay stable when $\xi < 0.48$, and enter into a chaotic state when $\xi > 0.68$; in the chaotic state, the average profits of the system (9) show a downward trend (see in Figure 10). Therefore, the oversize of ξ will make the decision making complicated and affect the profits of the manufacturer and retailer.

Moreover, the retailer in traditional channel makes the largest service level, but gets the smallest profit in channel competition, which causes a conflict between online and offline channels. However, the manufacturer and retailer build a profit distribution contract of BOPS channel, which moderates the conflict between the online and the offline channels.

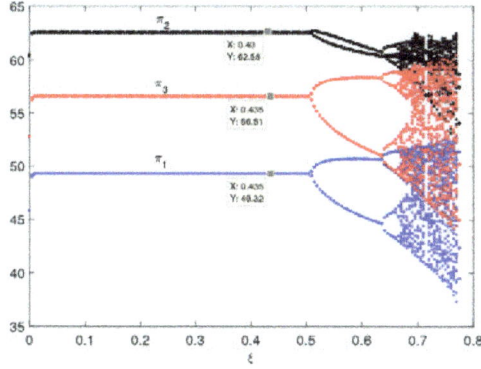

Figure 9. Profit diagram of system (9).

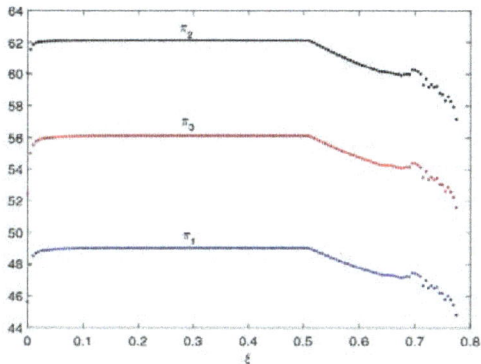

Figure 10. Average profit diagram of system (9).

5. Chaos Control

In a multichannel supply chain, all administrators want to achieve their own business goals easily and adjust their service strategies frequently to reduce uncertainty in market competition. Once the adjustment speeds of service levels carry out in an irrational state, the market will be out of order and fall into chaos, which is harmful to the manufacturer and retailer. The manufacturer and retailer need more information to eliminate the uncertainty of the system. Consequently, some methods should adopt to defer or remove the occurrence of bifurcation and chaos.

The above analysis shows that the change of ξ makes the multichannel supply chain system enter chaotic state gradually. In this section, the feedback control method is be used to delay or eliminate the chaotic behavior of the multichannel service game model, thus reduce the negative impact of chaos on the system (9). Wu and Ma [43] have used the feedback control method to control chaos in the product horizontal diversification supply chain.

The original dynamic system (9) is as follows

$$\begin{cases} s_1(t+1) = f_1[s_1(t), s_2(t), s_3(t)] \\ s_2(t+1) = f_2[s_1(t), s_2(t), s_3(t)] \\ s_3(t+1) = f_3[s_1(t), s_2(t), s_3(t)] \end{cases} \tag{12}$$

The controlled dynamic system can be expressed as follows

$$\begin{cases} s_1(t+1) = f_1[s_1(t), s_2(t), s_3(t)] \\ s_2(t+1) = f_2[s_1(t), s_2(t), s_3(t)] - Ps_2(t) \\ s_3(t+1) = f_3[s_1(t), s_2(t), s_3(t)] \end{cases} \tag{13}$$

where P represents the chaos control parameter. Selecting an appropriate value for P is essential to delay bifurcation, which can make the multichannel supply chain system return to the stable state from the chaotic state.

Figure 11 shows the bifurcation diagram and entropy diagram of the controlled system (13) with the change of P when $\beta = 0.3$ and $\xi = 0.72$. The controlled system (13) gradually enters the stable state from the chaotic state with the increase of P; when $0.73 \leq P \leq 1$, the controlled system (13) is in the stable state and has low entropy in which the market is in an orderly competitive state.

In the real market, the control parameter P can act as an external interference for decision-makers for the multichannel supply chain system. When the multichannel supply chain goes into a chaotic state with the increase in market uncertainty, the decision maker should actively intervene in market competition. In other words, the control parameter P can also regard as a decision-maker's learning and self-adaptive ability. The decision-makers can select appropriate values for P to achieve the multichannel supply chain back to the stable state.

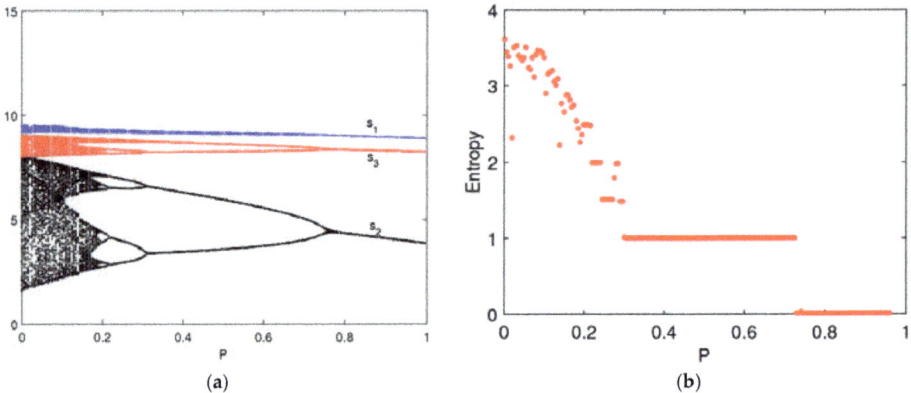

Figure 11. The evolution process of controlled system with P changing. (**a**) The bifurcation diagram and (**b**) the entropy diagram.

6. Conclusions

In this paper, we build a multichannel dynamic service game model where a retailer operates an offline channel and a manufacturer operates an online channel and offers customers the option to buy online and pick up in retailer's store (BOPS). The manufacturer and retailer maximize the channel profits as a business objective and make the channel service game under optimal pricing. The equilibrium solutions, the stable region, complexity entropy, and efficiency of the multichannel supply chain system are studied. The results show that the stability of the multichannel supply chain system will weaken with the service elasticity coefficient increasing and almost be unaffected by the feedback parameter adjustment of the retailer. If the manufacturer adjusts his service decision quickly, the system will go into a chaotic state, the average profits of the system will show a downward trend, and the system's entropy will increase. The BOPS channel strengthens the cooperation between the manufacturer and retailer and moderates the conflict between the online and the offline channels. In chaos, the service input is difficult to predict and the system entropy is large, the manufacturer and retailer need more information to familiarize themselves with the market environment. The chaotic system can be delayed or eliminated effectively using the method of feedback control, so that the manufacturer and retailer should cooperate with each other to make some measures to prevent and control chaos.

The conclusions of this paper involve the elucidation of a realistic guide for the manufacturer and retailer to make optimal service decisions to avoid chaos and profit loss. For example, the manufacturer should have sufficient information to analyze consumers' sensitivity to channel services. If the manufacturer blindly adjusts the channel service levels, it is easy to increase the uncertainty of the system and make the market go into chaos. In chaos, the manufacturer and retailer suffer unnecessary losses, and may even exit the market.

Nonetheless, some assumptions made in this paper limit the research results of the model. Loosing these assumptions can make the model close to market operation scenarios more. For instance, a model considered the interaction effect of service on channel demand will close to the actual situation. Second, other power structures between the manufacturer and retailer in the multichannel supply chain should take into account, as it may shed light on whether the current results will hold. Finally, the other decision objectives of the manufacturer and retailer should consider the future. We hope that the ideas and the model presented in this paper will lay the motivational ground for future research in these directions.

Author Contributions: Y.H. built the probabilistic selling game model and provided economic interpretation of the conclusions; X.C. carried out numerical simulation; Q.L. performed mathematical calculation; X.M. carried out writing-review & editing. All authors have read and approved the final manuscript.

Funding: The research was supported by the Henan Province Soft Science Research Plan Project (No: 182400410054); the Henan provincial government decision research tendering project (No: 2018B019); the Henan provincial social science planning decision consultation project (2018JC05); the Humanities and Social Science Foundation in Hubei Provincial Education Department (17Q088); and The Research Center of Enterprise Decision Support, Key Research Institute of Humanities and Social Sciences in Universities of Hubei Province (DSS20150205).

Acknowledgments: The authors thank the reviewers for their careful reading and providing some pertinent suggestions.

Conflicts of Interest: The authors declare no conflicts of interest.

References

1. Dan, B.; Qu, Z.; Zhang, H.; Liu, C.; Zhang, X. Mixed Channel Strategy for Manufacturers to Cope with Strong Retailers in Supply Chain. *Manag. Rev.* **2016**, *12*, 19.
2. Huang, S.; Yang, C.; Yang, J. Pricing and production decisions in dual-channel supply chains with demand and production cost disruptions. *Syst. Eng. Theory Pract.* **2014**, *34*, 1219–1229.
3. Li, B.; Zhu, M.; Jiang, Y.; Li, Z. Pricing policies of a competitive dual-channel green supply chain. *J. Clean. Prod.* **2016**, *112*, 2029–2042. [CrossRef]
4. Kouvelis, P.; Turcic, D.; Zhao, W.H. Supply chain contracting in environments with volatile input prices and frictions. *Manuf. Serv. Oper. Manag.* **2018**, *20*, 130–146. [CrossRef]
5. Li, B.; Hou, P.W.; Chen, P.; Li, Q.H. Pricing strategy and coordination in a dual channel supply chain with a risk-averse retailer. *Int. J. Prod. Econ.* **2016**, *178*, 154–168. [CrossRef]
6. Li, B.; Chen, P.; Li, Q.; Wang, W. Dual-channel supply chain pricing decisions with a risk-averse retailer. *Int. J. Prod. Res.* **2014**, *52*, 7132–7147. [CrossRef]
7. Radhi, M.; Zhang, G. Pricing policies for a dual-channel retailer with cross-channel returns. *Comput. Ind. Eng.* **2018**, *119*, 63–75. [CrossRef]

8. Li, W.; Chen, J.; Liang, G.; Chen, B. Money-back guarantee and personalized pricing in a Stackelberg manufacturer's dual-channel supply chain. *Int. J. Prod. Econ.* **2018**, *197*, 84–98. [CrossRef]
9. Ji, G.; Han, S.; Huatan, K. False failure returns: Optimal pricing and return policies in a dual-channel supply chain. *J. Syst. Sci. Syst. Eng.* **2018**, *27*, 292–321. [CrossRef]
10. Xie, J.; Zhang, W.; Liang, L.; Xia, Y.; Yin, J.; Yang, G. The revenue and cost sharing contract of pricing and servicing policies in a dual-channel closed-loop supply chain. *J. Clean. Prod.* **2018**, *191*, 361–383. [CrossRef]
11. Dumrongsiri, A. A supply chain model with direct and retail channels. *Eur. J. Oper. Res.* **2008**, *187*, 691–718. [CrossRef]
12. Yan, R.; Pei, Z. Retail services and firm profit in a dual-channel market. *J. Retail. Consum. Serv.* **2009**, *16*, 306–314. [CrossRef]
13. Mukhopadhyay, S.K.; Zhu, X.; Yue, X. Optimal Contract Design for Mixed Channels under Information Asymmetry. *Prod. Oper. Manag.* **2008**, *17*, 641–650. [CrossRef]
14. Ding, F.; Huo, J.-Z. A feasibility of service level coordination of dual channel supply chain effect of dual-channel supply chain. *Chin. Manag. Sci.* **2014**, *S1*, 485–490.
15. Giri, B.C.; Maiti, T. Service competition in a supply chain with two retailers under service level sensitive retail price and demand. *Int. J. Manag. Sci. Eng. Manag.* **2014**, *9*, 133–146. [CrossRef]
16. Ma, J.; Guo, Z. Research on the complex dynamic characteristics and RLS estimation's influence based on price and service game. *Math. Probl. Eng.* **2015**, *2015*. [CrossRef]
17. Ali, S.M.; Rahman, M.H.; Tumpa, T.J.; Rifat, A.A.M.; Paul, S.K. Examining price and service competition among retailers in a supply chain under potential demand disruption. *J. Retail. Consum. Serv.* **2018**, *40*, 40–47. [CrossRef]
18. Li, Q.H.; Li, B. Dual-channel supply chain equilibrium problems regarding retail services and fairness concerns. *Appl. Math. Model.* **2016**, *40*, 7349–7367. [CrossRef]
19. Zhou, Y.W.; Guo, J.; Zhou, W. Pricing/service strategies for a dual-channel supply chain with free riding and service-cost sharing. *Int. J. Prod. Econ.* **2018**, *196*, 198–210. [CrossRef]
20. Zhou, J.; Zhao, R. Dual-channel signaling strategy with channel competition. *Syst. Eng. Theory Pract.* **2018**, *38*, 414–428.
21. Chen, X.; Wang, X.; Jiang, X. The impact of power structure on the retail service supply chain with an O2O mixed channel. *J. Oper. Res. Soc.* **2016**, *67*, 294–301. [CrossRef]
22. Kong, L.; Liu, Z.; Pan, Y.; Xie, J.; Yang, G. Pricing and service decision of dual-channel operations in an O2O closed-loop supply chain. *Ind. Manag. Data Syst.* **2017**, *117*, 1567–1588. [CrossRef]
23. Cai, G.; Zhang, Z.G.; Zhang, M. Game theoretical perspectives on dual-channel supply chain competition with price discounts and pricing schemes. *Int. J. Prod. Econ.* **2009**, *117*, 80–96. [CrossRef]
24. Tao, J.T.; Li, B. Research on pricing policies in an O2O channel supply chain under the free-riding effect of online platform. *Ind. Eng. Manag.* **2018**, *1*, 38–44.
25. Zhao, J.; Hou, X.; Guo, Y.; Wei, J. Pricing policies for complementary products in a dual-channel supply chain. *Appl. Math. Model.* **2017**, *49*, 437–451. [CrossRef]
26. Chen, X.; Liu, Y.; Zhong, W. Optimal decision making for online and offline retailers under BOPS mode. *ANZIAM J.* **2016**, *58*, 187–208. [CrossRef]
27. Gao, F.; Su, X. Omnichannel retail operations with buy online and pickup in store. *Soc. Sci. Electron. Publ.* **2016**, *63*, 2478–2492.
28. Gallino, S.; Moreno, A. Integration of online and offline channels in retail: the impact of sharing reliable inventory availability information. *Soc. Sci. Electron. Publ.* **2014**, *60*, 1434–1451. [CrossRef]
29. Jin, M.; Li, G.; Cheng, T.C.E. Buy online and pick up in-store: Design of the service area. *Eur. J. Oper. Res.* **2018**, *268*, 613–623. [CrossRef]
30. Liu, Y.M.; Zhou, D. Is it always beneficial to implement BOPS? A comparative research with traditional dual channel. *Oper. Res. Manag. Sci.* **2018**, *2*, 23.
31. Moon, S.; Ji, W.; Moon, H.; Kim, D. A simulation of order resonance phenomenon in a supply chain triggered by reinforcing loop. *Int. J. Simul. Model.* **2018**, *17*, 231–244. [CrossRef]
32. Yan, B.; Jin, Z.; Wang, X.; Liu, S. Analyzing a mixed supply chain with a WeChat channel. *Electron. Commer. Res. Appl.* **2018**, *29*, 90–101. [CrossRef]
33. Matsui, K. When and what wholesale and retail prices should be set in multi-channel supply chains? *Eur. J. Oper. Res.* **2018**, *267*, 540–554. [CrossRef]

34. Zuo, Y.; Kajikawa, Y. Toward a theory of industrial supply networks: A multi-level perspective via network analysis. *Entropy* **2017**, *19*, 382. [CrossRef]
35. Wang, Z.; Soleimani, H.; Kannan, D.; Xu, L. Advanced cross-entropy in closed-loop supply chain planning. *J. Clean. Prod.* **2016**, *135*, 201–213. [CrossRef]
36. Kriheli, B.; Levner, E. Entropy-based algorithm for supply-chain complexity assessment. *Algorithms* **2018**, *11*, 35. [CrossRef]
37. Levner, E.; Ptuskin, A. Entropy-based model for the ripple effect: Managing environmental risks in supply chains. *Int. J. Prod. Res.* **2018**, *56*, 2539–2551. [CrossRef]
38. Lou, W.; Ma, J.; Zhan, X. Bullwhip entropy analysis and chaos control in the supply chain with sales game and consumer returns. *Entropy* **2017**, *19*, 64. [CrossRef]
39. Han, Z.; Ma, J.; Si, F.; Ren, W. Entropy complexity and stability of a nonlinear dynamic game model with two delays. *Entropy* **2016**, *18*, 317. [CrossRef]
40. Yao, D.Q.; Yue, X.; Liu, J. Vertical cost information sharing in a supply chain with value-adding retailers. *Omega* **2008**, *36*, 838–851. [CrossRef]
41. Dan, B.; Xu, G.; Liu, C. Pricing policies in a dual-channel supply chain with retail services. *Int. J. Prod. Econ.* **2012**, *139*, 312–320. [CrossRef]
42. Yang, J.Q.; Zhang, X.M.; Zhang, H.Y.; Liu, C. Cooperative inventory strategy in a dual-channel supply chain with transshipment consideration. *Int. J. Simul. Model.* **2016**, *15*, 365–376. [CrossRef]
43. Wu, F.; Ma, J. The equilibrium, complexity analysis and control in epiphytic supply chain with product horizontal diversification. *Nonlinear Dyn.* **2018**, *93*, 2145–2458. [CrossRef]

entropy

MDPI

Article

Dynamics and Complexity of a New 4D Chaotic Laser System

Hayder Natiq [1,2], Mohamad Rushdan Md Said [1,3,4,*], Nadia M. G. Al-Saidi [2] and Adem Kilicman [1,4]

[1]   Institute for Mathematical Research, Universiti Putra Malaysia, UPM Serdang 43000, Malaysia; haydernatiq86@gmail.com (H.N.); akilic@upm.edu.my (A.K.)
[2]   The Branch of Applied Mathematics, Applied Science Department, University of Technology, Baghdad 10075, Iraq; nadiamg08@gmail.com
[3]   Malaysia-Italy Centre of Excellence for Mathematical Science, Universiti Putra Malaysia, UPM Serdang 43000, Malaysia
[4]   Department of Mathematics, Universiti Putra Malaysia, UPM Serdang 43000, Malaysia
*   Correspondence: mrushdan@upm.edu.my

Received: 11 December 2018; Accepted: 2 January 2019; Published: 7 January 2019

Abstract: Derived from Lorenz-Haken equations, this paper presents a new 4D chaotic laser system with three equilibria and only two quadratic nonlinearities. Dynamics analysis, including stability of symmetric equilibria and the existence of coexisting multiple Hopf bifurcations on these equilibria, are investigated, and the complex coexisting behaviors of two and three attractors of stable point and chaotic are numerically revealed. Moreover, a conducted research on the complexity of the laser system reveals that the complexity of the system time series can locate and determine the parameters and initial values that show coexisting attractors. To investigate how much a chaotic system with multistability behavior is suitable for cryptographic applications, we generate a pseudo-random number generator (PRNG) based on the complexity results of the laser system. The randomness test results show that the generated PRNG from the multistability regions fail to pass most of the statistical tests.

Keywords: Hopf bifurcation; self-excited attractors; multistability; sample entropy; PRNG

1. Inroduction

The chaotic behavior as a rich nonlinear phenomenon has been detected in many non-natural and natural systems, and usually plays an important role in their performance [1,2]. Chaotic systems are complicated and have many interesting features, such as unpredictability, topological mixing, and high sensitivity to their initial conditions and parameters [3,4]. Therefore, chaotic systems have received significant attention from various fields including cryptography [5,6], secure communications [7,8], laser applications [9,10], biomedical engineering [11,12], and many others.

Existing chaotic systems can be classified into two categories: systems with self-excited attractors and systems with hidden attractors [13]. The chaotic system with self-excited attractors has a basin of attraction that is intersected with an unstable equilibrium, whereas the chaotic system with hidden attractors has a basin of attraction which does not intersect with any open neighborhoods of equilibria [14,15]. According to the above definition, most of the classical chaotic attractors are self-excited [16,17]. Meanwhile, it has been demonstrated that the attractors in dynamical systems with no equilibria [18,19], stable equilibria [20], lines of equilibria [21], and curves of equilibria [22] are hidden attractors.

However, with further investigation of chaos, it was unexpected to find that many systems with self-excited and hidden attractors have more than one attractor for a given set of parameters and different initial values. This phenomenon is known as multistability or coexisting attractors. The clear evidence of multistability was first experimentally manifested in a Q-switched gas laser [23], since then chaotic systems with multistability behaviors have been extensively reported. Munoz et al. presented a fractional-order chaotic system with multiple coexisting attractors [24]. Wang et al. established a 2D chaotic map with no-equilibria generating a pair of chaotic attractors [25]. Li et al. introduced a new method for constructing self-reproducing chaotic systems with extreme multistability [26]. In fact, multistability as a new research direction in chaos theory requires further research, especially, how to determine and locate this complicated nonlinear phenomenon in the chaotic systems.

Since the in-depth analysis of the local bifurcation is required to clarify the evolution of the chaotic state from the steady state, the scope of studying the bifurcation of the equilibria in the chaotic systems is of considerable interest [27]. Hopf bifurcation is one of an important local dynamic bifurcation, and is considered as the emergence of a limit cycle from an equilibrium point. Furthermore, the Hopf bifurcation plays a crucial role in analyzing the stability of the equilibria of the high-dimensional system [28,29]. Therefore, Hopf bifurcation is beneficial to analyzing the dynamic behavior of high-dimensional chaotic systems, as well as to the applications of controlling chaos [30].

Complexity of nonlinear dynamical systems has attracted attention in recent years due to its importance for measuring the predictability and randomness of the system time series [31,32]. The time series with high complexity led to a chaotic attractor, hence, the complexity is able to determine and locate the chaotic and periodic attractors in nonlinear systems [33,34]. Motivated by this observation, this paper applies Sample Entropy contour plot to determine multistability regions of a new 4D chaotic laser system, which is derived from Lorenz-Haken equations. The new chaotic system has one unstable equilibrium and symmetric stable equilibria, hence the chaotic attractor of the presented system is generally self-excited, meanwhile, the possible existence of a hidden chaotic attractor is an open problem.

The main contributions of this research work are as follows:

(i) We derive a new 4D chaotic laser system with three equilibria from Lorenz-Haken equations;
(ii) We investigate the stability of the symmetric equilibria, and the existence of coexisting multiple Hopf bifurcations on these equilibria;
(iii) We analyze the presence of complex coexisting behaviors in the laser system;
(iv) We use the complexity of the laser system time series to locate the regions of coexisting attractors when the parameters and initial values vary;
(v) Based on the complexity of the system time series, we study the randomness of multistability regions.

The rest of this paper is organized as follows: Section 2 introduces the new 4D chaotic laser system and studies its dynamical properties. Section 3 investigates the existence of Hopf bifurcation in the laser system. Section 4 provides the details about the multistability of the laser system. In Section 5, we use SamEn to locate the regions of the coexisting attractors, as well as to demonstrate the randomness of these regions. The conclusions are presented in Section 6.

2. A New 4D Chaotic Laser System From Lorenz-Haken Model

In this section, we discuss the dynamics of a new 4D chaotic laser system which is derived from the well-known Lorenz-Haken equations [35]. In the standard notation of Reference [36], the Lorenz-Haken equations is given by

$$\begin{cases} \dfrac{dx}{dt} = -\sigma(x-y) + iqx|x|^2, \\ \dfrac{dy}{dt} = -(1-i\delta)y + (r-z)x, \\ \dfrac{dz}{dt} = -bz + Re(x*y). \end{cases} \tag{1}$$

In the optical language, x is proportional to the electric field, y is proportional to the induced macroscopic polarization, $(r-z)$ denotes the inversion, $\sigma = \frac{\tau_P}{\tau_E}$, and $b = \frac{\tau_P}{\tau_N}$. Here, τ_E represents the optical field, τ_P is the induced polarization, and τ_N denotes the inversion parameter. Meanwhile, the parameter δ governs the coupling between amplitude and phase variations, and q is known as the linewidth enhancement factor.

Since both x and z can be chosen to be real [37], the dynamics of Equation (1) can be investigated by considering the following linear transformation

$$x = x_1, \quad y = x_2 + ix_3, \quad z = x_4.$$

Consequently, the new 4D chaotic laser system is defined as

$$\begin{cases} \dfrac{dx_1}{dt} = \sigma(x_2 - x_1), \\ \dfrac{dx_2}{dt} = -x_2 - \delta x_3 + (r - x_4)x_1, \\ \dfrac{dx_3}{dt} = \delta x_2 - x_3, \\ \dfrac{dx_4}{dt} = -bx_4 + x_1 x_2, \end{cases} \tag{2}$$

where x_i are state variables and σ, δ, r and b are parameters.

2.1. Chaotic Behavior Regions

To examine the dynamic characteristics of the system (2), Figure 1a,b depicts its bifurcation diagram and Lyapunov exponents, respectively, in which the parameters are set as $\sigma = 4$, $\delta = 0.5$, $r = 27$, and $0 \leq b \leq 2$. This figure clearly shows chaotic attractors for $b \in [0.15, 0.187] \cup [0.205, 2]$, quasi-periodic (when $b = 0.132$) and periodic attractors for $b \in [0, 0.15) \cup (0.187, 0.205)$. To demonstrate the chaotic behavior of the system (2), Figure 2 plots its phase portraits with $\sigma = 4$, $\delta = 0.5$, $r = 27$, $b = 2$ and for the initial values $(2, 1, 1, 2)$. As can be observed in Figure 2, the system (2) has a two-scroll chaotic attractor.

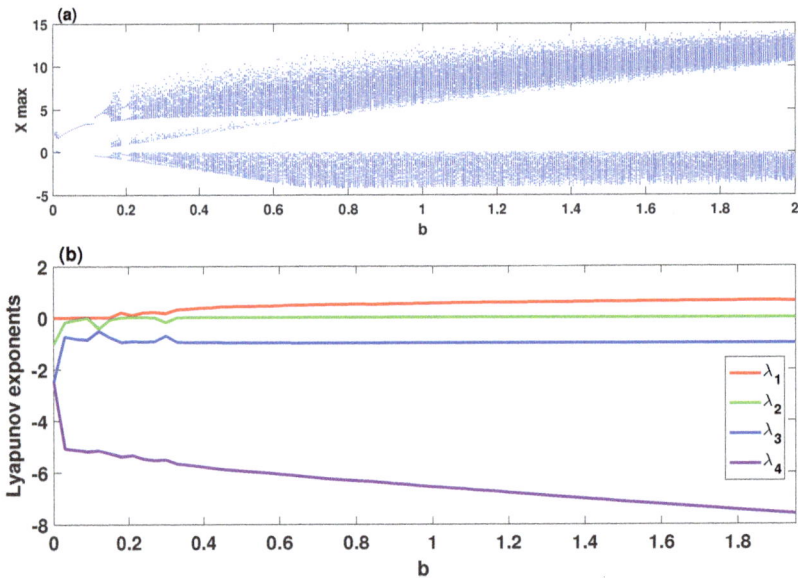

Figure 1. Dynamics of the system (2) versus the parameter b for the initial values $(2, 1, 1, 2)$ and with $\sigma = 4$, $\delta = 0.5$, $r = 27$: (**a**) bifurcation diagram; (**b**) Lyapunov exponents.

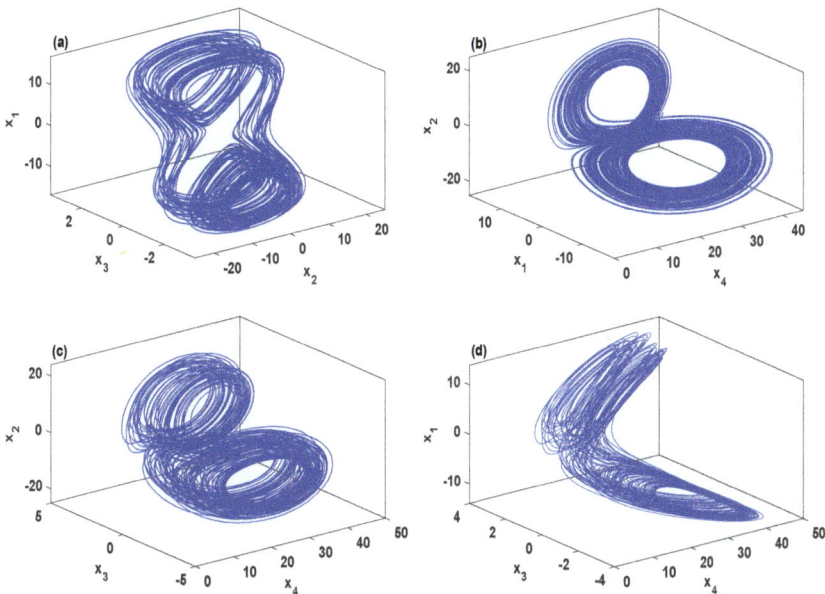

Figure 2. Different orientations on a two-scroll chaotic attractor of the system (2) for the initial values $(2, 1, 1, 2)$ and with the parameters $\sigma = 4$, $\delta = 0.5$, $r = 27$, $b = 2$. (**a**) (x_2, x_3, x_1) space; (**b**) (x_4, x_1, x_2) space; (**c**) (x_4, x_3, x_2) space; (**d**) (x_4, x_3, x_1) space.

2.2. Dissipation and Symmetry

The divergence of system (2) is defined as

$$\nabla V = \frac{\partial \dot{x}_1}{\partial x_1} + \frac{\partial \dot{x}_2}{\partial x_2} + \frac{\partial \dot{x}_3}{\partial x_3} + \frac{\partial \dot{x}_4}{\partial x_4} = -(\sigma + b + 2).$$

Thus, the system (2) becomes dissipative when $(\sigma + b + 2) > 0$. This means each volume element $V_0 e^{-(\sigma + b + 2)t}$ of system (2) shrinks to zero as $t \longrightarrow \infty$ at an exponential rate $(\sigma + b + 2)$.

Additionally, the system (2) has invariance under the coordinate transformation

$$(x_1, x_2, x_3, x_4) \longrightarrow (-x_1, -x_2, -x_3, x_4).$$

Consequently, the system (2) has rotational symmetry around the x_4-axis.

2.3. Equilibria and Stability

Suppose that the parameters $\sigma > 0$, $\delta > 0$, $r > 0$ and $b > 0$, then the equilibria of the system (2) can be calculated by solving the following equations

$$\begin{cases} \sigma(x_2 - x_1) = 0, \\ -x_2 - \delta x_3 + (r - x_4)x_1 = 0, \\ \delta x_2 - x_3 = 0, \\ -b x_4 + x_1 x_2 = 0. \end{cases}$$

From the above equations, it can be obtained that the equilibria of the system (2) have the following form:

$$E_i(k, \ k, \ \delta k, \ \frac{k^2}{b}),$$

where k is either 0 or $\pm\sqrt{b(r-(1+\delta^2))}$. The system (2) has one real equilibrium $E_1(0,0,0,0)$ when $r = 1 + \delta^2$, whereas it has three real equilibria if $r > 1 + \delta^2$

$$
\begin{cases}
E_1(0,0,0,0), \\
E_2\left(\sqrt{b(r-(1+\delta^2))}, \sqrt{b(r-(1+\delta^2))}, \delta\sqrt{b(r-(1+\delta^2))}, r-(1+\delta^2)\right), \\
E_3\left(-\sqrt{b(r-(1+\delta^2))}, -\sqrt{b(r-(1+\delta^2))}, -\delta\sqrt{b(r-(1+\delta^2))}, r-(1+\delta^2)\right).
\end{cases}
$$

Using the Jacobian matrix, the system (2) is linearized at the equilibrium E_i as follows

$$
J_{E_i} = \begin{pmatrix}
-\sigma & \sigma & 0 & 0 \\
r - \frac{k^2}{b} & -1 & -\delta & -k \\
0 & \delta & -1 & 0 \\
k & k & 0 & -b
\end{pmatrix}.
$$

Since the equilibria $E_{2,3}$ are symmetric about the x_4−axis, then they will have the same characteristics. Therefore, the characteristic equation of Jacobian matrix at the equilibrium E_2 with $b = 1$ can be written as

$$
f(\lambda) = (\lambda + 1) f_1(\lambda) = 0, \tag{3}
$$

where

$$
f_1(\lambda) = \lambda^3 + (2+\sigma)\lambda^2 + (1+2\sigma+\delta^2+k^2+\sigma k^2 - \sigma r)\lambda + (\sigma+\sigma\delta^2+3\sigma k^2 - \sigma r). \tag{4}
$$

It is obvious that Equation (3) always has one eigenvalue with negative real part which is $\lambda_1 = -1$, whereas the real parts of the other eigenvalues are not always negative. It is well-known that a system is asymptotically stable when all eigenvalues have negative real parts; otherwise, the system is unstable. By Routh–Hurwitz criterion, the real parts of all the eigenvalues of the system (2) are negative if and only if

$$
\begin{cases}
\Delta_1 = (2+\sigma) > 0, \\
\Delta_2 = \begin{vmatrix} (2+\sigma) & 1 \\ (\sigma+\sigma\delta^2+3\sigma k^2 - \sigma r) & (1+2\sigma+\delta^2+k^2+\sigma k^2 - \sigma r) \end{vmatrix} > 0, \\
\Delta_3 = (\sigma+\sigma\delta^2+3\sigma k^2 - \sigma r)\Delta_2 > 0.
\end{cases}
$$

By choosing the parameters $\sigma > 0$ and $\delta > 0$, these inequalities lead to the following condition:

$$
r < \frac{4\sigma + \sigma^2 - \sigma^2\delta^2}{\sigma - 2}. \tag{5}
$$

Thus, if the above conditions are satisfied, then the equilibrium E_2 is an asymptotically stable.

3. Local Bifurcation Analysis and Numerical Simulations

This section reviews the Hopf bifurcation using the bifurcation theories. In addition, the existence of coexisting symmetric Hopf bifurcations in the system (2) will be investigated with the variation of parameter $r \in R^+$.

3.1. Hopf Bifurcation

Hopf bifurcation is the source of a limit cycle, which usually appears when the stability of the equilibrium point changes at some critical parameter value. To illustrate the Hopf bifurcation of a dynamical system on the equilibrium point, consider a vector field as follows

$$\dot{x} = f(x, \zeta), \tag{6}$$

where $x \in R^4$ and $\zeta \in R^+$ represent the phase variables and the parameters, respectively. The vector field undergoes a Hopf bifurcation when the following conditions are satisfied simultaneously [38]:

(A)   nondegeneracy condition: the Jacobian matrix $J_{(x_0, \zeta_0)}$ has one pair of purely imaginary roots, and other roots have nonzero real parts;

(B)   transversality condition: the real part of differentiation characteristic equation with respect to the parameter ζ satisfy

$$Re(\frac{d\lambda}{d\zeta})\Big|_{\zeta=\zeta_0} \neq 0; \tag{7}$$

(C)   the first Lyapunov coefficient l_1 is nonzero.

In order to derive the first Lyapunov coefficient l_1, suppose that Equation (2) has an equilibrium point at $x = x_0$. By denoting $X = x - x_0$, we can write

$$F(X) = f(X, \zeta_0), \tag{8}$$

as

$$F(X) = AX + \frac{1}{2}B(X, X) + \frac{1}{6}C(X, X, X) + O(\| X \|^4), \tag{9}$$

where A is the Jacobian matrix, and B and C are symmetric multilinear vector functions which are defined as

$$\begin{cases} B_i(X, Y) = \sum\limits_{j,k=1}^{n} \frac{\partial^2 F_i(\eta)}{\partial \eta_j \partial \eta_k}\Big|_{\eta=0} X_j Y_k, & i = 1, 2, \ldots, n, \\ C_i(X, Y, Z) = \sum\limits_{j,k,l=1}^{n} \frac{\partial^3 F_i(\eta)}{\partial \eta_j \partial \eta_k \partial \eta_l}\Big|_{\eta=0} X_j Y_k Z_l, & i = 1, 2, \ldots, n. \end{cases} \tag{10}$$

Suppose that A possesses a pair of purely imaginary eigenvalues $\lambda_{1,2} = \pm i\omega$, meanwhile, the other eigenvalues have nonzero real part. Let p, q be an eigenvectors of A satisfying the following three conditions

$$\begin{cases} Aq = i\omega_0 q, \\ A^T p = -i\omega_0 p, \\ \langle p, q \rangle = \sum\limits_{i=1}^{n} \overline{p_i} \, q_i = 1. \end{cases} \tag{11}$$

By means of an immersion of the form $X = V(\mu, \overline{\mu})$, the 2D center manifold associated to the eigenvalues $\lambda_{1,2} = \pm i\omega$ is parameterized, where $V : C^2 \longrightarrow R^4$ has a Taylor expansion of the following form

$$V(\mu, \overline{\mu}) = \mu \, q + \overline{\mu} \, \overline{q} + \sum\limits_{2 \leq j+k \leq 3} \frac{1}{j!k!} v_{jk} \, \mu^j \, \overline{\mu}^k + O(|\mu|^4). \tag{12}$$

with $v_{jk} \in C^4$ and $v_{jk} = \bar{v}_{jk}$. By substituting Equation (12) into (8), one has

$$\frac{\partial V}{\partial \mu}\dot{\mu} + \frac{\partial V}{\partial \bar{\mu}}\dot{\bar{\mu}} = F(V(\mu, \bar{\mu})) \tag{13}$$

Defined by the coefficients $\mu^j \bar{\mu}^k$, the complex vectors v_{jk} can be obtained by solving Equation (13). On the chart μ for a center manifold, the system (13) can be written as

$$\dot{\mu} = i w_0 \mu + \frac{1}{2}G_{21}\mu|\mu|^2 + O(|\mu|^4). \tag{14}$$

Thus, the first Lyapunov coefficient can be defined as

$$l_1 = \frac{1}{2w_0}Re[\langle p, C(q, q, \bar{q})\rangle - 2\langle p, B(q, -v_{11})\rangle + \langle p, B(\bar{q}, v_{20})\rangle] \tag{15}$$

where $v_{11} = -A^{-1}B(q, \bar{q})$ and $v_{20} = (2iw_0I - A)^{-1}B(q, q)$.

3.2. Numerical Simulations

To investigate the existence of Hopf bifurcation in the system (2) at the equilibrium E_2, we will examine the conditions (A), (B) and (C) one by one.

Firstly, we assume that the characteristic Equation (3) has a pair of purely imaginary eigenvalues $\lambda_{1,2} = \pm i w_0$. By substituting $\lambda = i w_0$ into (4), one has

$$-i w_0^3 - (2 + \sigma)w_0^2 + (1 + 2\sigma + \delta^2 + k^2 + \sigma k^2 - \sigma r)i w_0 + (\sigma + \sigma\delta^2 + 3\sigma k^2 - \sigma r) = 0, \tag{16}$$

which leads to:

$$\begin{cases} -i w_0^3 + (1 + 2\sigma + \delta^2 + k^2 + \sigma k^2 - \sigma r)i w_0 = 0, \\ -(2 + \sigma)w_0^2 + (\sigma + \sigma\delta^2 + 3\sigma k^2 - \sigma r) = 0. \end{cases}$$

Thus, one can obtain that

$$\begin{cases} w_0 = \sqrt{\dfrac{\sigma + \sigma\delta^2 + 3\sigma k^2 - \sigma r}{2 + \sigma}}, \\ r = \dfrac{2 + 4\sigma + 2\delta^2 + 2k^2 + 2\sigma^2 + \sigma^2 k^2}{\sigma + \sigma^2}, \end{cases}$$

which are equivalent to

$$\begin{cases} w_0 = \sqrt{\dfrac{2\sigma k^2}{2 + \sigma}}, \\ r = \dfrac{4\sigma + \sigma^2 - \sigma^2\delta^2}{\sigma - 2}, \end{cases}$$

where $k = \sqrt{r - (1 + \delta^2)}$. It is worth noting that when $r = r_0 = \frac{2 + 4\sigma + 2\delta^2 + 2k^2 + 2\sigma^2 + \sigma^2 k^2}{\sigma + \sigma^2}$, the characteristic Equation (3) can be written as

$$f(\lambda) = (\lambda + 1)(\lambda + 2 + \sigma)\left(\lambda^2 + \frac{\sigma + \sigma\delta^2 + 3\sigma k^2 - \sigma r_0}{2 + \sigma}\right). \tag{17}$$

Therefore, the four eigenvalues of the system (2) are as follows

$$
\begin{cases}
\lambda_1 = -1, \\
\lambda_2 = -(2+\sigma), \\
\lambda_3 = i\sqrt{\dfrac{\sigma + \sigma\delta^2 + 3\sigma k^2 - \sigma r_0}{2+\sigma}} = i\omega_0, \\
\lambda_4 = -i\sqrt{\dfrac{\sigma + \sigma\delta^2 + 3\sigma k^2 - \sigma r_0}{2+\sigma}} = -i\omega_0.
\end{cases}
\tag{18}
$$

Consequently, the nondegeneracy condition (A) is satisfied when $r = r_0$.

Secondly, let $\lambda(r) = \pm i\omega_0(r)$, by substituting $\lambda(r)$ into Equation (10) and differentiate the both sides with respect to r, one obtains

$$
\frac{d\lambda(r)}{dr} = \frac{\sigma\lambda + \sigma}{3\lambda^2 + 2(2+\sigma)\lambda + (1+2\sigma+\delta^2+k^2+\sigma k^2 - \sigma r)},
\tag{19}
$$

which leads to:

$$
\frac{d\lambda(r)}{dr}\bigg|_{r=r_0,\lambda=i\omega_0} = \frac{\sigma(i\omega_0)+\sigma}{3(i\omega_0)^2 + 2(2+\sigma)i\omega_0 + (1+2\sigma+\delta^2+k^2+\sigma k^2 - \sigma r_0)},
\tag{20}
$$

Thus, one has

$$
Re(\lambda'(r=r_0)) = \frac{\sigma(1+2\sigma+\delta^2+k^2+\sigma k^2 + \omega_0^2 + 2\sigma\omega_0^2 - \sigma r_0)}{(1+2\sigma+\delta^2+k^2+\sigma k^2 - \sigma r - 3\omega_0^2)^2 + 4(2+\sigma)^2\omega_0^2} > 0,
\tag{21}
$$

where $\sigma = 4, \delta = 1.1, r_0 = \frac{4\sigma+\sigma^2-\sigma^2\delta^2}{\sigma-2} \approx 6.32$ and $\omega_0 = \sqrt{\frac{2\sigma k^2}{2+\sigma}} \approx 2.34$. Consequently, the transversality condition (B) is also verified.

At last, we will calculate the first Lyapunov coefficient l_1 under the above fixed parameters. The Jacobian matrix J on the equilibrium point E_2 is given by

$$
J_{E_2} = \begin{pmatrix} -4 & 4 & 0 & 0 \\ 2.21 & -1 & -1.1 & -2.0273 \\ 0 & 1.1 & -1 & 0 \\ 2.0273 & 2.0273 & 0 & -1 \end{pmatrix}.
\tag{22}
$$

The proper eigenvectors q and p are obtained by straightforward calculations

$$
q = \begin{pmatrix} 0.274 + 0.333i \\ 0.079 + 0.494i \\ 0.21 + 0.052i \\ 0.717 \end{pmatrix}, \quad p = \frac{1}{(0.078 - 0.891i)}\begin{pmatrix} -0.318 + 0.073i \\ -0.7 \\ 0.118 + 0.278i \\ 0.219 + 0.512i \end{pmatrix}
\tag{23}
$$

where the above eigenvectors q and p satisfy the three conditions (11), namely

$$
Aq = i\omega_0 q, \quad A^T p = -i\omega_0 p, \quad \langle p, q \rangle = \sum_{i=1}^{n} \overline{p_i}\, q_i = 1.
$$

From Equation (10), the multilinear vector functions of the system (2) are calculated as follows

$$B(x,y) = \begin{pmatrix} 0 \\ -x_1y_4 - x_4y_1 \\ 0 \\ x_1y_2 + x_2y_1 \end{pmatrix}, \quad C(x,y,z) = \begin{pmatrix} 0 \\ 0 \\ 0 \\ 0 \end{pmatrix}, \tag{24}$$

From (22)–(24), it follows that

$$B(q,q) = \begin{pmatrix} 0 \\ -0.393 - 0.477i \\ 0 \\ -0.285 + 0.323i \end{pmatrix}, \quad B(q,\bar{q}) = \begin{pmatrix} 0 \\ -0.393 \\ 0 \\ 0.372 \end{pmatrix},$$

$$\begin{cases} J_{E_2}^{-1} = \begin{pmatrix} -0.192 & -0.121 & 0.133 & 0.246 \\ 0.057 & -0.121 & 0.133 & 0.246 \\ 0.063 & -0.133 & -0.852 & 0.271 \\ -0.272 & -0.493 & 0.542 & 0 \end{pmatrix}, \\[4ex] (2i\omega_0 I - J_{E_2})^{-1} = \begin{pmatrix} 0.052 - 0.135i & -0.117 - 0.098i & 0.027 - 0.021i & 0.051 - 0.039i \\ -0.038 - 0.074i & -0.002 - 0.235i & 0.053 + 0.01i & 0.097 + 0.019i \\ -0.018 + 0.005i & -0.053 - 0.01i & 0.048 - 0.215i & 0.009 - 0.021i \\ -0.085 - 0.024i & -0.148 + 0.019i & 0.002 - 0.034i & 0.048 - 0.267i \end{pmatrix}. \end{cases}$$

Thus, one obtains

$$\begin{cases} v_{11} = [-0.139, -0.139, -0.153, -0.193]^T, \\ v_{20} = [-0.002 + 0.122i, -0.145 + 0.119i, 0.019 + 0.038i, 0.14 + 0.155i]^T. \end{cases} \tag{25}$$

By using (23)–(25), one gets

$$\begin{cases} \langle p, B(q, -v_{11}) \rangle = 0.066 - 0.191i, \\ \langle p, B(\bar{q}, v_{20}) \rangle = 0.064 - 0.130i, \\ \langle p, C(q, q, \bar{q}) \rangle = 0. \end{cases} \tag{26}$$

Consequently, the first Lyapunov is obtained by substituting (26) into (15)

$$l_1 = \frac{1}{2\omega_0} Re[\langle p, C(q,q,\bar{q}) \rangle - 2\langle p, B(q, -v_{11}) \rangle + \langle p, B(\bar{q}, v_{20}) \rangle] = -0.0145 < 0.$$

Therefore, the Hopf bifurcation of the system (2) at equilibrium point E_2 is nondegenerate and supercritical. Furthermore, the equilibria E_2 and E_3 are symmetric about the x_4−axis, hence, the system (2) should also undergo a Hopf bifurcation at E_3. Two numerical simulations are given in Figure 3. For $r = 5.5 < r_0$, the orbit of the system (2) with the initial values $(1.8, 1.8, 2, 4)$ is attracted to the stable equilibrium point E_2, whereas the orbit with the initial values $(-1.8, -1.8, -2, 4)$ is attracted to the other stable equilibrium point E_3, as illustrated in Figure 3a. In Figure 3b, by choosing $r = 6.5 > r_0$ with the initial values $(1.8, 1.8, 2, 4)$ and $(-1.8, -1.8, -2, 4)$, the orbits of the system are attracted to stable limit cycles emerging from E_2 and E_3, respectively.

According to Reference [39], $m = 2$, $\tau = 1$ and $r = 0.1 \sim 0.2$ times standard deviation (SD) of the time series. In our experiment, we fix $m = 2$, $\tau = 1$ and $r = 0.2 \times SD$.

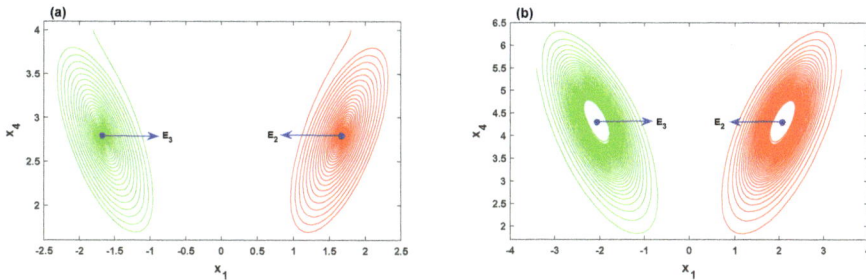

Figure 3. Hopf bifurcation of the system (2): (**a**) $r = 5.5 < r_0$, the orbit of the system is attracted to the stable symmetric equilibria E_2 and E_3; (**b**) $r = 6.5 > r_0$, the orbit of the system is attracted to a stable limit cycle emerging from the symmetric equilibria E_2 and E_3.

4. Multistability Behavior

A nonlinear dynamical system with multistability behavior can generate two or more attractors simultaneously depending on the initial values of the system. This section investigates the existence of multistability behavior in the system (2).

When we fix the parameters $\sigma = 2$, $\delta = 1.5$, $b = 0.7$ and select r as bifurcation parameter for over the range $r \in [7.5, 10]$, the coexisting bifurcation models of the state variable x_1 are depicted in Figure 4a. In this figure, the attractor colored in blue is initiated from $(-2, 1, 1, 1)$, meanwhile the attractor colored in red begins with the initial conditions $(1, 1, 1, 1)$. As can be observed in Figure 4a, the system (2) shows coexisting multiple chaotic attractors as well as the coexistence of multiple quasi-periodic attractors. To show the coexistence of multiple chaotic attractors visually, Figure 5 plots different orientations of the phase portraits of the system (2) when its parameters are set as $\sigma = 2$, $\delta = 1.5$, $b = 0.7$, and $r = 9.41$.

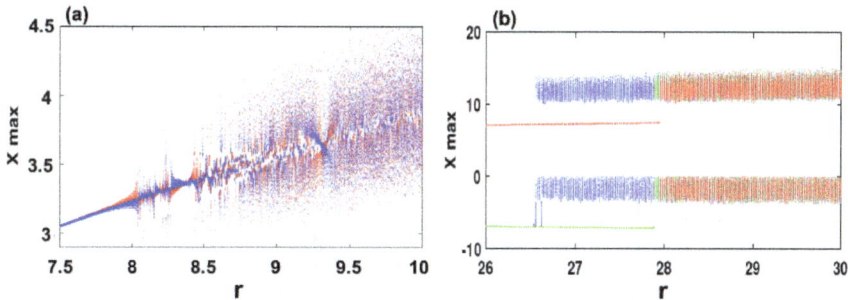

Figure 4. Bifurcation diagrams versus parameter r for illustrating the two and three coexisting attractors of the system (2): (**a**) $\sigma = 2$, $\delta = 1.5$, $b = 0.7$ for the initial values $(1, 1, 1, 1)$ (red) and $(-2, 1, 1, 1)$ (blue); (**b**) $\sigma = 4$, $\delta = 0.5$, $b = 2$ for the initial values $(2, 1, 1, 2)$ (blue), $(-2, 1, 1, -2)$ (red) and $(2, 1, 1, -2)$ (green).

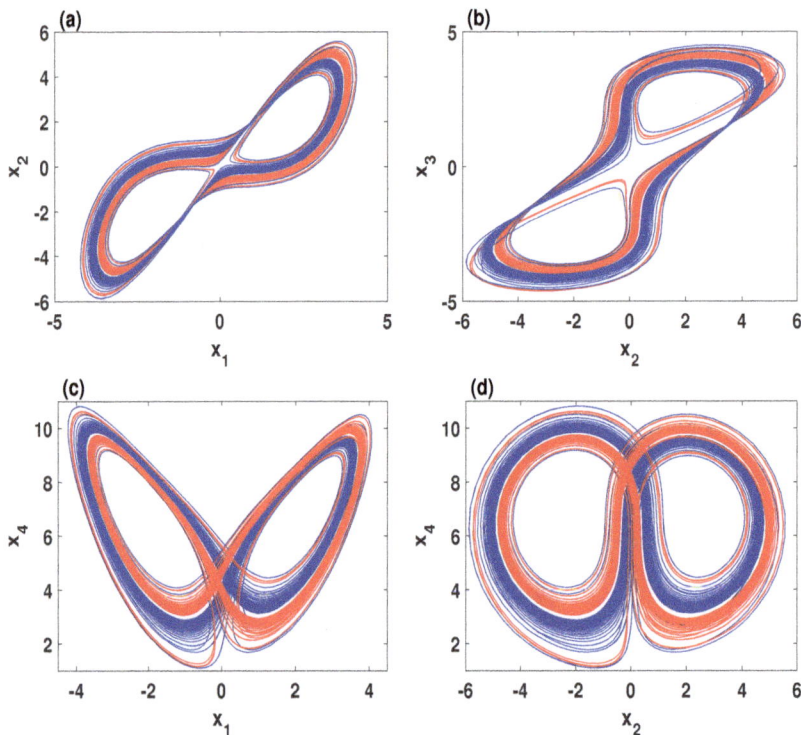

Figure 5. Multiple coexisting chaotic attractors of the system (2) when $\sigma = 2$, $\delta = 1.5$, $b = 0.7$, $r = 9.41$ for the initial values $(1, 1, 1, 1)$ (red) and $(-2, 1, 1, 1)$ (blue). (**a**) x_1–x_2 plane; (**b**) x_2–x_3 plane; (**c**) x_1–x_4 plane; (**d**) x_2–x_4 plane.

In addition, when we set $\sigma = 4$, $\delta = 0.5$, $b = 2$ with $26 \leq r \leq 30$, Figure 4b shows that the chaotic attractor with two stable fixed-point attractors coexist for the initial values $(\pm 2, 1, 1, \pm 2)$. For the orbit colored in blue, the evolution begins from attracting to the stable equilibrium E_3 within the range $26 \leq r \leq 26.7$, and then the system shows chaotic behavior when $r \geq 26.8$. For $(-2, 1, 1, -2)$ (red), the system converges to the stable equilibrium E_2 when $26 \leq r \leq 28$, and then exhibits chaotic behavior when $r \geq 28.1$. For the initial values $(2, 1, 1, -2)$ (green), the system attracts to the stable equilibrium E_3 when $26 \leq r \leq 27.8$, meanwhile the chaotic behavior is shown when $r \geq 27.9$. Selecting $r = 27$, an interesting dynamic is observed in the system (2) by plotting different orientations of the phase portraits with the corresponding time series, as shown in Figure 6. These portraits confirm the coexistence of three different attractors: (a) blue butterfly attractors surrounds the symmetric equilibria E_2 and E_3; (b) the red stable fixed-point attractor for E_2, and the green stable fixed-point attractor for E_3.

Through the above analysis, we can observe that the multistability behavior occurs in the system (2) with various kinds of coexisting attractors. Therefore, it can be concluded that the system (2) has high sensitivity to both initial values and parameters.

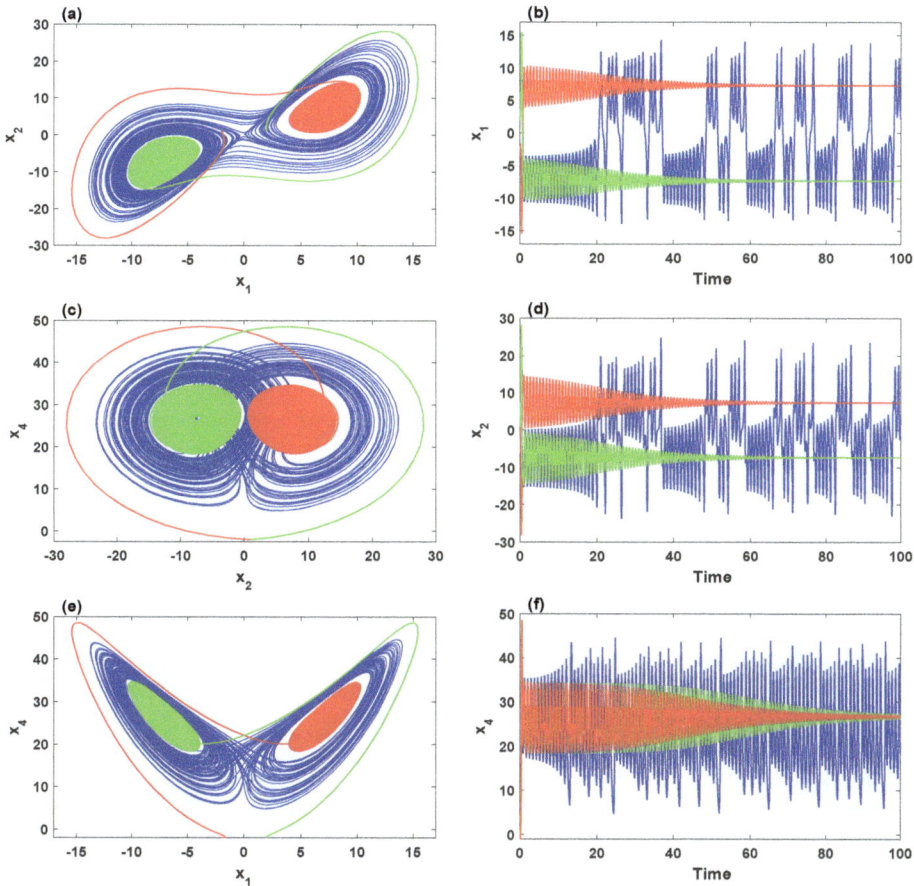

Figure 6. Three coexisting attractors with $\sigma = 4$, $\delta = 0.5$, $b = 2$, $r = 27$: (**a,c,e**) different perspectives on the coexistence of the chaotic and two stable fixed-point attractors for the initial values $(2, 1, 1, 2)$ (blue), $(-2, 1, 1, -2)$ (red) and $(2, 1, 1, -2)$ (green); (**b,d,f**) the corresponding time series of the state variables x_1, x_2 and x_4, respectively.

5. Complexity and Randomness of Multistability Regions

This section discusses determining and locating the parameters and initial values that show multistability behaviors, as well as investigates the randomness of the multistability regions.

5.1. Sample Entropy

Sample Entropy (SamEn) is a mathematical algorithm proposed by Richman [40]. It is used to provide a quantitative explanation about the complexity of nonlinear dynamical systems. Obviously, a system with bigger SamEn values indicates that it requires additional information to predict its attractor, hence, it is a chaotic system. Suppose that the time series $(y_i, i = 0, 1, 2, \ldots, M - 1)$ of a dynamical system with a length of M, then the SamEn algorithm can be calculated by the following steps:

(A) Reconstructing phase-space: for a given embedding dimension m and time delay τ, the reconstruction sequences are given by

$$Y_i = \{y_i, y_{i+\tau}, ..., y_{i+(m-1)\tau}\}, \quad y_i \in R^m \tag{27}$$

where $i = 1, 2, ..., M - m + \tau$.

(B)   Counting the vector pairs: let B_i be the number of vector Y_j such that

$$d[Y_i, Y_j] \leq r, \quad i \neq j \tag{28}$$

where r is the tolerance parameter, and $d[Y_i, Y_j]$ is the distance between Y_i and Y_j, which is defined by

$$d[Y_i, Y_j] = max\{|y(i+k) - y(j+k)| : 0 \leq k \leq m - 1\}. \tag{29}$$

(C)   Calculating probability: according to the obtained number of vector pairs, we can obtain

$$C_i^m(r) = \frac{B_i}{M - (m - 1)\tau}, \tag{30}$$

then calculate the probability by

$$\phi^m(r) = \frac{\sum_{i=1}^{M-(m-1)\tau} \ln C_i^m(r)}{[M - (m - 1)\tau]} \tag{31}$$

(D)   Calculating SamEn: repeating the above steps we can obtain $\phi^{m+1}(r)$, then SamEn is given by

$$SamEn(m, r, M) = \phi^m(r) - \phi^{m+1}(r). \tag{32}$$

According to Reference [39], $m = 2$, $\tau = 1$ and $r = 0.1 \sim 0.2$ times standard deviation (SD) of the time series. In our experiment, we fix $m = 2$, $\tau = 1$ and $r = 0.2 \times SD$.

It is well-known that the cross-section of the basins of attraction can determine the dynamical system behaviors when its initial values vary. However, it is interesting to ask if there is any technique that can determine the behaviors of a dynamical system when its initial values and parameters vary. Therefore, SamEn based contour plots are applied to locate the regions of chaotic and periodic state, and hence, to determine the parameters and initial values that show multistability behaviors. To locate those parameters and initial values in the system (2), we designed the following experiments: (1) consider r as bifurcation parameter and set $\sigma = 4$, $b = 2$ and $\delta = 0.5$; (2) let $(x_{10}, x_{20}, x_{30}, x_{40})$ be the initial values; (3) calculate SamEn versus varying the parameter r and one of an initial value; (4) calculate SamEn versus varying two of the initial values.

Figure 7 plots SamEn of the system (2) in a two-dimensional plane when $r \in (24, 30)$ and different initial values. It can be observed from Figure 7a–d that four cases are analyzed when the initial values are set as $(x_{10}, 1, 1, 2)$, $(2, x_{20}, 1, 2)$, $(2, 1, x_{30}, 2)$ and $(2, 1, 1, x_{40})$, respectively. From Figure 7, it can be seen that the parameter r and the initial values in the blue regions have smaller SamEn values, which means that the system (2) shows periodic state, whereas, those in the yellow and green regions lead to a chaotic state. Furthermore, Figure 8 shows the chaotic and periodic regions of system (2) when two of the initial values vary simultaneously.

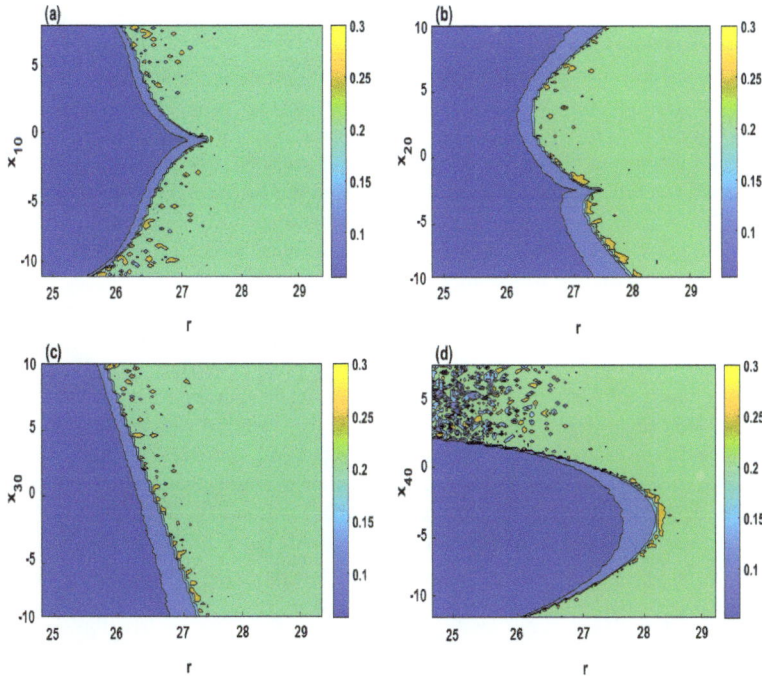

Figure 7. SamEn in the parameter r-initial value plane for $\sigma = 4$, $\delta = 0.5$, $b = 2$: (**a**) $r - x_{10}$ plane; (**b**) $r - x_{20}$ plane; (**c**) $r - x_{30}$ plane; (**d**) $r - x_{40}$ plane.

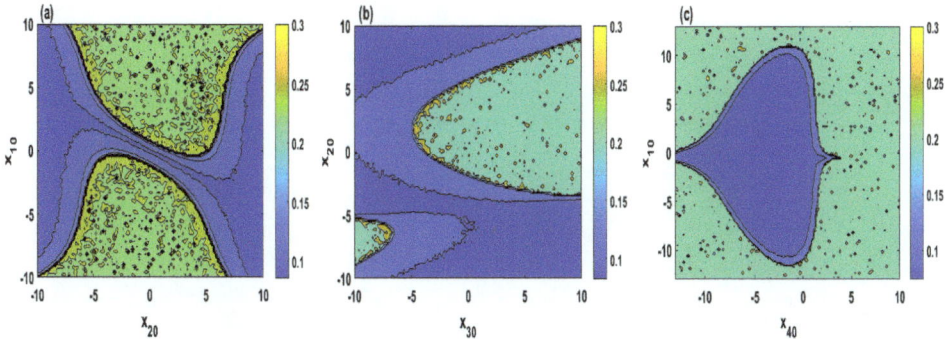

Figure 8. SamEn versus varying two of the initial values for $\sigma = 4$, $\delta = 0.5$, $b = 2$, $r = 27$: (**a**) $(x_{10}, x_{20}, 1, 2)$; (**b**) $(2, x_{20}, x_{30}, 2)$; (**c**) $(x_{10}, 1, 1, x_{40})$.

5.2. Chaos-Based PRNG

Many chaotic systems have been applied to generate pseudorandom number generator (PRNG). The need of PRNG arises in many cryptographic applications, e.g., common cryptosystems employ keys, data hiding, and auxiliary quantities used in generating digital signatures [41,42]. However, secret keys of most chaos-based cryptographic schemes are generated by parameters and initial values of the employed chaotic systems [43]. Those parameters and initial values might be from multistability regions; it is therefore important to investigate the randomness of the trajectories generating from multistability regions.

To investigate the randomness of blue-green regions (multistability behaviors) and green regions (chaotic), which is shown in Figure 7d, we use here a simple chaos-based PRNG as an example.

The generation procedures of the chaos-based PRNG are shown in Algorithm 1, for which x_1, x_2, x_3 and x_4 generates 1,000,000 bits binary string.

Several statistical tests can be employed to test the randomness of PRNG. Our experiment uses the highest standards of statistical packages which is NIST-800-22 [42]. The NIST-800-22 consists of 16 empirical statistical tests that provide true evaluation for the randomness of PRNG. Each test is developed to detect the non-random areas of a binary sequence from different sides, and then to derive a *p*-value. According to the recommendations in [24,44], we set the confidence level $\alpha = 0.01$, and we use a binary sequence with length of 1,000,000 bit as the testing input. Since the confidence level of each test in NIST is set to be 1%, then the sequence is considered to be random with a confidence of 99% when the obtained *p*-value is bigger than 0.01.

According to Algorithm 1, we can obtain four PRNG from the trajectory of x_1, x_2, x_3 and x_4 when the initial values are considered as input. For $\sigma = 4$, $\delta = 0.5$, $b = 2$ and $r \in [27, 29]$ with the initial values $(2, 1, 1, -2)$, the SamEn values of the selected parameters and initial values are within the blue-green regions (multistability), as shown in Figure 7d. The randomness of the corresponding PRNG that generated from the trajectory of x_1, x_2, x_3 and x_4 can be visually shown by depicting the NIST-800-22 test results, as seen in Figure 9. As can be observed from Figure 9, the four PRNG generating from multistability regions fail to pass most of the statistical tests. On the other hand, when $\sigma = 4$, $\delta = 0.5$, $b = 2$ and $r \in [27, 29]$ with the initial values $(2, 1, 1, 2)$, the SamEn values are within the green region (chaotic), as shown in Figure 7d. Table 1 lists the corresponding NIST-800-22 results for each of the four PRNG. It is obvious that the four PRNG can pass all the statistical tests.

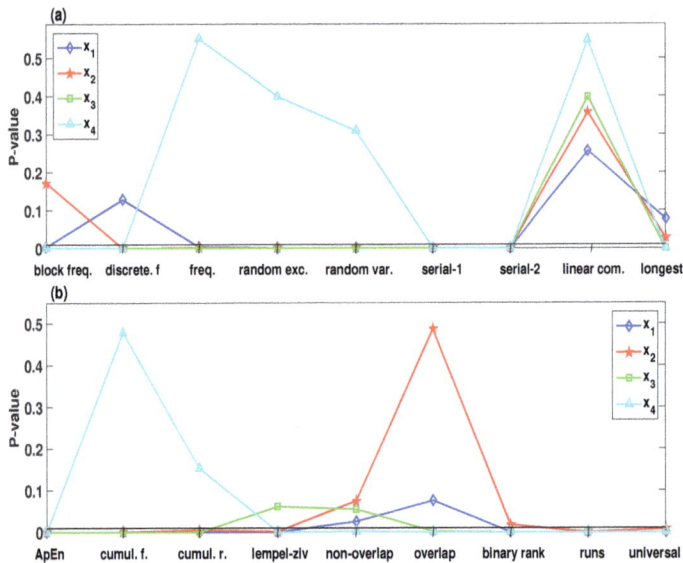

Figure 9. The statistical tests NIST SP800-22 of the pseudorandom number generator (PRNG) that generated by x_1, x_2, x_3, x_4 of the system (2) with $\sigma = 4$, $\delta = 0.5$, $b = 2$, $r \in [27, 29]$ and for the initial values $(2, 1, 1, -2)$. (**a**) Block-Frequency, Discrete Fourier Transform, Frequency (Monobit), Random Excursions, Random Excursions Variant, Serial-1, Serial-2, Linear Complexity, and Longest Run of Ones, respectively; (**b**) Approximate Entropy, Cumulative Sums (Forward), Cumulative Sums (Reverse), Lempel-ziv Compression, Non-overlapping Template, Overlapping Template, Binary Matrix Rank, Runs, and Universal Statistical.

Algorithm 1 The generation of chaos-based PRNG

Input: The initial values of system (2).

1: **for** $i = 1$ to 4 **do**

2:     **for** $r = 27$ to 29 **do**

3:         Truncate a chaotic sequence C_i from the trajectory of x_i;

4:         Convert the floating number C_i of x_i into a 32-bit binary using the IEEE-754-Standard;

5:         Fetch the last 16th digital number of the obtained binary string;

6:     **end for**

7: **end for**

Output: Four PRNG are generated from of x_1, x_2, x_3 and x_4.

Table 1. NIST-800-22 tests results of binary sequences generated by PRNG of x_1, x_2, x_3 and x_4 outputs.

	Each Sequence to be Tested Consists of 1,000,000 Bits				
NIST-800-22 Tests	***p*-Value (x_1)**	***p*-Value (x_2)**	***p*-Value (x_3)**	***p*-Value (x_4)**	**Result**
1. Block-Frequency (m = 128)	0.2116	0.8460	0.8313	0.0210	Random
2. Frequency (Monobit)	0.7611	0.0380	0.6570	0.3503	Random
3. Discrete Fourier Transform	0.3602	0.1792	0.1478	0.1225	Random
4. Approximate Entropy (m = 10)	0.9592	0.6512	0.6343	0.3659	Random
5. Cumulative Sums (Forward)	0.7617	0.0721	0.7280	0.5832	Random
Cumulative Sums (Reverse)	0.5578	0.0320	0.5106	0.1816	Random
6. Serial-1 (m = 16)	0.7937	0.2948	0.1635	0.9706	Random
Serial-2 (m = 16)	0.8885	0.7628	0.5357	0.9530	Random
7. Runs	0.9649	0.6196	0.4751	0.1530	Random
8. Longest Run of Ones	0.2568	0.0965	0.8242	0.2420	Random
9. Overlapping Template (m = 9)	0.7032	0.6461	0.5603	0.7085	Random
10. Non-overlapping Template (m = 9)	0.4960	0.5403	0.5150	0.5117	Random
11. Linear Complexity (m = 500)	0.4091	0.7263	0.1607	0.8582	Random
12. Binary Matrix Rank	0.2618	0.1029	0.2843	0.2376	Random
13. Lempel-ziv Compression	0.0769	0.2343	0.1411	0.9581	Random
14. Random Excursions	0.4628	0.2379	0.4787	0.3931	Random
15. Random Excursions Variant	0.6141	0.1814	0.3977	0.2865	Random
16. Universal Statistical	0.4931	0.7326	0.6056	0.1038	Random

6. Conclusions

This paper has introduced a new 4D chaotic laser system, which is derived from Lorenz-Haken equations. The new chaotic laser system has three equilibria and only two quadratic nonlinearities. The dynamics of the new system have been studied deeply, in which the system shows coexisting multiple Hopf bifurcations, and complex coexisting behaviors of two and three attractors. In addition, we applied SamEn contour plots for measuring the complexity of the system when its initial values and parameters vary. Simulation results have shown that multistability regions can be easily determined and located using SamEn contour plots. To examine the randomness of PRNG that generate from the multistability regions, we used the NIST-800-22 tests. Statistical test results demonstrate that the generated PRNG from multistability regions are non-random. This means that although the multistability behaviors indicate high sensitivity of chaotic systems, they might be unsuitable for cryptographic applications.

Author Contributions: H.N. designed the study and wrote the paper. M.R.M.S. and N.M.G.A-S. contributed to the methodology, simulations and algorithm design. A.K. undertook the theoretical analysis. All authors read and approved the manuscript.

Funding: This research was funded by Universiti Putra Malaysia Grant under the grant code (GP. 2018/9595200).

Acknowledgments: The authors would like to thank the Research Management Center (RMC) of Universiti Putra Malaysia for supporting this work.

Conflicts of Interest: The authors declare no conflict of interest.

References

1. Banerjee, S.; Rondoni, L.; Mukhopadhyay, S.; Misra, A.P. Synchronization of spatiotemporal semiconductor lasers and its application in color image encryption. *Opt. Commun.* **2011**, *284*, 2278–2291. [CrossRef]
2. Valli, D.; Banerjee, S.; Ganesan, K.; Muthuswamy, B.; Subramaniam, C.K. Chaotic time delay systems and field programmable gate array realization. In *Chaos, Complexity and Leadership 2012*; Springer: Dordrecht, The Netherlands, 2014; pp. 9–16.
3. Banerjee, S.; Saha, P.; Chowdhury, A.R. Chaotic scenario in the Stenflo equations. *Phys. Scr.* **2001**, *63*, 177. [CrossRef]
4. Natiq, H.; Banerjee, S.; He, S.; Said, M.R.M.; Kilicman, A. Designing an M-dimensional nonlinear model for producing hyperchaos. *Chaos Solitons Fractals* **2018**, *114*, 506–515. [CrossRef]
5. Ghosh, D.; Banerjee, S.; Chowdhury, A.R. Synchronization between variable time-delayed systems and cryptography. *Europhys. Lett.* **2007**, *80*, 30006. [CrossRef]
6. Banerjee, S. (Ed.) *Chaos Synchronization and Cryptography for Secure Communications: Applications for Encryption*; IGI Global: Hershey, PA, USA, 2010.
7. Saha, P.; Banerjee, S.; Chowdhury, A.R. Chaos, signal communication and parameter estimation. *Phys. Lett. A* **2004**, *326*, 133–139. [CrossRef]
8. Fataf, N.A.A.; Palit, S.K.; Mukherjee, S.; Said, M.R.M.; Son, D.H.; Banerjee, S. Communication scheme using a hyperchaotic semiconductor laser model: Chaos shift key revisited. *Eur. Phys. J. Plus* **2017**, *132*, 492. [CrossRef]
9. Banerjee, S.; Pizzi, M.; Rondoni, L. Modulation of output power in the spatio-temporal analysis of a semi conductor laser. *Opt. Commun.* **2012**, *285*, 1341–1346. [CrossRef]
10. Rondoni, L.; Ariffin, M.R.K.; Varatharajoo, R.; Mukherjee, S.; Palit, S.K.; Banerjee, S. Optical complexity in external cavity semiconductor laser. *Opt. Commun.* **2017**, *387*, 257–266. [CrossRef]
11. Mukherjee, S.; Palit, S.K.; Banerjee, S.; Ariffin, M.R.K.; Rondoni, L.; Bhattacharya, D.K. Can complexity decrease in congestive heart failure? *Phys. A Stat. Mech. Appl.* **2015**, *439*, 93–102. [CrossRef]
12. Banerjee, S.; Palit, S.K.; Mukherjee, S.; Ariffin, M.R.K.; Rondoni, L. Complexity in congestive heart failure: A time-frequency approach. *Chaos Interdiscip. J. Nonlinear Sci.* **2016**, *26*, 033105. [CrossRef]
13. Pham, V.-T.; Vaidyanathan, S.; Volos, C.K.; Jafari, S. Hidden attractors in a chaotic system with an exponential nonlinear term. *Eur. Phys. J. Spec. Top.* **2015**, *224*, 1507–1517. [CrossRef]
14. Leonov, G.A.; Kuznetsov, N.V. Hidden attractors in dynamical systems. From hidden oscillations in Hilbert–Kolmogorov, Aizerman, and Kalman problems to hidden chaotic attractor in Chua circuits. *Int. J. Bifurc. Chaos* **2013**, *23*, 1330002. [CrossRef]
15. Pham, V.T.; Volos, C.; Jafari, S.; Wang, X.; Vaidyanathan, S. Hidden hyperchaotic attractor in a novel simple memristive neural network. *Optoelectron. Adv. Mater. Rapid Commun.* **2014**, *8*, 1157–1163.
16. Dudkowski, D.; Jafari, S.; Kapitaniak, T.; Kuznetsov, N.V.; Leonov, G.A.; Prasad, A. Hidden attractors in dynamical systems. *Phys. Rep.* **2016**, *637*, 1–50. [CrossRef]
17. Tlelo-Cuautle, E.; de la Fraga, L.G.; Pham, V.T.; Volos, C.; Jafari, S.; de Jesus Quintas-Valles, A. Dynamics, FPGA realization and application of a chaotic system with an infinite number of equilibrium points. *Nonlinear Dyn.* **2017**, *89*, 1129–1139. [CrossRef]
18. Pham, V.-T.; Volos, C.; Gambuzza, L.V. A memristive hyperchaotic system without equilibrium. *Sci. World J.* **2014**, *2014*, 368986. [CrossRef] [PubMed]
19. Natiq, H.; Said, M.R.M.; Ariffin, M.R.K.; He, S.; Rondoni, L.; Banerjee, S. Self-excited and hidden attractors in a novel chaotic system with complicated multistability. *Eur. Phys. J. Plus* **2018**, *133*, 557. [CrossRef]
20. Wang, X.; Pham, V.T.; Jafari, S.; Volos, C.; Munoz-Pacheco, J.M.; Tlelo-Cuautle, E. A new chaotic system with stable equilibrium: From theoretical model to circuit implementation. *IEEE Access* **2017**, *5*, 8851–8858. [CrossRef]
21. Jafari, S.; Sprott, J.C. Simple chaotic flows with a line equilibrium. *Chaos Solitons Fractals* **2013**, *57*, 79–84. [CrossRef]

22. Pham, V.T.; Jafari, S.; Volos, C.; Giakoumis, A.; Vaidyanathan, S.; Kapitaniak, T. A chaotic system with equilibria located on the rounded square loop and its circuit implementation. *IEEE Trans. Circuits Syst. II Express Briefs* **2016**, *63*, 878–882. [CrossRef]
23. Arecchi, F.; Meucci, R.; Puccioni, G.; Tredicce, J. Experimental evidence of subharmonic bifurcations, multistability, and turbulence in a q-switched gas laser. *Phys. Rev. Lett.* **1982**, *49*, 1217. [CrossRef]
24. Munoz-Pacheco, J.; Zambrano-Serrano, E.; Volos, C.; Jafari, S.; Kengne, J.; Rajagopal, K. A new fractional-order chaotic system with different families of hidden and self-excited attractors. *Entropy* **2018**, *20*, 564. [CrossRef]
25. Wang, C.; Ding, Q. A New Two-Dimensional Map with Hidden Attractors. *Entropy* **2018**, *20*, 322. [CrossRef]
26. Li, C.; Sprott, J.C.; Hu, W.; Xu, Y. Infinite multistability in a self-reproducing chaotic system. *Int. J. Bifurc. Chaos* **2017**, *27*, 1750160. [CrossRef]
27. Sparrow, C. *The Lorenz Equations: Bifurcations, Chaos, and Strange Attractors*; Springer Science & Business Media: New York, NY, USA, 2012; Volume 41.
28. Pereira, U.; Coullet, P.; Tirapegui, E. The Bogdanov—Takens normal form: A minimal model for single neuron dynamics. *Entropy* **2015**, *17*, 7859–7874. [CrossRef]
29. Zhan, X.; Ma, J.; Ren, W. Research entropy complexity about the nonlinear dynamic delay game model. *Entropy* **2017**, *19*, 22. [CrossRef]
30. Han, Z.; Ma, J.; Si, F.; Ren, W. Entropy complexity and stability of a nonlinear dynamic game model with two delays. *Entropy* **2016**, *18*, 317. [CrossRef]
31. Dang, T.S.; Palit, S.K.; Mukherjee, S.; Hoang, T.M.; Banerjee, S. Complexity and synchronization in stochastic chaotic systems. *Eur. Phys. J. Spec. Top.* **2016**, *225*, 159–170. [CrossRef]
32. He, S.; Sun, K.; Wang, H. Complexity analysis and DSP implementation of the fractional-order Lorenz hyperchaotic system. *Entropy* **2015**, *17*, 8299–8311. [CrossRef]
33. Ma, J.; Ma, X.; Lou, W. Analysis of the Complexity Entropy and Chaos Control of the Bullwhip Effect Considering Price of Evolutionary Game between Two Retailers. *Entropy* **2016**, *18*, 416. [CrossRef]
34. He, S.; Li, C.; Sun, K.; Jafari, S. Multivariate Multiscale Complexity Analysis of Self-Reproducing Chaotic Systems. *Entropy* **2018**, *20*, 556. [CrossRef]
35. Haken, H. Analogy between higher instabilities in fluids and lasers. *Phys. Lett. A* **1975**, *53*, 77–78. [CrossRef]
36. Banerjee, S.; Saha, P.; Chowdhury, A.R. Chaotic aspects of lasers with host-induced nonlinearity and its control. *Phys. Lett. A* **2001**, *291*, 103–114. [CrossRef]
37. Van Tartwijk, G.H.M.; Agrawal, G.P. Nonlinear dynamics in the generalized Lorenz-Haken model. *Opt. Commun.* **1997**, *133*, 565–577. [CrossRef]
38. Kuznetsov, Y.A. Numerical Analysis of Bifurcations. In *Elements of Applied Bifurcation Theory*; Springer: New York, NY, USA, 2004; pp. 505–585.
39. Kaffashi, F.; Foglyano, R.; Wilson, C.G.; Loparo, K.A. The effect of time delay on approximate & sample entropy calculations. *Phys. D Nonlinear Phenom.* **2008**, *237*, 3069–3074.
40. Richman, J.S.; Moorman, J.R. Physiological time-series analysis using approximate entropy and sample entropy. *Am. J. Physiol. Heart Circ. Physiol.* **2000**, *278*, H2039–H2049. [CrossRef] [PubMed]
41. Volos, C.K.; Kyprianidis, I.M.; Stouboulos, I.N. Fingerprint images encryption process based on a chaotic true random bits generator. *Int. J. Multimedia Intell. Secur.* **2010**, *1*, 320–335. [CrossRef]
42. Rukhin, A.; Soto, J.; Nechvatal, J.; Smid, M.; Barker, E. *A Statistical Test Suite for Random and Pseudorandom Number Generators for Cryptographic Applications*; Booz-Allen and Hamilton Inc.: Mclean, VA, USA, 2001.
43. Natiq, H.; Al-Saidi, N.M.G.; Said, M.R.M.; Kilicman, A. A new hyperchaotic map and its application for image encryption. *Eur. Phys. J. Plus* **2018**, *133*, 6. [CrossRef]
44. Rodríguez-Orozco, E.; García-Guerrero, E.; Inzunza-Gonzalez, E.; López-Bonilla, O.; Flores-Vergara, A.; Cárdenas-Valdez, J.; Tlelo-Cuautle, E. FPGA-based Chaotic Cryptosystem by Using Voice Recognition as Access Key. *Electronics* **2018**, *7*, 414. [CrossRef]

![entropy logo] *entropy*

MDPI

Article

Entropy Analysis and Neural Network-Based Adaptive Control of a Non-Equilibrium Four-Dimensional Chaotic System with Hidden Attractors

Hadi Jahanshahi [1] , **Maryam Shahriari-Kahkeshi** [2], **Raúl Alcaraz** [3] , **Xiong Wang** [4], **Vijay P. Singh** [5] **and Viet-Thanh Pham** [6,*]

[1]   Department of Aerospace Engineering, Faculty of New Sciences and Technologies, University of Tehran, Tehran 14395-1561, Iran; hadi_jahanshahi@ut.ac.ir

[2]   Faculty of Engineering, Shahrekord University, Shahrekord 64165478, Iran; m.shahriyarikahkeshi@ec.iut.ac.ir

[3]   Research Group in Electronic, Biomedical and Telecommunication Engineering, University of Castilla-La Mancha (UCLM), 16071 Cuenca, Spain; raul.alcaraz@uclm.es

[4]   Institute for Advanced Study, Shenzhen University, Shenzhen 518060, China; wangxiong8686@szu.edu.cn

[5]   Department of Biological & Agricultural Engineering and Zachry Department of Civil Engineering, Texas A&M University, 2117 TAMU, College Station, TX 77843, USA; vsingh@tamu.edu

[6]   Nonlinear Systems and Applications, Faculty of Electrical and Electronics Engineering, Ton Duc Thang University, Ho Chi Minh City 700000, Vietnam

*   Correspondence: phamvietthanh@tdtu.edu.vn

Received: 21 January 2019; Accepted: 4 February 2019; Published: 7 February 2019

Abstract: Today, four-dimensional chaotic systems are attracting considerable attention because of their special characteristics. This paper presents a non-equilibrium four-dimensional chaotic system with hidden attractors and investigates its dynamical behavior using a bifurcation diagram, as well as three well-known entropy measures, such as approximate entropy, sample entropy, and Fuzzy entropy. In order to stabilize the proposed chaotic system, an adaptive radial-basis function neural network (RBF-NN)–based control method is proposed to represent the model of the uncertain nonlinear dynamics of the system. The Lyapunov direct method-based stability analysis of the proposed approach guarantees that all of the closed-loop signals are semi-globally uniformly ultimately bounded. Also, adaptive learning laws are proposed to tune the weight coefficients of the RBF-NN. The proposed adaptive control approach requires neither the prior information about the uncertain dynamics nor the parameters value of the considered system. Results of simulation validate the performance of the proposed control method.

Keywords: Non-equilibrium four-dimensional chaotic system; entropy measure; adaptive approximator-based control; neural network; uncertain dynamics

1. Introduction

A variety of chaotic systems with various features, such as multistability [1–3], extreme multistability [4,5], and multi-scroll attractors [6,7], have been introduced in recent years for investigating nonlinear dynamical systems. Dynamical systems can be categorized based on self-excited and hidden attractors [8]. From 1994, when the first non-equilibrium chaotic flow was reported in literature [9], almost 20 years have passed before another chaotic systems with non-equilibrium was introduced [10–15]. It can be easily concluded that the chaotic attractor in such systems is hidden. Given the fact that systems without equilibrium have unexpected responses to perturbations, these systems have become attractive systems for researchers.

However, all the aforementioned systems with no-equilibria are described by 3D differential equations. So, the question is if there is any 4D system with no-equilibria. The first 4D chaotic system was found by Rössler in 1979 [16], which was the first step in designing a 4D chaotic system. In the last few years, only a few works related with 4D chaotic dynamical systems with no-equilibria have been reported. In 2014, Wei et al. presented a new four-dimensional hyperchaotic system with no-equilibria developed by extension of the generalized diffusionless Lorenz equations [17]. In 2015, a no-equilibrium chaotic system with multiwing butterfly attractors constructed using a state feedback controller was proposed by Tahir et al. [18]. Motivated by complex dynamical behaviors of chaotic systems and unusual features of hidden attractors, a novel no-equilibrium chaotic system with an exponential nonlinearity was also proposed by Pham et al. in 2015 [19]. In 2016, Pham et al. introduced a novel four-dimensional continuous-time autonomous system with a cubic nonlinear term, which does not have equilibria [20]. In 2017, Bao et al. presented a memristive system, which does not display any equilibrium but can exhibit hyperchaotic, chaotic, and periodic dynamics as well as transient hyperchaos [21]. Furthermore, in 2018, Zhang et al. introduced a 4D chaotic composed of nine terms including only one constant term having also a line of equilibrium points or no equilibrium points [22].

In order to suppress the chaotic behavior of the nonlinear systems, several control methods have been implemented. Mobayen and Ma introduced a combination of finite-time robust-tracking theory and composite nonlinear feedback approach [23]. Shukla and Sharma designed a backstepping controller and analyzed the stability of the designed controller for a class of three-dimensional chaotic systems [24]. To name just a few, fuzzy controller [25–29], sliding mode controller [30–34], and hybrid controllers [35–39] are some other controllers that are implemented to control and synchronize the chaotic systems.

Artificial intelligence methods have been used widely to successfully solve a wide range of problems [40–44]. Designing the controllers based on the Neural network, as one of the most used artificial intelligence-based controllers (especially when dealing with complex nonlinear systems), is used extensively. Neural network-based control procedure can provide an efficient solution to the control of the complex, uncertain, and ill-defined systems. Some interesting results on using neural network to control and synchronize of complex systems have been studied in [45–48]. Yadmellat and Nikravesh have proposed a neural network-based output-feedback control method for nonlinear chaotic systems [49]. In another paper, Sarcheshmeh et al. designed two neural controllers to synchronize two master and slave chaotic satellites [50]. In order to suppress the disturbances in the chaotic systems, it is necessary to design an adaptive controller. In this regard, Fang et al. proposed a hybrid of an adaptive neural synchronization algorithm and a backstepping technique to synchronize a class of uncertain chaotic systems [51]. Shao et al. developed an adaptive neural network-based synchronization control strategy to stabilize a general form of unknown chaotic systems in the presence of unknown disturbances [52].

This paper focuses on the control of an uncertain four-dimensional chaotic system, which presents completely uncertain and chaotic nonlinear dynamics, such as an entropy analysis corroborates. Three well-known entropy-based metrics are computed from the time series generated by the system, thus highlighting different levels of complexity for different conditions. Since neural network is a universal approximator and it has a powerful tool for learning and approximating arbitrarily functions. Therefore, in this work, RBF-NN as a linear-in-parameter approximator has been chosen to approximate the uncertain nonlinear dynamics of the four-dimensional chaotic system. Moreover, no prior knowledge about system parameters is available. Then, the proposed indirect adaptive technique is proposed by using the developed RBF-NN-based model. Stability analysis shows that all of the closed-loop signals are semi-globally uniformly ultimately bounded and by proper choice of the design parameters the tracking error converges to the small vicinity of the origin. Also, weights of the RBF-NN are calibrated using the adaptive laws derived using the Lyapunov direct method. Simulation results verify the effectiveness of the proposed approach in control of the uncertain chaotic system with hidden attractors.

The paper is organized as follows. In Section 2, the four-dimensional chaotic system is described. In Section 3, the entropy analysis of the proposed system is presented. The RBF-NN and the design of the suggested control strategy are introduced in Section 4. In this section, the stability analysis of presented control algorithm is also discussed. The final section concludes the paper.

2. Four-dimensional Chaotic System

The general form of the proposed four-dimensional chaotic system is described as follows:
Let x, y, z, and w be the state variables of the system. Then,

$$\begin{aligned}
\dot{x} &= y \\
\dot{y} &= z \\
\dot{z} &= w \\
\dot{w} &= -aw + bx^2 - cy^2 + exy + fxz + g
\end{aligned} \tag{1}$$

where a, b, c, e, f, and g are system parameters. The behavior of the system depends on the numerical value of its parameters. The equilibrium states are found by setting the left-hand side of (1) to zero. Equation (1) gives $y = z = w = 0$, while $bx^2 = g$. If b and g are both nonzero with the same signs, then there are no equilibria. If $g = 0$, then Equation (1) gives $x = 0$, so there is the trivial equilibrium $(0,0,0,0)$. If $bg < 0$, there exist two equilibrium points $(\pm\sqrt{-g/b}, 0, 0, 0)$. The chaos of the dynamical system can be characterized by the Lyapunov exponent, which can be used to characterize the sensitivity of the system to the initial values. Considering Lyapunov exponents as L_1, L_2, L_3, and L_4 such that $L_1 > L_2 > L_3 > L_4$ and assuming $L_1 > 0$, $L_2 = 0$, $L_3 < 0$, and $L_4 < 0$, the dynamical behavior of the system (1) is chaotic. Taking $a = 1.05$, $b = 0.7$, $c = 0.19$, $e = 1.37$, $f = 1.79$, Figure 1 shows a bifurcation diagram which exhibits a periodic-doubling route to chaos of the peak of x (x max) of the system (1) versus parameter g, which is varied from -4 to 1.2. There are also some periodic windows in the chaotic region.

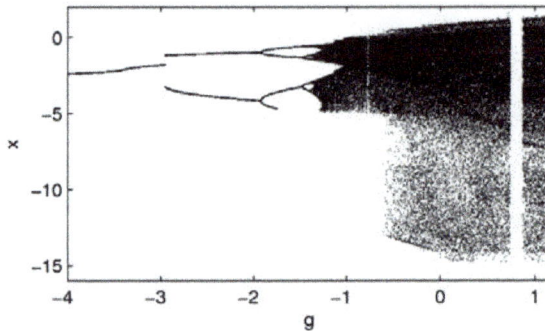

Figure 1. A bifurcation diagram exhibiting a periodic-doubling route to chaos of the peak of x (x max) of system (1) versus parameter g.

The system (1) exhibits periodic and chaotic behavior for different value of g. When $g = 1.15$, the Lyapunov dimension can be calculated by the Kaplan-Yorke dimension. In this case, by taking $a = 1.05$, $b = 0.7$, $c = 0.19$, $e = 1.37$, and $f = 1.79$, the Lyapunov exponent are as $L_1 = 0.185$, $L_2 = 0$, $L_3 = -0.195$, and $L_4 = -1.034$. So, the system shows a chaotic behavior. The phase portrait of the chaotic behavior of the system (1) is shown by Figure 2.

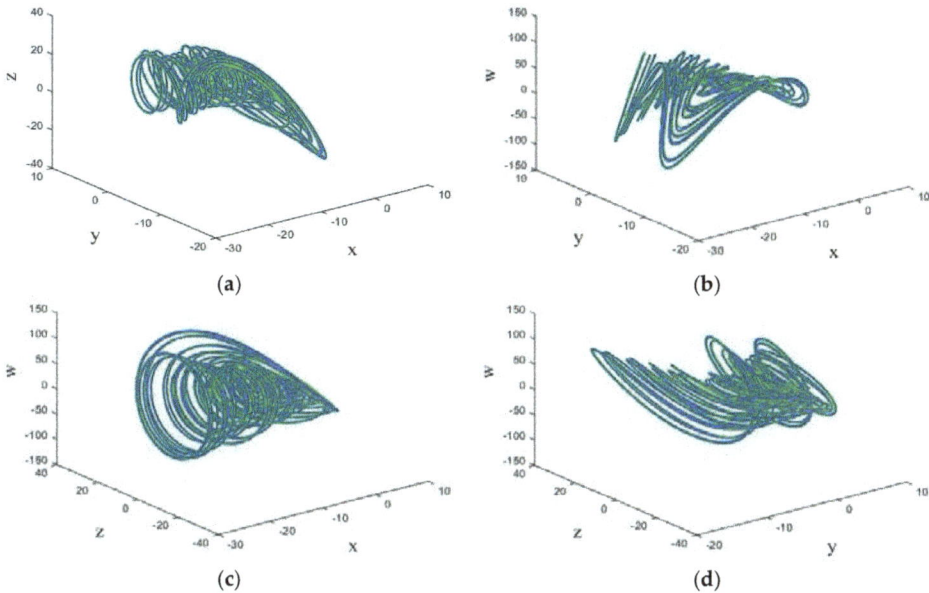

Figure 2. The three-dimensional (3D) chaotic portrait for system (1) in (**a**) *x-y-z* space, (**b**) *x-y-w* space, (**c**) *x-z-w* space, and (**d**) *y-z-w* space.

The largest Lyapunov exponent of the system (1) for $-4 < g < 1.2$, $a = 1.05$, $b = 0.7$, $c = 0.19$, $e = 1.37$, and $f = 1.79$ is shown by Figure 3.

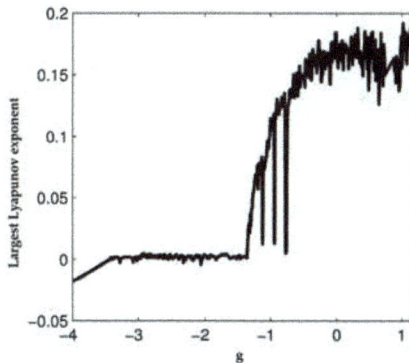

Figure 3. The largest Lyapunov exponent of the system (1).

Now, for a better understanding of the dynamic characteristics of system (1), its entropy has been analyzed by numerical simulation.

3. Entropy Analysis

As well as the positive largest Lyapunov exponent, entropy has been widely used to characterize chaotic systems [53]. This measure focuses on estimating seemingly unpredictable time evolution of chaotic systems and consequently tries to flesh out chaos in terms of randomness [54]. Thus, higher entropy indicates less predictability and a closer approach to stochastic behavior [55]. Although this information can be obtained through theoretical measures, such as Kolmogorov-Sinai entropy, they are often difficult to estimate from a finite data set [56]. Hence, some practical measures to estimate entropy of an underlying system from observed data have been developed in the last years, such as the well-established approximate entropy (ApEn) [57]. This metric has been widely used to characterize dynamical systems [58,59] because it is able to deal with short and noise data with outliers [60]. Briefly, ApEn quantifies times series regularity by computing repetitiveness of similar patterns and provides larger positive values for more irregular data. Hence, considering a N sample-length time series $x(n) = \{x(1), x(1), \ldots, x(N)\}$, this metric computation requires the following steps:

1. Form $N - m + 1$ m-sample length vectors, $\mathbf{X}_m(1), \ldots, \mathbf{X}_m(N - m + 1)$, defined by $\mathbf{X}_m(i) = \{x(i), x(i+1), \ldots, x(i+m-1)\}$, for $1 \le i \le N - m + 1$. Each vector contains m consecutive points from the ith sample.

2. Compute the Chebyshev distance for any pair of vectors $\mathbf{X}_m(i)$ and $\mathbf{X}_m(j)$. This distance is defined as the maximum absolute magnitude of the differences between coordinates, i.e.,

$$d_{ij}^m = \max_{k=0,\ldots,m-1} (|x(i+k) - x(j+k)|) \tag{2}$$

3. Estimate the number of pairs of vectors, $\mathbf{X}_m(j)$, whose distance with $\mathbf{X}_m(i)$ is less than or equal to r, i.e.,

$$C_i^m(r) = \frac{1}{N-m+1} \sum_{j=1}^{N-m+1} \Theta\left(r - d_{ij}^m\right) \tag{3}$$

$\Theta(z)$ being the Heaviside function, i.e., $\Theta(z) = 1$ for $z \ge 0$ and $\Theta(z) = 1$ for $z < 0$.

4. Calculate the global probability that any two sequences of size m present a distance lower than r, i.e.,

$$\varnothing^m(r) = \frac{1}{N-m+1} \sum_{i=1}^{N-m+1} \ln C_i^m(r) \tag{4}$$

5. Recompute the steps 1–4 for vectors with $m+1$ samples in length. In this case, Equations (3) and (4) should be replaced by

$$C_i^{m+1}(r) = \frac{1}{N-m} \sum_{j=1}^{N-m} \Theta\left(r - d_{ij}^{m+1}\right) \text{ and } \varnothing^{m+1}(r) = \frac{1}{N-m} \sum_{i=1}^{N-m} \ln C_i^{m+1}(r), \tag{5}$$

respectively.

6. Finally, ApEn can be computed by the difference

$$\text{ApEn}(m, r, n) = \varnothing^m(r) - \varnothing^{m+1}(r) \tag{6}$$

It is well known that this metric presents two limitations, such as it lacks relative consistency and is strongly dependent on the data length [61]. Indeed, when short times series are analyzed ApEn often provides lower values than expected [62]. These limitations have been overcome in its modified version proposed by Richman & Moorman and named sample entropy (SampEn) [61]. This new index presents two main differences from ApEn, i.e.,: (i) self-matches are excluded and (ii) a template-wise strategy is not used. Consequently, $N - m$ vectors of size m and $m+1$, for $1 \le i \le N - m$, are analyzed to compute SampEn, such that new Equations (3)–(5) can be expressed as

$$C_i^m(r) = \frac{1}{N-m-1} \sum_{j=1, j\neq i}^{N-m} \Theta\left(r - d_{ij}^m\right), \ \varnothing^m(r) = \frac{1}{N-m} \sum_{i=1}^{N-m} C_i^m(r), \tag{7}$$

$$C_i^{m+1}(r) = \frac{1}{N-m-1} \sum_{j=1, j\neq i}^{N-m} \Theta\left(r - d_{ij}^{m+1}\right), \text{ and } \varnothing^{m+1}(r) = \frac{1}{N-m} \sum_{i=1}^{N-m} C_i^{m+1}(r), \tag{8}$$

respectively. As a final step, SampEn can be estimated as

$$\text{SampEn}(m, r, N) = -\ln\left[\frac{\varnothing^{m+1}(r)}{\varnothing^m(r)}\right]. \tag{9}$$

Chen et al. [63] have proposed a modification of SampEn to avoid a poor statistical stability in some cases due to the binary classification of vectors achieved by the Heaviside function. This new index, named Fuzzy entropy (FuzzEn), considers a smoother definition of a vector match by using a family of exponential functions $D_{ij}^m(r, k) = \exp\left(-\left(d_{ij}^m / r\right)^k\right)$. To quantify the similarity degree among patterns. Thus, Equations (7) and (8) are redefined as

$$C_i^m(r, k) = \frac{1}{N - m - 1} \sum_{j=1, j\neq i}^{N-m} D_{ij}^m(r, k), \text{ and } C_i^{m+1}(r, k) = \frac{1}{N - m - 1} \sum_{j=1, j\neq i}^{N-m} D_{ij}^{m+1}(r, k), \tag{10}$$

respectively. Additionally, the mean from each vector $\mathbf{X}_m(i)$ is removed to highlight the local features of the data [63], thus resulting in

$$\mathbf{X}_m^*(i) = \{x(i), x(i+1), \dots, x(i+m-1)\} - \frac{1}{m}\sum_{l=0}^{m-1} x(i+1) \tag{11}$$

Clearly, the selection of parameters m and r has a strong impact on the entropy estimates obtained by these three indices. Although no widespread rules exist for their optimal choice, some previous works have recommended the use of $m = 1$ or 2 and r between 0.05 and 0.25 times the standard deviation of the data [57,61]. Thus, making use of $m = 2$, $r = 0.15$, and $k = 2$, the values of ApEn, SampEn, and FuzzEn computed from the times series $x(n)$ of the system (1) with length $N = 3000$ are displayed in Figure 4. As can be seen, the three entropy measures provided similar results. In fact, no perceptible differences can be noticed between ApEn and SampEn. Moreover, although FuzzEn revealed lower values than ApEn and SampEn, the same trend can be observed as a function of g. To this last respect, entropy shows low values when the system is in a stable state (i.e., for $g \leq -1.2$) and, contrarily, high values when the system is in a chaotic state (i.e., for $g > -1.2$). The higher the entropy, the higher the degree of uncertainty in the time series, thus requiring more level of information to keep system (1) in a stable state. Note that the large differences between values of ApEn/SampEn and FuzzEn are provoked by their different ways of estimating vector match. Thus, whereas all pairs of vectors presenting a distance larger than r do not contribute to entropy computation in ApEn/SampEn [61], FuzzEn always considers the degree of similarity between these patterns, thus obtaining more continuous and smooth entropy estimates [63].

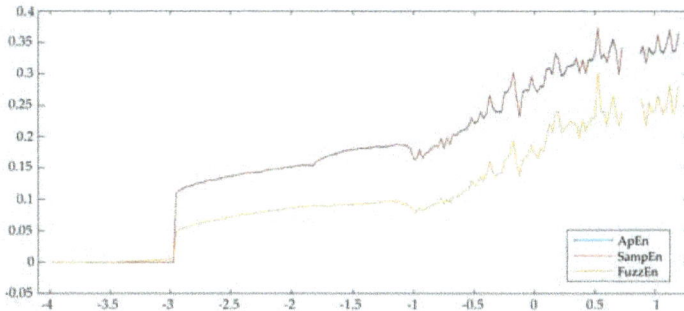

Figure 4. Values of ApEn, SampEn, and FuzzEn computed from $x(n)$ of the system (1) with respect to parameter g.

4. Brief Review of the RBF-NNs

The objective of control method is to derive the control input for stabilizing the four-dimensional chaotic system (1). Due to their inherent functional approximation and learning capabilities, RBF-NNs have recently received significant attention for approximation and modeling nonlinear functions [46,47]. According to the universal approximation property of the RBF-NN, it can approximate any continuous function $f(\mathbf{x}) : R^i \to R$ with an arbitrary accuracy δ in the following form:

$$f(\mathbf{x}) = \mathbf{\theta}^T \mathbf{\varphi}(\mathbf{x}) + \delta(\mathbf{x}) \qquad |\delta(\mathbf{x})| \leq \bar{\delta} \tag{12}$$

where $\mathbf{\theta} \in R^l$ represents the ideal weight vector, $\delta(\mathbf{x})$ denotes the approximation error, and l is the number of neurons. In (12), the ideal parameter vector $\mathbf{\theta} \in R^l$ satisfies

$$\mathbf{\theta} = \arg \min_{\hat{\mathbf{\theta}} \in R^l} \left\{ \sup_{\mathbf{x} \in \Omega} \left| f(\mathbf{x}) - \hat{\mathbf{\theta}}^T \mathbf{\varphi}(\mathbf{x}) \right| \right\} \tag{13}$$

where $\hat{\mathbf{\theta}} = \begin{bmatrix} \hat{\theta}_1 & \hat{\theta}_2 & \cdots & \hat{\theta}_l \end{bmatrix}^T \in R^l$ is the estimate of the ideal weight vector $\mathbf{\theta}$, and $\mathbf{\varphi}(\mathbf{x}) = \begin{bmatrix} \varphi_1(\mathbf{x}) & \varphi_2(\mathbf{x}) & \cdots & \varphi_l(\mathbf{x}) \end{bmatrix} \in R^l$ represents the vector of the basis functions.

It is worthwhile to note that the approximation error $\delta(\mathbf{x})$ is not known, but it is bounded, i.e., $|\delta(\mathbf{x})| \leq \bar{\delta}$.

In the RBF-NNs, the following well-known Gaussian functions are chosen as the basis functions $\varphi_j(\mathbf{x})$ for $j = 1, 2, \ldots, N$

$$\varphi_j(\mathbf{x}) = e^{-\left(\frac{(\mathbf{x} - \mathbf{c}_j)^T (\mathbf{x} - \mathbf{c}_j)}{\sigma_j^2} \right)} \tag{14}$$

where $\mathbf{c}_j = \begin{bmatrix} c_{j,1} & c_{j,2} & \cdots & c_{j,N} \end{bmatrix}^T$ and σ_j denote the center and width of the Gaussian functions, respectively. Figure 5 shows the architecture of the NN.

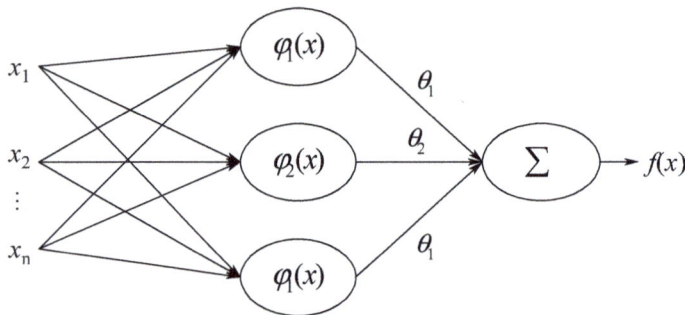

Figure 5. Architecture of the neural network.

Assumption 1. *This work assumes that the ideal weight vector has bounded norm, i.e., $\|\mathbf{\theta}\| \leq \bar{\theta}$. However, its bound is unknown.*

Remark 1. *Assumption 1 is only required for the stability analysis and design procedures of the control law does not need $\bar{\theta}$.*

4.1. Proposed Adaptive RBF-NN Controller

This section presents the proposed adaptive RBF-NN controller to suppress chaos in the considered system in (1). In the proposed method, all parameters of the system are as unknown as nonlinear dynamics and no prior knowledge about them is available. In order to handle the uncertain nonlinearity, the RBF-NN is invoked to model it. Then, the controller is designed by assuming that the RBF-NN-based model represents the true model of the system. Finally, adaptive learning laws based on the Lyapunov direct method are proposed to tune the adaptive parameters (weights coefficients) of the network.

Before designing the controller, let us rewrite the description of the four-dimensional chaotic system in (1) as follows:

$$\dot{\zeta} = A\zeta + \mathbf{b}(f(\zeta) + u) \tag{15}$$

where $\zeta = \begin{bmatrix} \zeta_1 & \zeta_2 & \zeta_3 & \zeta_4 \end{bmatrix}^T \in R^{4\times1}$ is the state vector, and $\zeta_1 = x$, $\zeta_2 = y$, $\zeta_3 = z$, and $\zeta_3 = w$; also, $f(\zeta) = -a\zeta_4 + b\zeta_1^2 - c\zeta_2^2 + e\zeta_1\zeta_2 + f\zeta_1\zeta_3 + g$ denotes the uncertain nonlinear dynamics, and $A \in R^{4\times4}$ and $\mathbf{b} \in R^{4\times1}$ are constant matrices as

$$A = \begin{bmatrix} 0 & 1 & 0 & 0 \\ 0 & 0 & 1 & 0 \\ 0 & 0 & 0 & 1 \\ 0 & 0 & 0 & 0 \end{bmatrix}, \quad \mathbf{b} = \begin{bmatrix} 0 \\ 0 \\ 0 \\ 1 \end{bmatrix} \tag{16}$$

Now, the control input is proposed as

$$u = -\hat{f}(\zeta) + y_d^{(4)} - \mathbf{k}^T \mathbf{e} \tag{17}$$

where $e = \zeta_1 - y_d$ is the tracking error, $\mathbf{e} = \begin{bmatrix} e & \dot{e} & \ddot{e} & \dddot{e} \end{bmatrix}^T \in R^{4\times1}$ represent the error vector and $\mathbf{k} = [k_4, k_3, k_2, k_1]^T \in R^{4\times1}$ denotes the design parameters that are selected such that all roots of the characteristic polynomial $\Delta(s) = s^4 + k_1 s^3 + k_2 s^2 + k_3 s_3 + k_4$ are in the open left-half of the complex plane.

Now substituting (6) and (7) in (5), we will have

$$\begin{aligned} \zeta^{(4)} &= f(\zeta) - \hat{f}(\zeta) + y_d^{(4)} - k^T \mathbf{e} \\ &= \tilde{\theta}^T \varphi(\zeta) + y_d^{(4)} - k^T \mathbf{e} + \delta \end{aligned} \tag{18}$$

where $\tilde{\theta} = \theta - \hat{\theta}$ denotes the parameter approximation error, and adaptive parameters θ are tuned by using the proposed adaptive laws as follows:

$$\dot{\hat{\theta}} = \gamma \mathbf{e}^T P \mathbf{b} \varphi(\zeta) \tag{19}$$

where $\gamma > 0$ is the learning rate, and $P \in R^{4\times4}$ represents a positive definite/semi definite matrix which satisfies the following Riccati-like equation:

$$A_c^T P + P A_c + \sigma P^T P + Q = 0 \tag{20}$$

where $Q \in R^{4\times4}$ is a positive definite matrix, and $\sigma > 0$ is a design parameter.

Before presenting stability analysis, the error dynamics is obtained by considering (15) and (18) as

$$\dot{\mathbf{e}} = A_c \mathbf{e} + \mathbf{b}\tilde{\theta}^T \varphi(\zeta + \delta) \tag{21}$$

where

$$A_c = \begin{bmatrix} 0 & 1 & 0 & 0 \\ 0 & 0 & 1 & 0 \\ 0 & 0 & 0 & 1 \\ k_4 & k_3 & k_2 & k_1 \end{bmatrix} \tag{22}$$

Now, stability analysis of the proposed controller is presented by considering the following Lyapunov function:

$$V = \frac{1}{2}\mathbf{e}^T P\mathbf{e} + \frac{1}{2\gamma}\tilde{\theta}^T\tilde{\theta} \tag{23}$$

Differentiating (21) with respect to time, results in

$$\dot{V} = \frac{1}{2}\dot{\mathbf{e}}^T P\mathbf{e} + \frac{1}{2}\mathbf{e}^T P\dot{\mathbf{e}} - \frac{1}{\gamma}\tilde{\theta}^T\dot{\theta} \tag{24}$$

Substitution of (19) in (23), results in

$$\begin{aligned}
\dot{V} &= \frac{1}{2}\left(\mathbf{e}^T A_c^T + \boldsymbol{\varphi}(\zeta)^T\tilde{\theta}\mathbf{b}^T + \delta\right)P\mathbf{e} + \frac{1}{2}\mathbf{e}^T P\left(A_c\mathbf{e} + \mathbf{b}\tilde{\theta}^T\boldsymbol{\varphi}(\zeta) + \delta\right) - \frac{1}{\gamma}\tilde{\theta}^T\dot{\theta} \\
&= \frac{1}{2}\mathbf{e}^T\left(A_c^T P + PA_c\right)\mathbf{e} - \frac{1}{\gamma}\tilde{\theta}^T\left(\dot{\theta} - \gamma\mathbf{e}^T P\mathbf{b}\boldsymbol{\varphi}(\zeta)\right) + \delta P\mathbf{e}
\end{aligned} \tag{25}$$

Again, substituting the proposed adaptive learning law (19) in (25), yields

$$\begin{aligned}
\dot{V} &= \frac{1}{2}\mathbf{e}^T\left(A_c^T P + PA_c\right)\mathbf{e} + \delta P\mathbf{e} \\
&\leq \frac{1}{2}\mathbf{e}^T\left(A_c^T P + PA_c + \sigma P^T P\right)\mathbf{e} + \frac{1}{2\sigma}\bar{\delta}^2 \\
&\leq -\frac{1}{2}\mathbf{e}^T Q\mathbf{e} + \frac{1}{2\sigma}\bar{\delta}^2 \leq -\frac{1}{2}\underline{\lambda}(Q)\mathbf{e}^T\mathbf{e} + \frac{1}{2\sigma}\bar{\delta}^2
\end{aligned} \tag{26}$$

where $\underline{\lambda}(Q)$ denotes to the minimum eigenvalue of matrix Q. As it is obtained from (26), the condition $\|\mathbf{e}\| \leq \bar{\delta}^2/\sigma\underline{\lambda}$ results in $\dot{V} \leq 0$. This inequality shows that all of the closed-loop signals (i.e., $\mathbf{e}$ and $\tilde{\theta}$) are semi-globally uniformly ultimately bounded [48].

Remark 2. *The design parameter σ in the Riccati-like Equation (20) has been proposed to attenuate the inevitable effects of the approximation error on $\dot{V}$.*

Remark 3. *It should be noted that the proposed controller does not require any off-line learning phase.*

4.2. Simulation Results

This section presents some simulation results to investigate the effectiveness of the proposed adaptive RBF-NN-based controller. A typical chaotic behavior of the uncontrolled system was discussed in Section 2. Now, the control objective is to stabilize the considered unknown chaotic system in (1) and to derive it to the equilibrium point.

To design the proposed controller, one RBF-NN composed of 50 neurons was constructed. The center of the membership functions and initial weights of the network were set at 1. For simulation, σ_i and γ were set to 0.01, and 0.5, respectively, and the initial conditions were chosen as $\zeta(0) = \begin{bmatrix} 0 & -1 & 0 & -1.5 \end{bmatrix}^T$. As mentioned before, the proposed approach does not require any training data and any off-line learning phase. After the construction of the RBF-NN, it is used to model the uncertain function $f(\zeta)$ and then the control input (17) is applied. The design parameters k_1, k_2, k_3 and k_4 in the control input (17) are chosen such that the all roots of the characteristic polynomial $\Delta(s)$ remain in the open left-half of the complex plane. For simulation, these parameters were chosen as $k_1 = 20$, $k_2 = 24$, $k_3 = 25$, and $k_4 = 22$. Also, by solving the Riccati-like Equation (20), the following matrix P was obtained:

$$P = \begin{bmatrix} 5 & 0 & 0 & 0 \\ 0 & 5 & 0 & 0 \\ 0 & 0 & 5 & 0 \\ 15 & 25 & 20 & 10 \end{bmatrix} \tag{27}$$

Also, adjustable parameters $\hat{\theta} \in R^{50}$ was adjusted based on the proposed adaptive learning law in (19).

Figures 6–10 depicts the simulation results. To highlight the performance of the proposed approach, at first the control input was set as zero, then after $t = 50$ s the proposed control method was activated. As obtained from the depicted results in Figure 6, before the activation of the proposed controller, the system has chaotic behavior but after the activation of it, the chaos was suppressed, and the desired behavior is obtained.

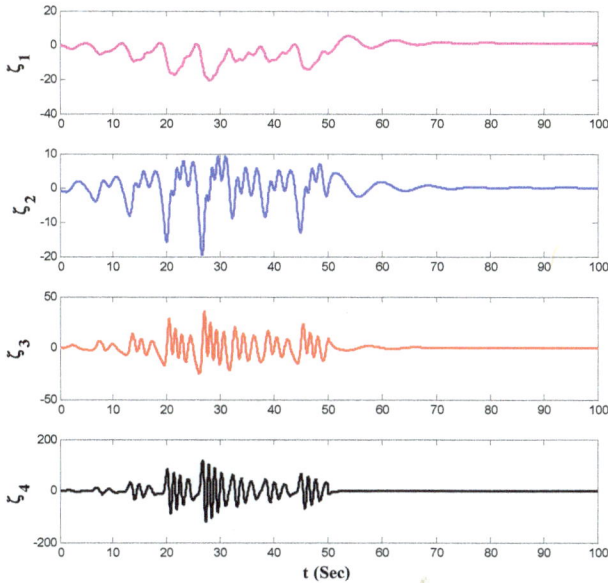

Figure 6. The state variables when the proposed control input is activated at $t = 50$ s.

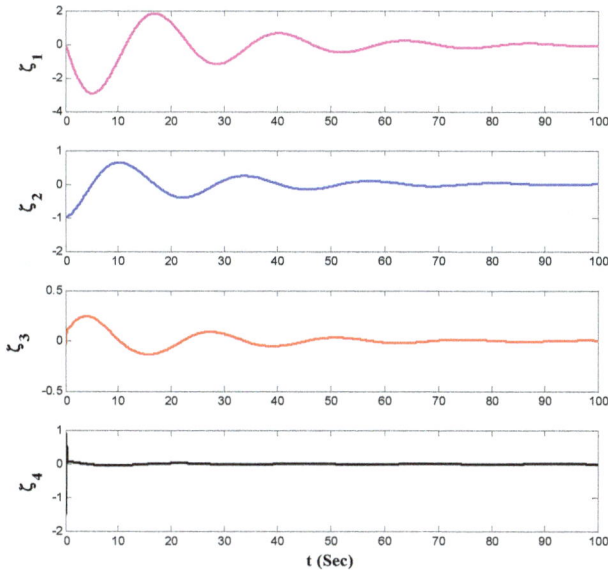

Figure 7. The state variables in the presence of the proposed control method.

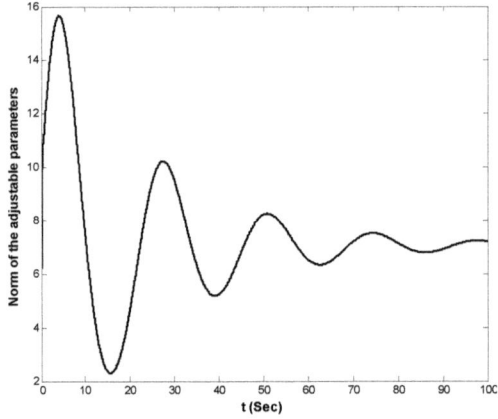

Figure 8. Norm of the weights of the RBF-NN.

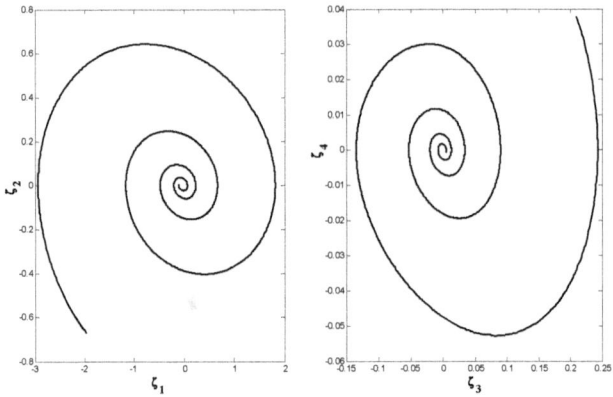

Figure 9. Phase portraits of the controlled system.

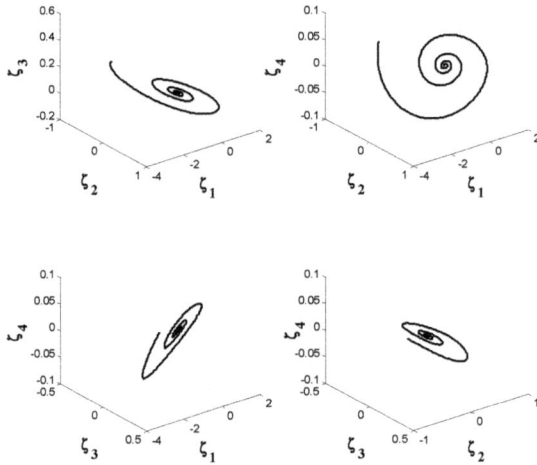

Figure 10. The 3-D behavior of the controlled system.

The state variables of the system by using the proposed controller are shown in Figure 7. Also, norm of the estimated weight coefficients is shown in Figure 8. The obtained result in Figure 8 shows that the norm of the adjustable parameters is bounded. Figures 9 and 10 depict the phase portraits and the three-dimensional behavior of the controlled system, respectively. The reported results demonstrate the ability of the proposed approach to stabilize the considered non-equilibrium four-dimensional chaotic system with hidden attractors.

5. Conclusions

In this study, a new adaptive radial basis function-neural network-based control scheme was proposed to stabilize a specific four-dimensional chaotic system, which shows a periodic-double and low-entropy route preceding high-entropy chaotic states. The proposed controller design requires neither any initial information about the dynamics of the chaotic system nor its parameters. The uncertain dynamics of the considered four-dimensional system is approximated by using the RBF-NN, and then the proposed indirect adaptive control law is proposed based on the developed model. Stability analysis is presented, and adaptive learning law is derived for calibrating weights of the RBF-NN. Simulation results verify the acceptable performance of the proposed method for stabilizing the considered chaotic system.

Author Contributions: Conceptualization, H.J.; Investigation, H.J., M.S.-K., and R.A.; Methodology, V.P.S.; Software, M.S.-K. and X.W.; Supervision, X.W.; Validation, V.-T.P.; Writing—original draft, H.J., M.S.-K., and R.A.; Writing—review & editing, V.P.S., and V.-T.P.

Funding: The author Xiong Wang was supported by the National Natural Science Foundation of China (No. 61601306) and the Shenzhen Overseas High Level Talent Peacock Project Fund (No. 20150215145C).

Conflicts of Interest: The authors declare no conflict of interest.

References

1. Lai, Q.; Chen, S. Generating multiple chaotic attractors from Sprott B system. *Int. J. Bifurc. Chaos* **2016**, *26*, 1650177. [CrossRef]
2. Sharma, P.R.; Shrimali, M.D.; Prasad, A.; Kuznetsov, N.V.; Leonov, G.A. Control of multistability in hidden attractors. *Eur. Phys. J. Spec. Top.* **2015**, *224*, 1485–1491. [CrossRef]
3. Sprott, J.C.; Jafari, S.; Khalaf, A.J.M.; Kapitaniak, T. Megastability: Coexistence of a countable infinity of nested attractors in a periodically-forced oscillator with spatially-periodic damping. *Eur. Phys. J. Spec. Top.* **2017**, *226*, 1979–1985. [CrossRef]
4. Bao, B.; Jiang, T.; Xu, Q.; Chen, M.; Wu, H.; Hu, Y. Coexisting infinitely many attractors in active band-pass filter-based memristive circuit. *Nonlinear Dyn.* **2016**, *86*, 1711–1723. [CrossRef]
5. Bao, B.-C.; Xu, Q.; Bao, H.; Chen, M. Extreme multistability in a memristive circuit. *Electron. Lett.* **2016**, *52*, 1008–1010. [CrossRef]
6. Munoz-Pacheco, J.M.; Tlelo-Cuautle, E.; Toxqui-Toxqui, I.; Sanchez-Lopez, C.; Trejo-Guerra, R. Frequency limitations in generating multi-scroll chaotic attractors using CFOAs. *Int. J. Electron.* **2014**, *101*, 1559–1569. [CrossRef]
7. Tlelo-Cuautle, E.; Rangel-Magdaleno, J.J.; Pano-Azucena, A.D.; Obeso-Rodelo, P.J.; Nuñez-Perez, J.C. FPGA realization of multi-scroll chaotic oscillators. *Commun. Nonlinear. Sci. Numer. Simul.* **2015**, *27*, 66–80. [CrossRef]
8. Leonov, G.A.; Kuznetsov, N.V. Hidden attractors in dynamical systems. From hidden oscillations in Hilbert–Kolmogorov, Aizerman, and Kalman problems to hidden chaotic attractor in Chua circuits. *Int. J. Bifurc. Chaos* **2013**, *23*, 1330002. [CrossRef]
9. Sprott, J.C. Some simple chaotic flows. *Phys. Rev. E Stat. Nonlin. Soft Matter Phys.* **1994**, *50*, R647. [CrossRef]
10. Pham, V.-T.; Volos, C.; Jafari, S.; Kapitaniak, T. Coexistence of hidden chaotic attractors in a novel no-equilibrium system. *Nonlinear Dyn.* **2017**, *87*, 2001–2010. [CrossRef]
11. Pham, V.-T.; Jafari, S.; Volos, C.; Gotthans, T.; Wang, X.; Hoang, D.V. A chaotic system with rounded square equilibrium and with no-equilibrium. *OPTIK* **2017**, *130*, 365–371. [CrossRef]

12. Jafari, S.; Sprott, J.C.; Golpayegani, S.M.R.H. Elementary quadratic chaotic flows with no equilibria. *Phys. Lett. A* **2013**, *377*, 699–702. [CrossRef]
13. Wei, Z. Dynamical behaviors of a chaotic system with no equilibria. *Phys. Lett. A* **2011**, *376*, 102–108. [CrossRef]
14. Pham, V.-T.; Akgul, A.; Volos, C.; Jafari, S.; Kapitaniak, T. Dynamics and circuit realization of a no-equilibrium chaotic system with a boostable variable. *AEU Int. J. Electron. C.* **2017**, *78*, 134–140. [CrossRef]
15. Ren, S.; Panahi, S.; Rajagopal, K.; Akgul, A.; Pham, V.-T.; Jafari, S. A new chaotic flow with hidden attractor: The first hyperjerk system with no equilibrium. *Z. Naturforsch. A* **2018**, *73*, 239–249. [CrossRef]
16. Rossler, O.E. An equation for hyperchaos. *Phys. Lett. A* **1979**, *71*, 155–157. [CrossRef]
17. Wei, Z.; Wang, R.; Liu, A. A new finding of the existence of hidden hyperchaotic attractors with no equilibria. *Math. Comput. Simul.* **2014**, *100*, 13–23. [CrossRef]
18. Tahir, F.R.; Jafari, S.; Pham, V.-T.; Volos, C.; Wang, X. A novel no-equilibrium chaotic system with multiwing butterfly attractors. *Int. J. Bifurc. Chaos* **2015**, *25*, 1550056. [CrossRef]
19. Pham, V.T.; Vaidyanathan, S.; Volos, C.K.; Jafari, S. Hidden attractors in a chaotic system with an exponential nonlinear term. *Eur. Phys. J. Spec. Top.* **2015**, *224*, 1507–1517. [CrossRef]
20. Pham, V.-T.; Vaidyanathan, S.; Volos, C.; Jafari, S.; Kingni, S.T. A no-equilibrium hyperchaotic system with a cubic nonlinear term. *OPTIK* **2016**, *127*, 3259–3265. [CrossRef]
21. Bao, B.C.; Bao, H.; Wang, N.; Chen, M.; Xu, Q. Hidden extreme multistability in memristive hyperchaotic system. *Chaos Soliton. Fract.* **2017**, *94*, 102–111. [CrossRef]
22. Zhang, S.; Zeng, Y.; Li, Z.; Wang, M.; Xiong, L. Generating one to four-wing hidden attractors in a novel 4D no-equilibrium chaotic system with extreme multistability. *Chaos* **2018**, *28*, 013113. [CrossRef] [PubMed]
23. Mobayen, S.; Ma, J. Robust finite-time composite nonlinear feedback control for synchronization of uncertain chaotic systems with nonlinearity and time-delay. *Chaos Soliton. Fract.* **2018**, *114*, 46–54. [CrossRef]
24. Shukla, M.K.; Sharma, B.B. Stabilization of a class of fractional order chaotic systems via backstepping approach. *Chaos Soliton. Fract.* **2017**, *98*, 56–62. [CrossRef]
25. Wang, Y.; Yu, H. Fuzzy synchronization of chaotic systems via intermittent control. *Chaos Soliton. Fract.* **2018**, *106*, 154–160. [CrossRef]
26. Hsiao, F.-H. Robust H∞ fuzzy control of dithered chaotic systems. *Neurocomputing* **2013**, *99*, 509–520. [CrossRef]
27. Lin, C.-M.; Huynh, T.-T. Function-Link Fuzzy Cerebellar Model Articulation Controller Design for Nonlinear Chaotic Systems Using TOPSIS Multiple Attribute Decision-Making Method. *Int. J. Fuzzy Syst.* **2018**, *20*, 1839–1856. [CrossRef]
28. Zhang, X.; Li, D.; Zhang, X. Adaptive fuzzy impulsive synchronization of chaotic systems with random parameters. *Chaos Soliton. Fract.* **2017**, *104*, 77–83. [CrossRef]
29. Rajagopal, K.; Jahanshahi, H.; Varan, M.; Bayır, I.; Pham, V.-T.; Jafari, S.; Karthikeyan, A. A hyperchaotic memristor oscillator with fuzzy based chaos control and LQR based chaos synchronization. *AEU Int. J. Electron. C* **2018**, *94*, 55–68. [CrossRef]
30. Mobayen, S. Chaos synchronization of uncertain chaotic systems using composite nonlinear feedback based integral sliding mode control. *ISA Trans.* **2018**, *77*, 100–111. [CrossRef]
31. Chen, X.; Park, J.H.; Cao, J.; Qiu, J. Adaptive synchronization of multiple uncertain coupled chaotic systems via sliding mode control. *Neurocomputing* **2018**, *273*, 9–21. [CrossRef]
32. Deepika, D.; Kaur, S.; Narayan, S. Uncertainty and disturbance estimator based robust synchronization for a class of uncertain fractional chaotic system via fractional order sliding mode control. *Chaos Soliton. Fract.* **2018**, *115*, 196–203. [CrossRef]
33. Sun, Z. Synchronization of fractional-order chaotic systems with non-identical orders, unknown parameters and disturbances via sliding mode control. *Chin. J. Phys.* **2018**, *56*, 2553–2559. [CrossRef]
34. Liu, H.; Yang, J. Sliding-mode synchronization control for uncertain fractional-order chaotic systems with time delay. *Entropy* **2015**, *17*, 4202–4214. [CrossRef]
35. Shieh, C.-S.; Hung, R.-T. Hybrid control for synchronizing a chaotic system. *Appl. Math. Model.* **2011**, *35*, 3751–3758. [CrossRef]
36. Tsai, J.S.-H.; Fang, J.-S.; Yan, J.-J.; Dai, M.-C.; Guo, S.-M.; Shieh, L.-S. Hybrid robust discrete sliding mode control for generalized continuous chaotic systems subject to external disturbances. *Nonlinear Anal. Hybrid Syst.* **2018**, *29*, 74–84. [CrossRef]

37. Cai, P.; Yuan, Z.Z. Hopf bifurcation and chaos control in a new chaotic system via hybrid control strategy. *Chin. J. Phys.* **2017**, *55*, 64–70. [CrossRef]

38. Jahanshahi, H. Smooth control of HIV/AIDS infection using a robust adaptive scheme with decoupled sliding mode supervision. *Eur. Phys. J. Spec. Top.* **2018**, *227*, 707–718. [CrossRef]

39. Jahanshahi, H.; Rajagopal, K.; Akgul, A.; Sari, N.N.; Namazi, H.; Jafari, S. Complete analysis and engineering applications of a megastable nonlinear oscillator. *Int. J. Non Linear Mech.* **2018**, *107*, 126–136. [CrossRef]

40. Najafizadeh Sari, N.; Jahanshahi, H.; Fakoor, M. Adaptive Fuzzy PID Control Strategy for Spacecraft Attitude Control. *Int. J. Fuzzy Syst.* **2019**. [CrossRef]

41. Mahmoodabadi, M.J.; Jahanshahi, H. Multi-objective optimized fuzzy-PID controllers for fourth order nonlinear systems. *Eng. Sci. Technol.* **2016**, *19*, 1084–1098. [CrossRef]

42. Kosari, A.; Jahanshahi, H.; Razavi, S.A. Optimal FPID control approach for a docking maneuver of two spacecraft: Translational motion. *J. Aerospace Eng.* **2017**, *30*, 04017011. [CrossRef]

43. Hou, R.; Wang, L.; Gao, Q.; Hou, Y.; Wang, C. Indirect adaptive fuzzy wavelet neural network with self-recurrent consequent part for AC servo system. *ISA Trans.* **2017**, *70*, 298–307. [CrossRef] [PubMed]

44. Solgi, Y.; Ganjefar, S. Variable structure fuzzy wavelet neural network controller for complex nonlinear systems. *Appl. Soft Comput.* **2018**, *64*, 674–685. [CrossRef]

45. Ahn, C.K. Neural network $\mathcal{H}_{\infty}$ chaos synchronization. *Nonlinear Dyn.* **2010**, *60*, 295–302. [CrossRef]

46. Hsu, C.-F. Hermite-neural-network-based adaptive control for a coupled nonlinear chaotic system. *Neural Comput. Appl.* **2013**, *22*, 421–433. [CrossRef]

47. Gokce, K.; Uyaroglu, Y. An Adaptive Neural Network Control Scheme for Stabilizing Chaos to the Stable Fixed Point. *Inf. Technol. Control* **2017**, *46*, 219–227. [CrossRef]

48. Zouari, F.; Boulkroune, A.; Ibeas, A. Neural adaptive quantized output-feedback control-based synchronization of uncertain time-delay incommensurate fractional-order chaotic systems with input nonlinearities. *Neurocomputing* **2017**, *237*, 200–225. [CrossRef]

49. Yadmellat, P.; Nikravesh, S.K.Y. A recursive delayed output-feedback control to stabilize chaotic systems using linear-in-parameter neural networks. *Commun. Nonlinear Sci. Numer. Simul.* **2011**, *16*, 383–394. [CrossRef]

50. Sarcheshmeh, S.F.; Esmaelzadeh, R.; Afshari, M. Chaotic satellite synchronization using neural and nonlinear controllers. *Chaos Soliton. Fract.* **2017**, *97*, 19–27. [CrossRef]

51. Fang, L.; Li, T.; Wang, X.; Gao, X. Adaptive synchronization of uncertain chaotic systems via neural network-based dynamic surface control design. In Proceedings of the 10th International Symposium on Neural Networks (2013 ISNN), Dalian, China, 4–6 July 2013; pp. 104–111.

52. Shao, S.; Chen, M.; Yan, X. Prescribed performance synchronization for uncertain chaotic systems with input saturation based on neural networks. *Neural Comput. Appl.* **2018**, *29*, 1349–1361. [CrossRef]

53. Gomez, I.S.; Losada, M.; Lombardi, O. About the Concept of Quantum Chaos. *Entropy* **2017**, *19*, 205. [CrossRef]

54. Frigg, R. In what sense is the Kolmogorov-Sinai entropy a measure for chaotic behaviour?—bridging the gap between dynamical systems theory and communication theory. *Br. J. Philos. Sci.* **2004**, *55*, 411–434. [CrossRef]

55. Young, L.-S. Entropy in dynamical systems. In *Entropy*; Princeton University Press: Princeton, NJ, USA, 2003; pp. 313–328

56. Grassberger, P.; Procaccia, I. Estimation of the Kolmogorov entropy from a chaotic signal. *Phys. Rev. A* **1983**, *28*, 2591. [CrossRef]

57. Pincus, S.M. Approximate entropy as a measure of system complexity. *Proc. Natl. Acad. Sci. USA* **1991**, *88*, 2297–2301. [CrossRef]

58. Wang, C.; Ding, Q. A New Two-Dimensional Map with Hidden Attractors. *Entropy* **2018**, *20*, 322. [CrossRef]

59. Xu, G.; Shekofteh, Y.; Akgül, A.; Li, C.; Panahi, S. A new chaotic system with a self-excited attractor: Entropy measurement, signal encryption, and parameter estimation. *Entropy* **2018**, *20*, 86. [CrossRef]

60. Pincus, S. Approximate entropy (ApEn) as a complexity measure. *Chaos* **1995**, *5*, 110–117. [CrossRef]

61. Richman, J.S.; Moorman, J.R. Physiological time-series analysis using approximate entropy and sample entropy. *Am. J. Physiol. Heart Circ. Physiol.* **2000**, *278*, H2039–H2049. [CrossRef]

62. Borowska, M. Entropy-based algorithms in the analysis of biomedical signals. *Stud. Logic Grammar Rhetoric* **2015**, *43*, 21–32. [CrossRef]
63. Chen, W.; Wang, Z.; Xie, H.; Yu, W. Characterization of surface EMG signal based on fuzzy entropy. *IEEE Trans. Neural Syst. Rehabil. Eng.* **2007**, *15*, 266–272. [CrossRef]

entropy

MDPI

Article

Adaptive Synchronization of Fractional-Order Complex Chaotic system with Unknown Complex Parameters

Ruoxun Zhang [1,2,*], Yongli Liu [1] and Shiping Yang [2]

[1]   College of Primary Education, Xingtai University, Xingtai 054001, China; liuyongliliu@126.com
[2]   College of Physics Science and Information Engineering, Hebei Normal University,
    Shijiazhuang 050016, China; yangship@hebtu.edu.cn
*   Correspondence: xtzhrx@126.com

Received: 27 December 2018; Accepted: 19 February 2019; Published: 21 February 2019

Abstract: This paper investigates the problem of synchronization of fractional-order complex-variable chaotic systems (FOCCS) with unknown complex parameters. Based on the complex-variable inequality and stability theory for fractional-order complex-valued system, a new scheme is presented for adaptive synchronization of FOCCS with unknown complex parameters. The proposed scheme not only provides a new method to analyze fractional-order complex-valued system but also significantly reduces the complexity of computation and analysis. Theoretical proof and simulation results substantiate the effectiveness of the presented synchronization scheme.

Keywords: synchronization; fractional-order; complex-variable chaotic system; unknown complex parameters

1. Introduction

In the past 20 years, fractional-order chaotic systems have been extensively studied due to their wide applications in the fields of secure communication, control engineering, finance, physical and mathematical science, entropy, encryption and signal processing [1–4]. Meanwhile, synchronization of such systems has aroused tremendous attention of many researchers. Lots of excellent results were obtained and some methods of synchronization have been presented [5–17]. In various synchronization methods, the adaptive control approach is an effective method to realize the synchronization of uncertain systems.

The aforementioned works mainly investigated the fractional-order systems with real variables, not involving complex variables. Because complex variables that double the number of variables can generate complicated dynamical behaviors, enhance anti-attack ability and achieve higher transmission efficiency [18–20], many researchers have taken complex variables into the fractional-order systems and investigated dynamics behavior, stability, stabilization and synchronization of FOCCS in recent years. In [21–23], fractional-order complex-variable Chen system, T system and Lorenz system have been investigated, respectively. Recently, Zhang et al. [24] have investigated synchronization of fractional-order complex-valued delayed neural networks. Li et al. [25] presented adaptive synchronization scheme for fractional-order complex-valued neural networks with discrete and distributed delays. Sun et al. [26] proposed real combination synchronization of three fractional-order complex-variable chaotic systems, Yadav et al. [27] studied Dual function projective synchronization of fractional order complex chaotic systems, Nian et al. [28] realized synchronization of fractional-order complex chaotic system with parametric and external disturbances via sliding mode control method and Jiang et al. have studied complex modified projective synchronization (CMPS) for FOCCS in [29]. However, in these papers [24–29], the parameters of the FOCCS are exactly known in priori. In fact, in many practical engineering situations, most of system parameters cannot be accurately determined in advance and chaos synchronization will be destroyed with these uncertain factors. Hence, it is an important problem to realize synchronization of FOCCS with unknown complex parameters.

Inspired by the above discussions, the synchronization problem of FOCCS with unknown complex parameters was investigated in this paper. Using the inequality of the fractional derivative containing complex variable and the stability theory for fractional-order complex-valued system, we realized synchronization of such systems by constructing a suitable response system. It should be noted that we deal with the synchronization problem of fractional-order uncertain complex-variable system in complex-valued domain. That is to say, it is not necessary to separate the complex-variable system into its real and imaginary parts. This greatly reduces the complexity of computation and the difficulty of theoretical analysis.

Notation: $\mathbb{C}^n$ denotes complex n-dimensional space. For $z \in \mathbb{C}^n$, z^r, z^i, $\bar{z}$, z^T, z^H and $||z||$ are the real part, imaginary part, conjugate, transpose, conjugate transpose and l_2-norm of z, respectively. For a matrix $A \in \mathbb{C}^{n \times n}$, A^H denotes its conjugate transpose.

2. Preliminaries

Definition 1 [30]. *The fractional integral of order α for a function f is defined as:*

$$I^\alpha f(t) = {}_{t_0}D_t^{-\alpha} f(t) = \frac{1}{\Gamma(\alpha)} \int_{t_0}^t (t-\tau)^{\alpha-1} f(\tau) d\tau \tag{1}$$

where $t \geq t_0$ and $\alpha > 0$.

Definition 2 [30]. *Caputo's fractional derivative of order α for a function $f \in \mathbb{R}^n$ is defined by:*

$$_{t_0}^{C}D_t^\alpha f(t) = \frac{1}{\Gamma(n-\alpha)} \int_{t_0}^t \frac{f^{(n)}(\tau)}{(t-\tau)^{\alpha-n+1}} d\tau \tag{2}$$

where $t \geq t_0$ and n is a positive integer such that $n - < \alpha < 1$.

Lemma 1 [31]. *Let $x(t) \in \mathbb{R}^n$ be a continuous and derivable vector function. Then, for any time instant $t \geq t_0$ and $\alpha \in (0,1)$:*

$$\frac{1}{2}{}_{t_0}^{C}D_t^\alpha [x^T(t)x(t)] \leq x^T(t){}_{t_0}^{C}D_t^\alpha x(t) \tag{3}$$

Corollary 1. *For a scalar derivable function $\varphi(t)$ and a constants C, we have:*

$$\frac{1}{2}{}_{t_0}^{C}D_t^\alpha (\varphi(t) - C)^2 \leq (\varphi(t) - C){}_{t_0}^{C}D_t^\alpha \varphi(t) \tag{4}$$

Lemma 2 [32] . *Let $z \in \mathbb{C}^n$ be a differentiable complex-valued vector. Then, $\forall t \geq t_0$ and $\alpha \in (0, 1]$, the following inequality holds:*

$$_{t_0}^{C}D_t^\alpha z^H(t)Pz(t) \leq z^H(t)P{}_{t_0}^{C}D_t^\alpha z(t) + \left({}_{t_0}^{C}D_t^\alpha z(t)\right)^H Pz(t) \tag{5}$$

where $P \in \mathbb{C}^{n \times n}$ is a constant positive definite Hermitian matrix.

Lemma 3. *For the fractional-order complex-variable systems:*

$$_{0}^{C}D_t^\alpha z(t) = h(z(t)) \tag{6}$$

where $0 < \alpha < 1$, $z(t) = (z_1, z_2, \cdots, z_n)^T \in \mathbb{C}^n$ *is the system complex state vector,* $h \in \mathbb{C}^n$ *is a continuous nonlinear function complex vector, which satisfies the globally Lipschitz continuity condition in the complex domain. Let* $z(t) = 0$ *be an equilibrium point of system (1) and let* $V_1(t) = z^H(t)z(t)$ *and* $V_2(z(t)) \geq 0$ *are continuously differentiable functions. If:*

$$V(t) = V_1(t) + V_2(z(t)) \tag{7}$$

and:

$$_0^t D_t^\alpha V(t) \leq -\theta V_1(t) \tag{8}$$

where θ *is a positive constant. Then* $z(t) = 0$ *is asymptotically stable.*

Proof. See the Appendix A. It was pointed out [33–35] that a similar theorem is obtained for the real systems. □

Remark 1. *Using Lemmas 2.2–2.3, one can directly analyze fractional order complex-variable system in the complex space.*

3. Main Results

We considered a kind of FOCCS described by:

$$_0^C D_t^\alpha z(t) = Az(t) + f(z(t)) \tag{9}$$

where $0 < \alpha < 1$, $z(t) = (z_1, z_2, \cdots, z_n)^T \in \mathbb{C}^n$ is the system complex state vector, $f \in \mathbb{C}^n$ is a continuous nonlinear function vector, which satisfies the globally Lipschitz continuity condition in the complex domain and $A \in \mathbb{C}^{n \times n}$ is unknown (complex or real) parameter matrix. Furthermore, Equation (10) can be rewritten as:

$$_0^C D_t^\alpha z(t) = g(z(t))\theta + f(z(t)) \tag{10}$$

where $g : \mathbb{C}^n \to \mathbb{C}^{n \times m}$ is a complex function matrix, and $\theta = (\theta_1, \theta_2, \cdots, \theta_m)^T$ is the system unknown complex parameter vector. For system (9) or (10), there are two propositions as follows.

Proposition 1. *There exists a positive constant* l_1 *such that the following inequality holds:*

$$(z-w)^H[g(z) - g(w)]\theta + \{[g(z) - g(w)]\theta\}^H(z-w) \leq l_1(z-w)^H(z-w) \tag{11}$$

Proof. Given that $g(z(t))\theta = Az(t)$ results in:

$$\begin{aligned}
&(z-w)^H[g(z) - g(w)]\theta + \{[g(z) - g(w)]\theta\}^H(z-w) \\
&= (z-w)^H A(z-w) + (z-w)^H A^H(z-w) \\
&= (z-w)^H(A + A^H)(z-w)
\end{aligned}$$

Since $A + A^H$ is Hermitian Matrix:

$$\lambda_m(z-w)^H(z-w) \leq (z-w)^H(A + A^H)(z-w) \leq \lambda_M(z-w)^H(z-w)$$

where λ_m and λ_M are the minimum and maximum eigenvalue of $A + A^H$, respectively [36,37].
Let $l_1 = \max(|\lambda_m|, |\lambda_M|)$, then, one has:

$$(z-w)^H[g(z) - g(w)]\theta + \{[g(z) - g(w)]\theta\}^H(z-w) = (z-w)^H(A + A^H)(z-w) \leq l_1(z-w)^H(z-w)$$

□

Proposition 2 [38] . For the Lipschitz continuous function $f \in \mathbb{C}^n$, there exists a positive constant l_2 such that the following inequality holds:

$$(z - w)^H[f(z) - f(w)] + [f(z) - f(w)]^H(z - w) \leq l_2(z - w)^H(z - w) \tag{12}$$

Proof. For $f \in \mathbb{C}^n$, Lipschitz is continuous, then $||f(z) - f(w)|| \leq L||z - w||$, where $L \geq 0$ is a constant. It follows:

$$\begin{aligned}
(z - w)^H[f(z) - f(w)] + [f(z) - f(w)]^H(z - w) &= 2\mathrm{Re}\{(z - w)^H[f(z) - f(w)]\} \\
&\leq 2|z - w|_T|f(z) - f(w)| \leq |z - w|_T|z - w| + |f(z) - f(w)|_T|f(z) - f(w)| \\
&= (z - w)^H(z - w) + \left\Vert f(z) - f(w) \right\Vert_2 \leq (z - w)^H(z - w) + L^2 \left\Vert z - w \right\Vert^2 \\
&= (1 + L^2)(z - w)^H(z - w) = l_2(z - w)^H(z - w)
\end{aligned}$$

where $l_2 = L^2 + 1$, $\left| f(z) - f(w) \right| = (|f_1(z) - f_1(w)|, |f_2(z) - f_2(w)|, \cdots, |f_n(z) - f_n(w)|)^T$ and $\left| z - w \right| = (|z_1 - w_1|, |z_2 - w_2|, \cdots, |z_n - w_n|)^T$. □

Remark 2. *It is easy to check that many typical FOCCSs, such as the fractional-order complex-variable Chen system [21], T system [22] and Lorenz system [23] all satisfy Propositions 1 and 2.*

Choose system (11) as the master system, then the controlled response system is given by:

$$ {}_0^C D_t^\alpha w(t) = g(w(t))\hat{\theta} + f(w(t)) + u(t) \tag{13}$$

where $w(t) = (w_1, w_2, \cdots, w_n)^T$ is the complex state vector, $\hat{\theta} \in \mathbb{C}^m$ represents the estimate vector of unknown vector θ, and $u(t) = (u_1(t), u_2(t), \cdots, u_n(t))^T$ is controller to be determined.

Theorem 1. *Asymptotically synchronization and parameter identification of systems (13) and (11) can be achieved under adaptive controller:*

$$u(t) = -ke(t) \tag{14}$$

and the complex update laws:

$$ {}_0^C D_t^\alpha k = \sigma e^H(t)e(t) \tag{15}$$

$$ {}_0^C D_t^\alpha e_\theta = {}_0^C D_t^\alpha \hat{\theta} = -\eta g^H(w(t))e(t) \tag{16}$$

where $e(t) = w(t) - z(t)$ is the error vector, $e_\theta = \hat{\theta} - \theta$ is the parameter error, σ, η are two arbitrary positive constants.

Proof. From the error vector and systems (11) and (13), it yields:

$$\begin{aligned}
{}_0^C D_t^\alpha e(t) &= g(w(t))\hat{\theta} + f(w(t)) - g(z(t))\theta + f(z(t)) + u(t) \\
&= g(w(t))e_\theta + [g(w(t)) - g(z(t))]\theta + f(w(t)) - f(z(t)) + u(t)
\end{aligned}$$

Let us present the Lyapunov function:

$$V(t, e(t)) = e^H(t)e(t) + \frac{1}{\sigma}(k - k^*)^2 + \frac{1}{\eta}e_\theta^H(t)e_\theta(t) \tag{17}$$

where k^* is to be determined. □

Using Lemma 2.1, Corollary 2.1 and Lemma 2.2, we have:

$$
\begin{aligned}
{}_0^C D_t^\alpha V(t,e(t)) &= {}_0^C D_t^\alpha [e^H(t)e(t) + \tfrac{1}{\sigma}(k-k^*)^2 + \tfrac{1}{\eta}e_\theta^H(t)e_\theta(t)] \\
&\le e^H(t)\,{}_0^C D_t^\alpha e(t) + [{}_0^C D_t^\alpha e(t)]^H e(t) + \tfrac{2}{\sigma}(k-k^*)\,{}_0^C D_t^\alpha k + \tfrac{1}{\eta}e_\theta^H(t)\,{}_0^C D_t^\alpha e_\theta(t) + \tfrac{1}{\eta}[{}_0^C D_t^\alpha e_\theta(t)]^H e_\theta(t) \\
&\le e^H(t)\{g(w(t))e_\theta + [g(w(t))-g(z(t))]\theta + f(w(t)) - f(z(t)) - ke(t)\} + \{g(w(t))e_\theta \\
&\quad + [g(w(t)) - g(z(t))]\theta + f(w(t)) - f(z(t)) - ke(t)\}^H e + \tfrac{2}{\sigma}(k-k^*)\,{}_0^C D_t^\alpha k \\
&\quad + \tfrac{1}{\eta}e_\theta^H(t)\,{}_0^C D_t^\alpha e_\theta(t) + \tfrac{1}{\eta}[{}_0^C D_t^\alpha e_\theta(t)]^H e_\theta(t) \\
&\le e^H(t)\{g(w(t))e_\theta + [g(w(t))-g(z(t))]\theta + f(w(t)) - f(z(t)) - ke(t)\} + \{g(w(t))e_\theta \\
&\quad + [g(w(t)) - g(z(t))]\theta + f(w(t)) - f(z(t)) - ke(t)\}^H e + \tfrac{2}{\sigma}(k-k^*)\,{}_0^C D_t^\alpha k \\
&\quad + \tfrac{1}{\eta}e_\theta^H(t)\,{}_0^C D_t^\alpha e_\theta(t) + \tfrac{1}{\eta}[{}_0^C D_t^\alpha e_\theta(t)]^H e_\theta(t) \\
&\le e^H(t)\{g(w(t))e_\theta + [g(w(t))-g(z(t))]\theta + f(w(t)) - f(z(t)) - ke(t)\} + \{g(w(t))e_\theta \\
&\quad + [g(w(t)) - g(z(t))]\theta + f(w(t)) - f(z(t)) - ke(t)\}^H e + \tfrac{2}{\sigma}(k-k^*)\,{}_0^C D_t^\alpha k \\
&\quad + \tfrac{1}{\eta}e_\theta^H(t)\,{}_0^C D_t^\alpha e_\theta(t) + \tfrac{1}{\eta}[{}_0^C D_t^\alpha e_\theta(t)]^H e_\theta(t)
\end{aligned}
$$

Substitute Equations (15) and (16) into the inequality above, we further have:

$$
\begin{aligned}
{}_0^C D_t^\alpha V(t,e(t)) &\le e^H(t)\{g(w(t))e_\theta + [g(w(t))-g(z(t))]\theta + f(w(t)) - f(z(t)) - ke(t)\} \\
&\quad + \{g(w(t))e_\theta + [g(w(t))-g(z(t))]\theta + f(w(t)) - f(z(t)) - ke(t)\}^H e \\
&\quad + 2(k-k^*)e^H(t)e(t) - e_\theta^H(t)g^H(w(t))e(t) - [g^H(w(t))e(t)]^H e_\theta(t) \\
&\le e^H(t)[g(w(t)-g(z(t))]\theta + \big\{[g(w(t)-g(z(t))]\theta\big\}^H e(t) \\
&\quad + e^H(t)[f(w(t)-f(z(t))] + [f(w(t)-f(z(t))]^H e(t) - 2k^*e^H(t)e(t)
\end{aligned}
$$

From Propositions 1 and 2, we can obtain:

$$
e^H(t)[g(w(t))-g(z(t))]\theta + \{[g(w(t))-g(z(t)]\theta\}^H e(t) \le l_1 e^H(t)e(t)
$$

and:

$$
e^H(t)[f(w(t))-f(z(t)] + [f(w(t))-f(z(t)]^H e(t) \le l_2 e^H(t)e(t)
$$

then, one has:

$$
{}_0^C D_t^\alpha V(t,e(t)) \le e^H(t)[l_1 + (l_2 - 2k^*)I]e(t)
$$

Let $2k^* = l_1 + l_2 + 1$, then:

$$
{}_0^C D_t^\alpha V(t,e(t)) \le -e^H(t)e(t) \tag{18}
$$

According to Lemma 2.3, one has $\lim\limits_{t\to\infty} e^H(t)e(t) = 0$, which implies $\lim\limits_{t\to\infty} e(t) = 0$, which shows that the systems (11) and (13) can obtain asymptotically synchronization. Meanwhile, according to Remark 1 of Theorem 1 in [39], parameter identification is achieved.

Remark 3. *In previous work [24–29], the common method to analyze fractional complex-valued systems is to separate into two real-valued systems according to their real and imaginary parts, and then the criteria on synchronization were obtained by investigating these real-valued systems. However, there are two problems with this approach. One is that the dimension of the real-valued system is twice that of the original complex-valued system, which adds the complexity of computation and analysis. The other is that this method requires that complex-valued functions be explicitly separated into real and imaginary parts. However, this separation is not always expressible in an analytical form. Unlike from previous works, in our proposed method, the entire analysis process is performed in the complex-valued domain, and the complex function theory is used to derive synchronization conditions without separating the original complex-valued chaotic system into two real-valued systems, which reduces the complexity of analysis and computation. Moreover, the proposed method can be applied to other complex-valued systems, such as complex networks with fractional-order complex-variable dynamics and fractional-order complex-valued neural network systems.*

Remark 4. *If the system parameters are known, the update law will be reduced to (15) only.*

4. Numerical Simulations

In this section, in order to show the effectiveness of the proposed scheme in preceding section, numerical example on fractional-order complex chaotic system will be provided. When numerically solving such systems, we first adopt the predictor–corrector method [40] by MATLAB. Lyapunov exponents of the systems are calculated by adopting the Wolf et al. algorithm [41] with some changes.

Consider the Lorenz-like fractional-order complex chaotic system with commensurate order:

$$
\begin{pmatrix} {}^C_0 D_t^\alpha z_1 \\ {}^C_0 D_t^\alpha z_2 \\ {}^C_0 D_t^\alpha z_3 \end{pmatrix} = \begin{pmatrix} a(z_2 + z_1) \\ -cz_2 - z_1 z_3 \\ \bar{z}_1 z_1 - bz_3 \end{pmatrix} = \begin{pmatrix} z_2 + z_1 & 0 & 0 \\ 0 & -z_2 & 0 \\ 0 & 0 & -z_3 \end{pmatrix} \begin{pmatrix} a \\ c \\ b \end{pmatrix} + \begin{pmatrix} 0 \\ -z_1 z_3 \\ \bar{z}_1 z_1 \end{pmatrix} \quad (19)
$$

where z_1, z_2, z_3 are the complex state variables and a, b, c are system parameters; let $a = 10 + i$, $b = 3$, $c = 16 + 0.3i$. The maximum Lyapunov exponent (MLE) spectrum is depicted in Figure 1a, and the bifurcation diagram is presented in Figure 1b. Figure 1a,b shows that system (19) is chaotic with fractional order $\alpha \in [0.91, 0.98] \cup [0.985, 1]$ and parameters $a = 10 + i$, $b = 3$, $c = 16 + 0.3i$. When the fractional-order $\alpha = 0.95$, the attractor trajectories are illustrated in Figure 2.

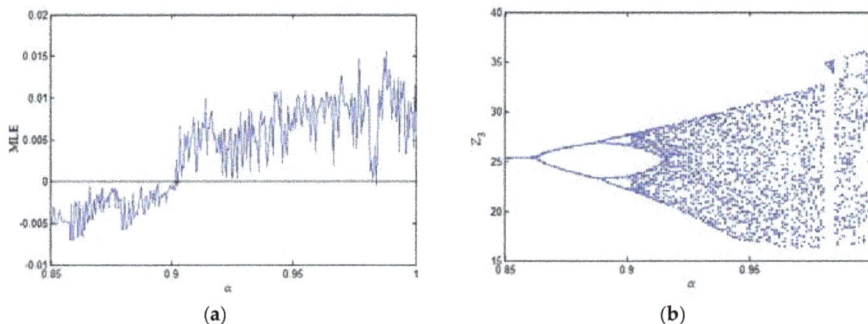

(a) (b)

Figure 1. Dynamic behaviors of the fractional-order complex Lorenz-like System with commensurate order α ($a = 10 + i$, $b = 3$, $c = 16 + 0.3i$). (a) maximal Lyapunov exponent; (b) bifurcation diagram.

Recently, ref. [42] described how to perform a successful simulation and optimization, and how to synthesize the mathematical models using CMOS technology. The application of metaheuristics to optimize MLE by varying the parameters of the oscillators was discussed.

Field-programmable gate array (FPGA)-based implementation of chaotic oscillators has demonstrated its usefulness in the development of engineering applications in a wide variety of fields, such as: random number generators, robotics and chaotic secure communication systems, signal processing. Very recently, Pano-Azucena et al. [43] implemented the chaotic system using a field-programmable gate array (FPGA) based on trigonometric polynomials. Reference [44] detailed the FPGA-based implementation of all the fractional order chaotic oscillators applying Grünwald-Letnikov(G-L) method. Their work proved experimentally that applying G-L method with 256 elements of memory; it can observe different families of Fractional-order chaotic attractors having working frequencies between 77.59 MHz and 84.9 MHz. This is very beneficial to the development of fractional-order chaos in engineering applications. For the FPGA-based implementation of FOCCS, the FOCCS was first separated into two real-variable systems according to their real and imaginary parts, and then these real-variable systems can be implemented using FPGA by the method proposed by. In order to find much better behavior and characteristics of the FOCCS in the complex domain, we used the G-L method to numerically solve the system (19) again. The MLE spectrum with varying parameter a^i (the imaginary of a) is depicted in Figure 3a, the bifurcation diagram is presented in Figure 3b, the state trajectories are illustrated in Figure 4.

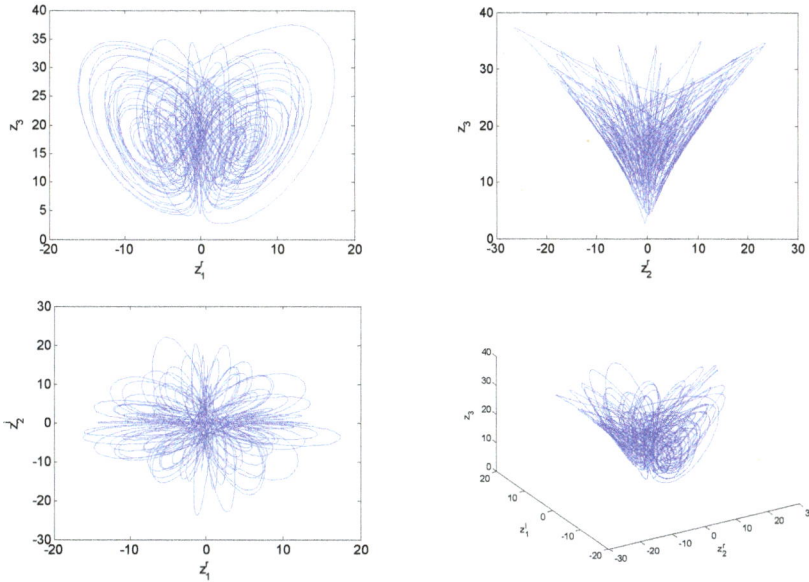

Figure 2. Chaotic attractors of fractional-order complex Lorenz-like system with $a = 10 + i$, $b = 3$, $c = 16 + 0.3i$ and commensurate order $\alpha = 0.95$.

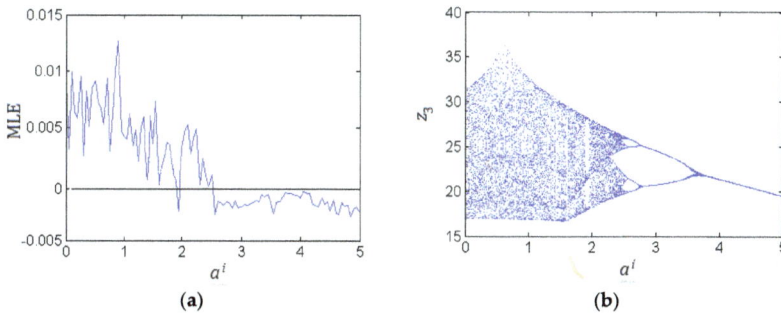

(a)   **(b)**

Figure 3. Dynamic behaviors of the fractional-order complex Lorenz-like System with commensurate order 0.95 ($a^r = 10$, $b = 3$, $c = 16 + 0.3i$). (**a**) maximal Lyapunov exponent; (**b**) bifurcation diagram.

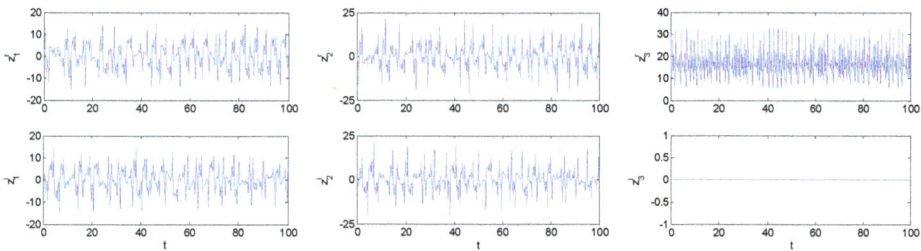

Figure 4. The state trajectories of fractional-order complex Lorenz-like system with $a = 10 + i$, $b = 3$, $c = 16 + 0.3i$ and commensurate order $\alpha = 0.95$.

Taking the system (19) as master system, and assuming the parameters a, b and c are unknown, the response system is given as follows:

$$\begin{pmatrix} {}^C_0 D^\alpha_t w_1 \\ {}^C_0 D^\alpha_t w_2 \\ {}^C_0 D^\alpha_t w_3 \end{pmatrix} = \begin{pmatrix} w_2 + w_1 & 0 & 0 \\ 0 & -w_2 & 0 \\ 0 & 0 & -w_3 \end{pmatrix} \begin{pmatrix} \hat{a} \\ \hat{c} \\ \hat{b} \end{pmatrix} + \begin{pmatrix} 0 \\ -w_1 w_3 \\ \overline{w}_1 w_1 \end{pmatrix} + \begin{pmatrix} u_1 \\ u_2 \\ u_3 \end{pmatrix} \qquad (20)$$

where $\hat{a}$, $\hat{b}$, $\hat{c}$ are parameter estimations. u_1, u_2, u_3 are the controller.

According to Theorem 1, the controllers and the update rules are selected as:

$$u_1 = -ke_1,\ u_2 = -ke_2,\ u_3 = -ke_3$$
$$ {}^C_0 D^\alpha_t k = \sigma e^H e = \sigma(\overline{e}_1 e_1 + \overline{e}_2 e_2 + \overline{e}_3 e_3),\ (\sigma > 0) \qquad (21)$$

$$\begin{pmatrix} {}^C_0 D^\alpha_t \hat{a} \\ {}^C_0 D^\alpha_t \hat{c} \\ {}^C_0 D^\alpha_t \hat{b} \end{pmatrix} = \begin{pmatrix} w_1 + w_2 & 0 & 0 \\ 0 & -w_2 & 0 \\ 0 & 0 & -w_3 \end{pmatrix}^H \begin{pmatrix} e_1 \\ e_2 \\ e_3 \end{pmatrix} = \begin{pmatrix} -(w_1 + w_2)e_1 \\ \overline{w}_2 e_2 \\ \overline{w}_3 e_3 \end{pmatrix} \qquad (22)$$

In the simulation, let $\alpha = 0.95$, $(a, b, c) = (30 + i, 3, 26 + 0.6i)$, the initial conditions $z(0) = (1 + i, -2 - i, 6)^T$, $w(0) = (-1 + i, -3 + i, 10)^T$, $(\hat{a}(0), \hat{b}(0), \hat{c}(0)) = (20, 2, 20)$, $k(0) = 0$ and $\sigma = 6$. Two systems can achieve synchronization and the parameters are identified, as shown in Figures 5 and 6.

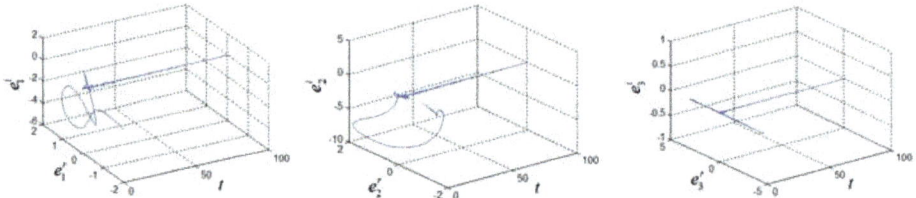

Figure 5. Synchronization errors e_1, e_2, e_3 of fractional-order complex Lorenz-like chaotic system.

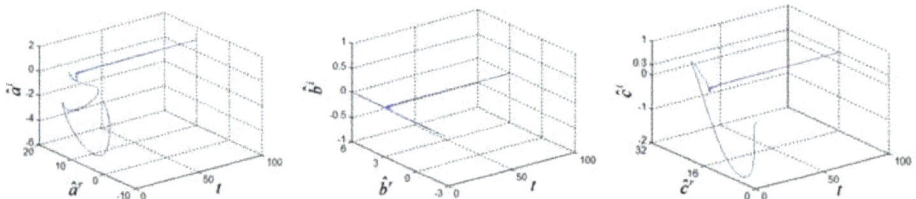

Figure 6. Estimated complex parameters of fractional-order complex Lorenz-like chaotic system.

5. Conclusions

We studied the adaptive synchronization of FOCCS with unknown complex parameters, and proposed a method for analyzing FOCCS without separating system into real and imaginary parts. By this method, the constructed response system can be asymptotically synchronized to an uncertain drive system with a desired complex scaling diagonal matrix. The proposed synchronization scheme retains the complex nature of fractional-order complex chaotic system. It not only provides a new method of analyzing FOCCS, but also significantly decreases the complexity of computation and analysis. We hope that the work performed will be helpful to further research of nonlinear fractional order complex-variable systems.

Author Contributions: Conceptualization, R.Z.; software, Y.L.; supervision, S.Y.

Entropy **2019**, *21*, 207

Funding: The author R.Z was supported by the Natural Science Foundation of Hebei province, China (No.A2015108010), and science and technology support program of Xingtai, China (No. 2016ZC191).

Conflicts of Interest: The authors declare no conflict of interest.

Appendix A

Proof of Lemma 3. By α-integrating (8), we have:

$$V(t) - V(0) \le -\theta I^\alpha V_1(t) = -\theta I^\alpha z^H(t)z(t) \le 0 \qquad (A1)$$

Thus, $V(t) \le V(0)$, $t \ge 0$. From (6), we can obtain that $V_1(t) = z^H(t)z(t)$ is bounded. Furthermore, since $h(z(t))$ satisfies the globally Lipschitz continuity condition, from (6), one has, $||{}_0^C D_t^\alpha z(t)||$
$= ||h(z(t))|| \le l||z^H(t)z(t))||$, i.e., $[\sum\limits_{i=1}^{n} (\overline{{}_0^C D_t^\alpha z_i(t)}\, {}_0^C D_t^\alpha z_i(t))]^{1/2} \le l[\sum\limits_{i=1}^{n} (\overline{z_i(t)}\, z_i(t))]^{1/2}$, where l is positive constant. Given that $V_1(t) = z^H(t)z(t)$ is bounded, we have $|z_i|$ is bounded, and then $|{}_0^C D_t^\alpha z_i(t)|$ is bounded. Thus, there exists a constant $M > 0$, such that $|{}_0^C D_t^\alpha z_i(t)| \le M$. For $0 \le t_1 < t_2$ and any $\varepsilon > 0$, if $|t_2 - t_1| < \delta(\varepsilon) = [\frac{\varepsilon \Gamma(\alpha+1)}{2M}]^{\frac{1}{\alpha}}$, one can get:

$$
\begin{aligned}
|z_i(t_1) - z_i(t_2)| &= |{}_0 D_t^{-\alpha}\, {}_0^C D_t^\alpha z_i(t_1) + z_i(0) - [{}_0 D_t^{-\alpha}\, {}_0^C D_t^\alpha z_i(t_2) + z_i(0)]| \\
&= |{}_0 D_t^{-\alpha}\, {}_0^C D_t^\alpha z_i(t_1) - {}_0 D_t^{-\alpha}\, {}_0^C D_t^\alpha z_i(t_2)| \\
&= \frac{1}{\Gamma(\alpha)} |\int_0^{t_1} [(t_1 - \tau)^{\alpha-1} - (t_2 - \tau)^{\alpha-1}]\, {}_0^C D_\tau^\alpha z_i(\tau) d\tau - \int_{t_1}^{t_2} (t_2 - \tau)^{\alpha-1}\, {}_0^C D_\tau^\alpha z_i(\tau) d\tau| \\
&\le \frac{M}{\Gamma(\alpha)} \{ \int_0^{t_1} [(t_1 - \tau)^{\alpha-1} - (t_2 - \tau)^{\alpha-1}] d\tau + \int_{t_1}^{t_2} (t_2 - \tau)^{\alpha-1} d\tau \} \\
&= \frac{M}{\Gamma(\alpha+1)} [(t_1^\alpha - t_2^\alpha) + 2(t_2 - t_1)^\alpha] \\
&\le \frac{2M}{\Gamma(\alpha+1)} (t_2 - t_1)^\alpha \le \varepsilon
\end{aligned}
$$

Hence, $z_i(t)$ is uniformly continuous. Therefore, $V_1(t) = z^H(t)z(t)$ is uniformly continuous. $\square$

Next, we adopt contradiction to prove $\lim\limits_{t\to\infty} V_1(t) = \lim\limits_{t\to\infty} z^H(t)z(t) = 0$ using the idea of Theorem 1 in Reference [45].

Suppose that $V_1(t) \ne 0$ as $t \to \infty$. Then there exists a monotone increasing sequence $(t_k)_{k\in N^+}$ ($t_k \to \infty$ as $k \to \infty$) and a positive constant $\varepsilon > 0$ such that $V_1(t_k) > \varepsilon$. Since the uniform continuity of $V_1(t)$, for the given ε, $\exists \delta > 0$ ($\delta \le \inf_{j\in N^+} \{t_{j+1} - t_j\}$, which implies that the intervals $[t_k, t_k + \delta]$ are nonoverlapping) such that $|V_1(t_k) - V_1(t)| \le \varepsilon/2$. Then, for any $t \in [t_k, t_k + \delta]$ it follows that $V_1(t) = V_1(t_k) - V_1(t_k) + V_1(t) \ge |V_1(t_k)| - |V_1(t_k) - V_1(t)| > \varepsilon/2$. Thus, for any $k = 1, 2, 3, \ldots$, from (A1), we have:

$$
\begin{aligned}
V(t_k + \delta) - V(0) &\le \frac{-\theta}{\Gamma(\alpha)} \int_0^{t_k+\delta} \frac{V_1(\tau)}{(t_k+\delta-\tau)^{1-\alpha}} d\tau \\
&= \frac{-\theta}{\Gamma(\alpha)} [\int_0^{t_1} + \int_{t_1}^{t_1+\delta} + \int_{t_1+\delta}^{t_2} + \int_{t_2}^{t_2+\delta} + \int_{t_2+\delta}^{t_3} + \int_{t_3}^{t_3+\delta} + \cdots + \int_{t_k}^{t_k+\delta}] \frac{V_1(\tau)}{(t_k+\delta-\tau)^{1-\alpha}} d\tau \\
&\le \frac{-\theta}{\Gamma(\alpha)} [\int_{t_1}^{t_1+\delta} + \int_{t_2}^{t_2+\delta} + \int_{t_3}^{t_3+\delta} + \cdots + \int_{t_k}^{t_k+\delta}] \frac{V_1(\tau)}{(t_k+\delta-\tau)^{1-\alpha}} d\tau \\
&\le \frac{-\theta\varepsilon}{2\Gamma(\alpha)} [\int_{t_1}^{t_1+\delta} + \int_{t_2}^{t_2+\delta} + \int_{t_3}^{t_3+\delta} + \cdots + \int_{t_k}^{t_k+\delta}] (t_k + \delta - \tau)^{\alpha-1} d\tau \\
&= \frac{-\theta\varepsilon}{2\Gamma(\alpha)} \sum_{j=1}^{k} \int_{t_j}^{t_j+\delta} (t_k + \delta - \tau)^{\alpha-1} d\tau
\end{aligned}
$$

Given that $(t_k + \delta - \tau)^{\alpha-1} \geq (t_k + \delta - t_j)^{\alpha-1}$ for all $\tau \in [t_j, t_j + \delta]$ results in:

$$V(t_k + \delta) - V(0) \leq \frac{-\theta\varepsilon}{2\Gamma(\alpha)} \sum_{j=1}^{k} \int_{t_j}^{t_j+\delta} (t_k + \delta - \tau)^{\alpha-1}\, d\tau$$

$$\leq \frac{-\theta\varepsilon}{2\Gamma(\alpha)} \sum_{j=1}^{k} \int_{t_j}^{t_j+\delta} (t_k + \delta - t_j)^{\alpha-1}\, d\tau$$

$$= \frac{-\theta\varepsilon\delta}{2\Gamma(\alpha)} \sum_{j=1}^{k} (t_k + \delta - t_j)^{\alpha-1}$$

$$\leq \frac{-\theta\varepsilon\delta}{2\Gamma(\alpha)} \sum_{j=1}^{k} (t_k + \delta - t_1)^{\alpha-1}$$

$$= \frac{-\theta\varepsilon\delta}{2\Gamma(\alpha)} \frac{k}{(t_k + \delta - t_1)^{1-\alpha}}$$

$$\leq \frac{-\theta\varepsilon\delta}{2\Gamma(\alpha)} \frac{k}{(kd)^{1-\alpha}} = \frac{-\theta\varepsilon\delta}{2\Gamma(\alpha)} \frac{k^{\alpha}}{(d)^{1-\alpha}}$$

where $d = \sup_{j \in N, 2 \leq j \leq k} \{t_j - t_{j-1}\}$ (since $V_1(t)$ is a uniformly continuous function and assumed $V_1(t) \neq 0$, as $t \to \infty$, d is bounded). Obviously, $V(t_k + \delta) - V(0) \leq \frac{-\theta\varepsilon\delta}{2\Gamma(\alpha)} \frac{k^{\alpha}}{(d)^{1-\alpha}} \to -\infty$ as $k \to \infty$, which contradict with $V(t) \geq 0$. Therefore, $\lim_{t\to\infty} V_1(t) = \lim_{t\to\infty} z^H(t)z(t) = 0$. i.e., $z(t) = 0$ is asymptotically stable. This completes the proof for Lemma 3.

References

1. Gorenlo, R.; Mainardi, F. *Fractional Calculus*; Springer: Berlin, Germany, 1997.
2. Caponetto, R. *Fractional Order Systems: Modeling and Control Applications*; World Scientiic: Singapore, 2010.
3. Magin, R.L. *Fractional Calculus in Bioengineering*; Begell House Redding: New York, NY, USA, 2006.
4. Radwan, A.G.; Shamim, A.; Salama, K.N. Theory of fractional order elements based impedance matching networks. *IEEE Microwave Wirel. Comp. Lett.* **2011**, *21*, 120–122. [CrossRef]
5. Radwan, A.G.; Abd-El-Haiz, S.K.; Abdelhaleem, S.H. Image encryption in the fractional-order domain. In Proceedings of the 1st International Conference on Engineering and Technology, Cairo, Egypt, 10–11 October 2012.
6. Kiani-B, A.; Fallah, K.; Pariz, N.; Leung, H. A chaotic secure communication scheme using fractional chaotic systems based on an extended fractional Kalman filter. *Commun. Nonlinear Sci. Numer. Simul.* **2009**, *14*, 863–879. [CrossRef]
7. Li, G.J.; Liu, H. Stability analysis and synchronization for a class of fractional-order neural networks. *Entropy* **2016**, *18*, 55. [CrossRef]
8. Yin, C.; Cheng, Y.; Zhong, S.M.; Bai, Z. Fractional-order switching type control law design for adaptive sliding mode technique of 3D fractional-order nonlinear systems. *Complexity* **2016**, *21*, 363–373. [CrossRef]
9. Liu, H.; Li, S.G.; Wang, H.X.; Huo, Y.H.; Luo, J.H. Adaptive synchronization for a class of uncertain fractional-order neural networks. *Entropy* **2015**, *17*, 7185–7200. [CrossRef]
10. Lu, J.G. Chaotic dynamics of the fractional-order Lü system and its synchronization. *Phys. Lett. A* **2006**, *354*, 305–311. [CrossRef]
11. Li, C.G.; Chen, G.R. Chaos in the fractional-order Chen system and it's control. *Chaos Solitons Fractals* **2004**, *22*, 549–554. [CrossRef]
12. Agrawal, S.K.; Srivastava, M.; Das, S. Synchronization of fractional order chaotic systems using active control method. *Chaos Solitons Fractals* **2012**, *45*, 737–752. [CrossRef]
13. Chen, L.P.; Chai, Y.; Wu, R.C. Lag projective synchronization in fractional-order chaotic (hyperchaotic) systems. *Phys. Lett. A* **2011**, *375*, 2099–2110. [CrossRef]
14. Zhang, Y.; Sun, J. Chaotic synchronization and anti-synchronization based on suitable separation. *Phys. Lett. A* **2004**, *330*, 442–447. [CrossRef]

15. Zhang, R.X.; Yang, S.P. Robust synchronization of two different fractional-order chaotic systems with unknown parameters using adaptive sliding mode approach. *Nonlinear Dyn.* **2013**, *71*, 269–278. [CrossRef]
16. Zhang, R.X.; Yang, S.P. Adaptive synchronization of fractional-order chaotic systems via a single driving variable. *Nonlinear Dyn.* **2011**, *66*, 831–837. [CrossRef]
17. Zhang, R.X.; Yang, S.P. Robust chaos synchronization of fractional-order chaotic systems with unknown parameters and uncertain perturbations. *Nonlinear Dyn.* **2012**, *69*, 983–992. [CrossRef]
18. Wang, S.; Wang, X.; Han, B. Complex generalized synchronization and parameter identification of nonidentical nonlinear complex systems. *PLoS ONE* **2016**, *11*, e0152099. [CrossRef] [PubMed]
19. Jiang, C.; Zhang, F.; Li, T. Synchronization and anti-synchronization of N-coupled fractional-order complex chaotic systems with ring connection. *Math. Meth. Appl. Sci.* **2018**, *41*, 2625–2638. [CrossRef]
20. Zhang, F.F.; Liu, S. Self-time-delay synchronization of time-delay coupled complex chaotic system and its applications to communication. *Int. J. Mod. Phys. C* **2014**, *25*, 559–583. [CrossRef]
21. Luo, C.; Wang, X.Y. Chaos in the fractional-order complex Lorenz system and its synchronization. *Nonlinear Dyn.* **2013**, *71*, 241–257. [CrossRef]
22. Luo, C.; Wang, X. Chaos generated from the fractional-order complex Chen system and its application to digital secure communication. *Int. J. Mod. Phys. C* **2013**, *24*, 72–77. [CrossRef]
23. Liu, X.J.; Hong, L.; Yang, L.X. Fractional-order complex T system: Bifurcations, chaos control, and synchronization. *Nonlinear Dyn.* **2014**, *75*, 589–602. [CrossRef]
24. Zhang, W.W.; Cao, J.D.; Chen, D.Y.; Alsaadi, F.E. Synchronization in fractional-order complex-valued delayed neural networks. *Entropy* **2018**, *20*, 54. [CrossRef]
25. Li, L.; Wang, Z.; Lu, J.W.; Li, Y.X. Adaptive synchronization of fractional-order complex-valued neural networks with discrete and distributed delays. *Entropy* **2018**, *20*, 124. [CrossRef]
26. Sun, J.; Deng, W.; Cui, G.Z.; Wang, Y.F. Real combination synchronization of three fractional-order complex-variable chaotic systems. *Optik* **2016**, *127*, 11460–11468. [CrossRef]
27. Yadav, V.K.; Srikanth, N.; Das, S. Dual function projective synchronization of fractional order complex chaotic systems. *Optik* **2016**, *127*, 10527–10538. [CrossRef]
28. Nian, F.Z.; Liu, X.M.; Zhang, Y. Sliding mode synchronization of fractional-order complex chaotic system with parametric and external disturbances. *Chaos Solitons Fractals* **2018**, *116*, 22–28. [CrossRef]
29. Jiang, C.M.; Liu, S.T.; Zhang, F.F. Complex modified projective synchronization for fractional-order chaotic complex systems. *Int. J. Auto Comput.* **2017**, *15*, 603–615. [CrossRef]
30. Podlubny, I. *Fractional Differential Equations*; Academic Press: New York, NY, USA, 1999.
31. Aguila-Camacho, N.; Duarte-Mermoud, M.A.; Gallegos, J.A. Lyapunov functions for fractional order systems. *Commun. Nonlinear Sci. Numer. Simulat.* **2014**, *19*, 2951–2957. [CrossRef]
32. Quan, X.; Zhuang, S.; Liu, S.; Xiao, J. Decentralized adaptive coupling synchronization of fractional-order complex-variable dynamical networks. *Neurocomputing* **2016**, *186*, 119–126.
33. Zhang, R.X.; Liu, Y. A new Barbalat's Lemma and Lyapunov stability theorem for fractional order systems. In Proceedings of the 29th Chinese Control and Decision Conference, Chongqing, China, 28–30 May 2017; pp. 3676–3681.
34. Wang, F.; Yang, Y.Q. Fractional order Barbalat's Lemma and its applications in the stability of fractional order nonlinear systems. *Math. Model. Anal.* **2017**, *22*, 503–513. [CrossRef]
35. Wang, C. Adaptive fuzzy control for uncertain fractional-order financial chaotic systems subjected to input saturation. *PLoS ONE* **2016**, *11*, e0164791. [CrossRef]
36. Xu, Q.; Xu, X.H.; Zhuang, S.X.; Xiao, J.X.; Song, C.H. New complex projective synchronization strategies for drive-response networks with fractional complex-variable dynamics. *Appl. Math. Comput.* **2018**, *338*, 552–566. [CrossRef]
37. Tao, F.; Sun, J.T. Stability of complex-valued impulsive system with delay. *Appl. Math. Comput.* **2014**, *240*, 102–108.
38. Wu, Z.Y.; Chen, G.R.; Fu, X.C. Synchronization of a network coupled with complex-variable chaotic systems. *Chaos* **2012**, *22*, 023127. [CrossRef]
39. Gu, Y.; Yu, Y.; Wang, H. Synchronization-based parameter estimation of fractional-order neural networks. *Physica A Stat. Mech. Appl.* **2017**, *483*, 351–361. [CrossRef]
40. Diethelm, K.; Neville, J.F.; Freed, A.D. A predictor-corrector approach for the numerical solution of fractional differential equations. *Nonlinear Dyn.* **2002**, *29*, 1–4. [CrossRef]

41. Wolf, A.; Swift, J.B.; Swinney, H.L.; Vastano, J.A. Determining Lyapunov exponents from a time series. *Physica D* **1985**, *16*, 285–317. [CrossRef]

42. Carbajal-Gomez, V.H.; Tlelo-Cuautle, E.; Muñoz-Pacheco, J.M.; de la Fraga, L.G.; Sanchez-Lopez, C.; Fernandez-Fernandez, F.V. Optimization and CMOS design of chaotic oscillators robust to PVT variations: INVITED (In Press). *Integration* **2018**. [CrossRef]

43. Pano-Azucena, A.D.; Tlelo-Cuautle, E.; Rodriguez-Gomez, G. de la Fraga, L.G. FPGA-based implementation of chaotic oscillators by applying the numerical method based on trigonometric polynomials. *AIP Adv.* **2018**, *8*, 075217. [CrossRef]

44. Pano-Azucena, A.D.; Ovilla-Martinez, B.; Tlelo-Cuautle, E.; Muñoz-Pacheco, J.M.; de la Fraga, L.G. FPGA-based implementation of different families of fractional-order chaotic oscillators applying Grounwald-Letnikov method. *Commun. Nonlinear Sci. Numer. Simulat.* **2019**, *72*, 516–527. [CrossRef]

45. Bao, H.; Park, J.H.; Cao, J. Adaptive synchronization of fractional-order memristor-based neural networks with time delay. *Nonlinear Dyn.* **2015**, *82*, 1343–1354. [CrossRef]

![entropy]

Article

Chaotic Map with No Fixed Points: Entropy, Implementation and Control

Van Van Huynh [1], Adel Ouannas [2], Xiong Wang [3], Viet-Thanh Pham [4,*], Xuan Quynh Nguyen [5] and Fawaz E. Alsaadi [6]

[1] Modeling Evolutionary Algorithms Simulation and Artificial Intelligence, Faculty of Electrical and Electronics Engineering, Ton Duc Thang University, Ho Chi Minh City, Vietnam; huynhvanvan@tdtu.edu.vn
[2] Department of Mathematics and Computer Science, University of Larbi Tebessi, Tebessa 12002, Algeria; adel.ouannas@yahoo.com
[3] Institute for Advanced Study, Shenzhen University, Shenzhen 518060, China; wangxiong8686@szu.edu.cn
[4] Nonlinear Systems and Applications, Faculty of Electrical and Electronics Engineering, Ton Duc Thang University, Ho Chi Minh City, Vietnam
[5] National Council for Science and Technology Policy, Hanoi, Vietnam; Quynhnx@hactech.edu.vn
[6] Department of Information Technology, Faculty of Computing and IT, King Abdulaziz University, Jeddah 21589, Saudi Arabia; fawazkau@gmail.com
* Correspondence: phamvietthanh@tdtu.edu.vn

Received: 29 January 2019; Accepted: 12 March 2019; Published: 14 March 2019

Abstract: A map without equilibrium has been proposed and studied in this paper. The proposed map has no fixed point and exhibits chaos. We have investigated its dynamics and shown its chaotic behavior using tools such as return map, bifurcation diagram and Lyapunov exponents' diagram. Entropy of this new map has been calculated. Using an open micro-controller platform, the map is implemented, and experimental observation is presented. In addition, two control schemes have been proposed to stabilize and synchronize the chaotic map.

Keywords: chaotic map; fixed point; chaos; approximate entropy; implementation

1. Introduction

Discrete maps have attracted significant attention in the study of dynamical systems [1–4]. Discrete maps appear in various disciplines including physiology, chemistry, physics, ecology, social sciences and engineering [3,5–7]. It has previously been observed that simple first-order nonlinear maps can generate complex dynamical behavior including chaos [8]. Chaotic maps such as Hénon map [9], Logistic map [8], Lozi map [10], and zigzag map [11] are found. When investigating chaotic maps, the stability of fixed point plays a vital role. The authors tried to find fixed points and studied the behavior of orbits near fixed points. Relation of fixed points and critical transitions is illustrated in [12]. Previous studies have established that conventional chaotic maps often have unstable fixed points.

More recent studies have focused on chaotic maps related to the hidden attractor category [13–15]. Hidden attractors in chaotic maps are reported in [16], where a 1D map with no fixed point is extended from Logistic map. Jiang et al. introduced a list of two-dimensional maps with no fixed point [17]. These maps are inspired by Hénon map. By applying a Jerk-like structure, a gallery of 3D maps having hidden dynamics is investigated [17]. Ouannas proposed a fractional map having no fixed point [18]. Xu et al. found hidden dynamics of a two-dimensional map based on Arnold's cat map [19]. The authors built a hardware implementation of the map using Field-programmable gate array (FPGA). However, detailed investigation of chaotic maps without fixed point should be examined further.

Our work discovers a new no equilibrium map with chaos. In Section 2, the map's model is introduced, and its dynamics is reported. Realization of the map in an Arduino Uno board is presented in Section 3. In Section 4, control approaches for such a map are designed. Section 5 summarizes our work.

2. Chaotic Map

By using nonlinear functions, we construct a map described by:

$$\begin{cases} x\,(n+1) = x\,(n) + y\,(n)\,, \\ y\,(n+1) = y\,(n) - a\,|y\,(n)| - x\,(n)\,y\,(n) + b(x\,(n))^2 - c(y\,(n))^2 + d, \end{cases} \tag{1}$$

where a, b, c, and d are positive parameters.

The fixed points $E(x,y)$ of the map can be found by solving

$$\begin{cases} x = x + y, \\ y = y - a\,|y| - xy + bx^2 - cy^2 + d. \end{cases} \tag{2}$$

By rewriting Equation (2), we have

$$bx^2 + d = 0. \tag{3}$$

Therefore, there is no any fixed point in the map in Equation (1) for such positive parameters b and d.

We observe chaos in the map for $a = 0.01$, $b = 0.1$, $c = 2$, $d = 0.1$ and the initial conditions $(x(0), y(0)) = (1.5, 0.5)$ (see Figure 1). Similar to the reported map in [18], the map in Equation (1) belongs to a class of maps without fixed point. Compared with the map reported in [18], the map in Equation (1) is not a fractional one.

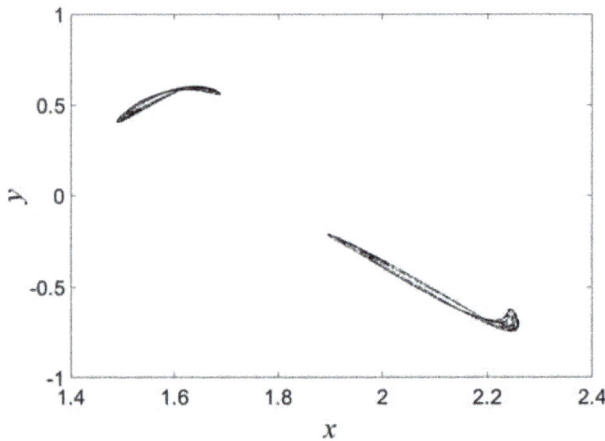

Figure 1. Strange attractor of the map for $a = 0.01$, $b = 0.1$, $c = 2$, $d = 0.1$ and $(x(0), y(0)) = (1.5, 0.5)$.

2.1. Dynamics of the Map

Dynamics of the proposed map were studied. It was found that the map displays interesting dynamics when varying the parameter c and keeping $a = 0.01$, $b = 0.1$, $d = 0.1$ and $(x(0), y(0)) = (1.5, 0.5)$. Note that, since we wanted to keep the system NE (no equilibrium), we have frizzed the parameters b and d. Changing parameter a as bifurcation parameter did not show a proper route to chaos and in some values resulted in unbounded solutions. Thus, we chose c as the bifurcation parameter. In addition, note that the initial condition used in our simulations was not dominant and affected only the initial transient regime. As seen in the bifurcation diagram (Figure 2a) and the finite-time local Lyapunov exponents (Figure 2b), the map in Equation (1) displays a period doubling route to chaos. The time interval for calculating finite-time local Lyapunov exponents [20] is 10,000. Since it has no equilibrium, it has no period-1 cycle. The bifurcation starts from a period-2 cycle. Then, it continues with period-doubling until chaos is born a little before $c = 2$.

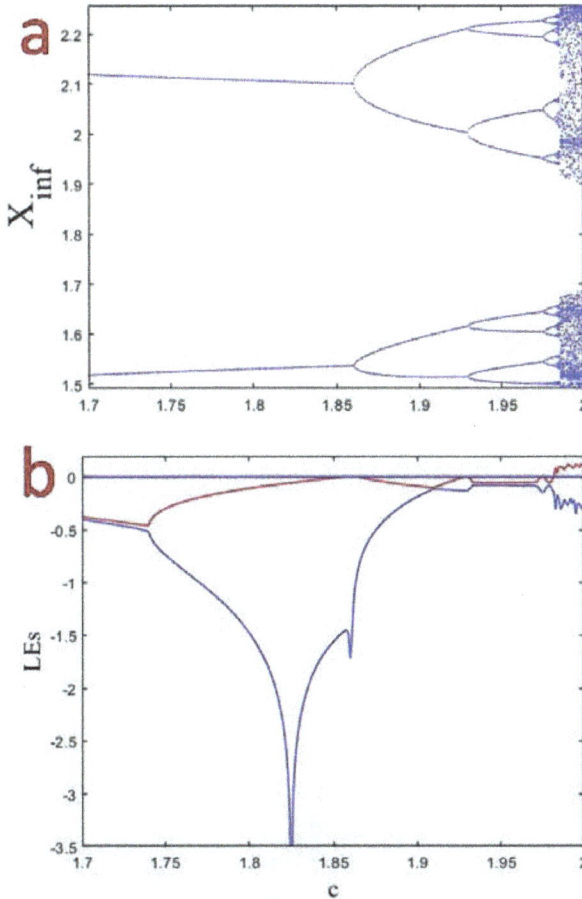

Figure 2. Bifurcation diagram (**a**); and Lyapunov exponents (**b**) when varying c for $a = 0.01$, $b = 0.1$, $d = 0.1$ and $(x(0), y(0)) = (1.5, 0.5)$.

2.2. Entropy of the New Map

Previous research has established that entropy is an effective index for estimating information in a particular system [21–23]. The authors applied entropy measurement to consider the complexity/chaos of dynamical systems [24–27]. In particular, approximate entropy (ApEn) [28,29] is useful to study chaotic systems [19,30]. It is noted that there is no reported threshold to be achieve in the ApEn in order to exhibit chaos [28,29]. Xu et al. reported the ApEn of a new system with chaos [19]. Their values of ApEn ranged from 0 to 0.12. Wang and Ding presented a table of AnEn test for four chaotic maps [30]. Here, calculated approximate entropy (ApEn) for the proposed the map in Equation (1) is reported in Table 1. Obtained entropy in Table 1 illustrates the complexity of the map when it exhibits chaos.

Table 1. Calculated approximate entropy of the map in Equation (1) for $a = 0.01$, $b = 0.1$, $d = 0.1$ and $(x(0), y(0)) = (1.5, 0.5)$.

Case	c	ApEn
1	1.985	0.0306
2	1.99	0.2142
3	1.995	0.2184
4	2	0.2525

3. Implementation of the Map Using Microcontroller

Chaotic maps are useful for designing pseudorandom number generators [31–34], building S-Box [35], proposing color image encryption [36] or constructing secure communication [37]. Therefore, implementation of chaotic maps is a practical topic in the literature. Some approaches have been used to realize theoretical models of chaotic maps. Valtierra et al employed a skew-tent map in switched-capacitor circuits [6]. Bernoulli shift map, Borujeni maps, zigzag, and tent are done with a field-programmable gate array architecture [7]. Wang and Ding introduced FPGA hardware implementation of a map with hidden attractors [30]. It is worth noting that using microcontroller is an effective approach to implement chaotic maps [37,38]. The open-source platform named Arduino provides a reasonable development tool because of its free development software [39–41]. In our work, we used an Arduino Uno board based on microcontroller to realize the proposed map in Equation (1), as shown in Figure 3. Pins 9 and 10 of the Arduino Uno board are configured as two digital outputs. However, we could choose different pins for digital outputs because Arduino Uno board has 14 digital pins. We wrote a program for the map in the Arduino development environment. It is noted that the algorithm steps and program structure used in our implementation are similar to those reported in [38]. The output pin 9 was activated when $x > 1.8$ while the output pin 10 was activated when $y > 0$. Figure 4 displays the experimental waveforms at pins 9 and 10.

Figure 3. Arduino Uno board for implementing chaotic the map in Equation (1).

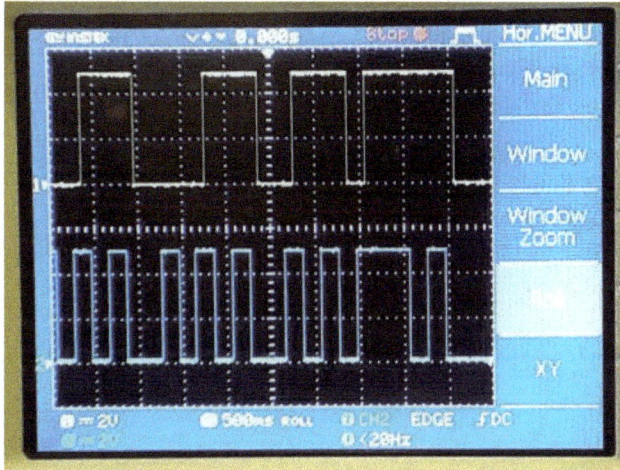

Figure 4. Captured waveforms at pins 9 and 10 of the Arduino Uno board.

4. Control Schemes for the Proposed Map

When investigating chaotic maps, stabilization and synchronization are vital aspects. Two control laws for stabilizing and synchronizing the proposed non-fixed-point map are introduced in this section.

4.1. Stabilization

The aim of stabilizing the proposed map is to devise an adaptive control law such that all system states are stabilized to 0. The controlled map is

$$
\begin{cases}
x(n+1) = x(n) + y(n) + u_x, \\
y(n+1) = y(n) - a|y(n)| - x(n)y(n) + bx^2(n) - cy^2(n) + d + u_y,
\end{cases}
\tag{4}
$$

where u_x and u_y are controllers to be determined.

The map in Equation (4) can be stabilized with the control law in Equation (5)

$$
\begin{cases}
u_x = -\frac{1}{2}x(n), \\
u_y = -\frac{1}{2}y(n) + a|y(n)| + x(n)y(n) - bx^2(n) + cy^2(n) - d
\end{cases}
\tag{5}
$$

Substituting the control law in Equation (5) into Equation (4), we get

$$
\begin{cases}
x(n+1) = \frac{1}{2}x(n) + y(n), \\
y(n+1) = \frac{1}{2}y(n).
\end{cases}
\tag{6}
$$

The written form of the error system in Equation (6) is

$$
(x(n+1), y(n+1))^T = \mathbf{M} \times (x(n), y(n))^T,
\tag{7}
$$

where

$$
\mathbf{M} = \begin{pmatrix} \frac{1}{2} & 1 \\ 0 & \frac{1}{2} \end{pmatrix}.
\tag{8}
$$

Therefore, the map in Equation (1) is stabilized.

We illustrated the result by selecting parameters $(a, b, c, d) = (0.01, 0.1, 2, 0.1)$ and $(x(0), y(0)) = (1.5, 0.5)$. In Figure 5, the evolution of states verifies the control law.

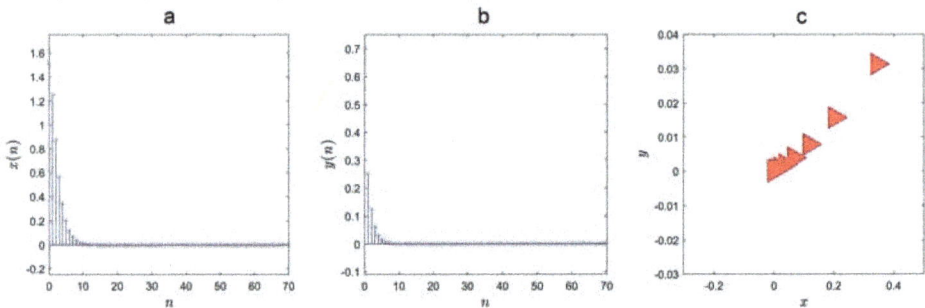

Figure 5. Stabilization when applying the proposed control law: (**a**) $x(n)$, (**b**) $y(n)$, and (**c**) $x - y$ plane.

4.2. Synchronization

Researchers have discovered synchronization of discrete systems [42–44]. We consider the drive system in Equation (9)

$$\begin{cases} x_m(n+1) = y_m(n), \\ y_m(n+1) = x_m(n) + a_1 x_m^2(n) + a_2 y_m^2(n) - a_3 x_m(n) y_m(n) - a_4, \end{cases} \tag{9}$$

It has been shown in [17] that the map in Equation (9) exhibits chaotic behaviors with no fixed points. The map in Equation (9) is one of the first example of discrete-time systems without fixed points, i.e, the map in Equation (9) has hidden attractors. The map in Equation (9) is inspired by the well-known Hénon map.

The subscript s denotes the response system's states. The response is given by

$$\begin{cases} x_s(n+1) = x_s(n) + y_s(n), \\ y_s(n+1) = y_s(n) - a|y_s(n)| - x_s(n) y_s(n) + b x_s^2(n) - c y_s^2(n) + d, \end{cases} \tag{10}$$

where $u_i(t)$ $(i = 1, 2)$ are synchronization controllers.

The error system is

$$\begin{aligned} e_1(n) &= x_s(n) - x_m(n), \\ e_2(n) &= y_s(n) - y_m(n), \end{aligned} \tag{11}$$

We find the controllers u_1 and u_2 based on Theorem 1.

Theorem 1. *By selecting*

$$\begin{cases} u_1 = -\frac{1}{2}x_s(n) - \frac{1}{2}x_m(n) - \frac{2}{3}y_s(n) + \frac{2}{3}y_m(n), \\ u_2 = \frac{1}{3}x_s(n) - \frac{2}{3}x_m(n) - \frac{3}{2}y_s(n) + \frac{1}{2}y_m(n) \\ \qquad a|y_s(n)| + x_s(n) y_s(n) - b x_s^2(n) + c y_s^2(n) - d \\ \qquad + a_1 x_m^2(n) + a_2 y_m^2(n) - a_3 x_m(n) y_m(n) - a_4, \end{cases} \tag{12}$$

the drive system in Equation (9) and the response system in Equation (10) are synchronized.

Proof. The error system in Equation (11) is rewritten as

$$\begin{cases} e_1\,(n+1) = x_s\,(n) + y_s\,(n) - y_m\,(n) + u_1, \\ e_2\,(n+1) = y_s\,(n) - a\,|y_s\,(n)| - x_s\,(n)\,y_s\,(n) + bx_s^2\,(n) - cy_s^2\,(n) + d \\ \qquad - x_m\,(n) - a_1 x_m^2\,(n) - a_2 y_m^2\,(n) + a_3 x_m\,(n)\,y_m\,(n) + a_4 + u_2, \end{cases} \tag{13}$$

Substituting the control law in Equation (12) into Equation (13) yields the reduced dynamics

$$\begin{cases} e_1\,(n+1) = \tfrac{1}{2}e_1\,(n) + \tfrac{1}{3}e_2\,(n), \\ e_2\,(n+1) = \tfrac{1}{3}e_1\,(n) - \tfrac{1}{2}e_2\,(n). \end{cases} \tag{14}$$

The Lyapunov function is $V\,(e_1(n), e_2(n)) = e_1^2(n) + e_2^2(n)$,

$$\begin{aligned} \Delta V &= V\,(e_1(n+1), e_2(n+1)) - V\,(e_1(n), e_2(n)) \\ &= \frac{1}{4}e_1^2\,(n) + \frac{1}{3}e_1\,(n)\,e_2\,(n) + \frac{1}{9}e_2^2\,(n) \\ &\quad \frac{1}{4}e_1^2\,(n) - \frac{1}{3}e_1\,(n)\,e_2\,(n) + \frac{1}{9}e_2^2\,(n) - e_1^2\,(n) - e_2^2\,(n) \\ &= -\frac{1}{2}e_1^2\,(n) - \frac{7}{9}e_2^2\,(n) < 0. \end{aligned}$$

By means of Lyapunov stability theory, the maps in Equations (9) and (10) are synchronized. □

Figure 6 depicts the time evolution of states of systems in Equations (9) and (10) after control. As reported in Figure 7, synchronization is obtained.

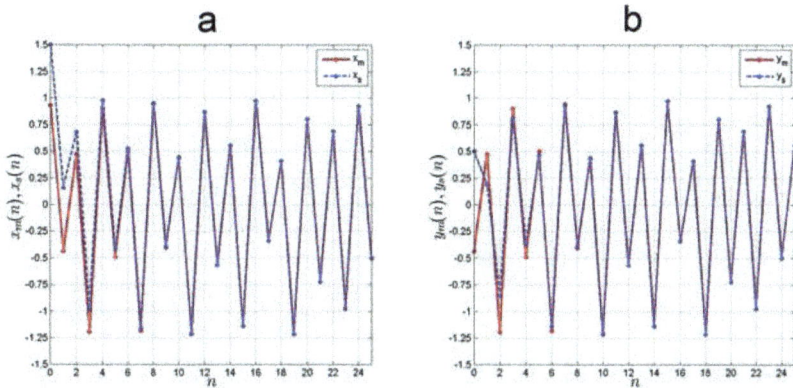

Figure 6. Evolution of states when applying the control: (**a**) $x_m(n)$, $x_s(n)$ and (**b**) $y_m(n)$, $y_s(n)$.

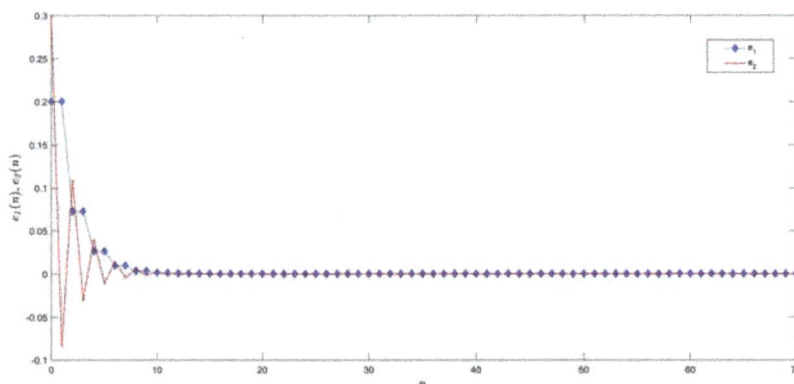

Figure 7. Synchronization errors.

5. Conclusions

This work has introduced a new chaotic map, which can be considered as a system with hidden attractor. Having no fixed point is a notable feature of the proposed map. Chaos in the map is observed and confirmed by positive Lyapunov exponent. Realization of the map using an open-source electronic platform is given to illustrate its feasibility. Experimental results are recorded and displayed by oscilloscope. Approximate entropy is calculated to determine the complexity of the map. We have also presented stabilization and synchronization for the map. In future research, this map will be embedded into practical applications such as data encryption, signal transmission or motion planning.

Author Contributions: Conceptualization, V.V.H.; Formal analysis, A.O.; Investigation, V.V.H. and X.Q.N.; Methodology, A.O. and V.-T.P.; Project administration, V.-T.P.; Resources, X.W.; Software, X.W. and F.E.A.; Supervision, V.-T.P. and F.E.A.; Validation, V.V.H.; Visualization, A.O. and X.Q.N.; Writing—original draft, X.W. and X.Q.N.; and Writing—review and editing, F.E.A.

Funding: Xiong Wang was supported by the National Natural Science Foundation of China (No. 61601306) and Shenzhen Overseas High Level Talent Peacock Project Fund (No. 20150215145C).

Conflicts of Interest: The authors declare no conflict of interest.

References

1. Pierre, C.; Jean-Pierre, E. *Iterated Map on the Interval as Dynamical Systems*; Springer: Berlin, Germany, 1980.
2. Bahi, J.M.; Guyeux, C. *Iterated Map on the Interval as Dynamical Systems*; CRC Press: Boca Raton, FL, USA, 2013.
3. Elaydi, S.N. *Discrete Chaos: With Applications in Science and Engineering*, 2nd ed.; Chapman and Hall/CRC: Boca Raton, FL, USA, 2007.
4. Gibson, W.T.; Wilson, W.C. Individual-based chaos: Extensions of the discrete logistic model. *J. Theor. Biol.* **2013**, *339*, 84–92. [CrossRef] [PubMed]
5. Borujeni, S.; Ehsani, M. Modified logistic maps for cryptographic application. *Appl. Math.* **2015**, *6*, 773–782. [CrossRef]
6. Valtierra, J.L.; Tlelo-Cuautle, E.; Rodriguez-Vazquez, A. A switched-capacitor skew-tent map implementation for random number generation. *Int. J. Circuit Theor. Appl.* **2017**, *45*, 305–315. [CrossRef]
7. De la Fraga, L.G.; Torres-Perez, E.; Tlelo-Cuautle, E.; Mancillas-Lopez, C. Hardware implementation of pseudo-random number generators based on chaotic maps. *Nonlinear Dyn.* **2017**, *90*, 1661–1670. [CrossRef]
8. May, R.M. Simple mathematical models with very complicated dynamics. *Nature* **1976**, *261*, 459–467. [CrossRef]
9. Hénon, M.A. A two-dimensional mapping with a strange attractor. *Commun. Math. Phys.* **1976**, *50*, 69–77. [CrossRef]

10. Lozi, R. Un atracteur étrange du type attracteur de Hénon. *J. Phys.* **1978**, *39*, 9–10.

11. Nejati, H.; Beirami, A.; Ali, W. Discrete-time chaotic-map truly random number generators: Design, implementation, and variability analysis of the zigzag map. *Analog Integr. Circuits Signal Process.* **2012**, *73*, 363–374. [CrossRef]

12. Scheffer, M.; Bascompte, J.; Brock, W.A.; Brovkin, V.; Carpenter, S.R.; Dakos, V.; Held, H.; van Nes, E.H.; Rietkerk, M.; Sugihara, G. Early-warning singals for critical transitions. *Nature* **2009**, *461*, 53–59. [CrossRef]

13. Leonov, G.A.; Kuznetsov, N.V.; Kuznetsova, O.A.; Seldedzhi, S.M.; Vagaitsev, V.I. Hidden oscillations in dynamical systems. *Trans. Syst. Control* **2011**, *6*, 54–67.

14. Leonov, G.A.; Kuznetsov, N.V. Hidden attractors in dynamical systems: From hidden oscillation in Hilbert-Kolmogorov, Aizerman and Kalman problems to hidden chaotic attractor in Chua circuits. *Int. J. Bifurc. Chaos* **2013**, *23*, 1330002. [CrossRef]

15. Dudkowski, D.; Jafari, S.; Kapitaniak, T.; Kuznetsov, N.; Leonov, G.; Prasad, A. Hidden attractors in dynamical systems. *Phys. Rep.* **2016**, *637*, 1–50. [CrossRef]

16. Jafari, S.; Pham, V.T.; Moghtadaei, M.; Kingni, S.T. The relationship between chaotic maps and some chaotic systems with hidden attractors. *Int. J. Bifurc. Chaos* **2016**, *26*, 1650211. [CrossRef]

17. Jiang, H.; Liu, Y.; Wei, Z.; Zhang, L. Hidden chaotic attractors in a class of two-dimensional maps. *Nonlinear Dyn.* **2016**, *85*, 2719–2727. [CrossRef]

18. Ouannas, A.; Wang, X.; Khennaoui, A.A.; Bendoukha, S.; Pham, V.T.; Alsaadi, F. Fractional form of a chaotic map without fixed points: Chaos, entropy and control. *Entropy* **2018**, *20*, 720. [CrossRef]

19. Xu, G.; Shekofteh, Y.; Akgul, A.; Li, C.; Panahi, S. New chaotic system with a self-excited attractor: Entropy measurement, signal encryption, and parameter estimation. *Entropy* **2018**, *20*, 86. [CrossRef]

20. Kuznetsov, N.V.; Leonov, G.A.; Mokaev, T.N.; Prasad, A.; Shrimali, M.D. Finite–time Lyapunov dimension and hidden attractor of the Rabinovich system. *Nonlinear Dyn.* **2018**, *92*, 267–285. [CrossRef]

21. Borda, M. *Fundamentals in Information Theory and Coding*; Springer: Berlin, Germany, 2011.

22. Gray, R.M. *Entropy and Information Theory*; Springer: Berlin, Germany, 2011.

23. Bossomaier, T.; Barnett, L. *An Introduction to Transfer Entropy: Information Flow in Complex Systems*; Springer: Berlin, Germany, 2016.

24. Eckmann, J.; Ruelle, D. Ergodic theory of chaos and strange attractors. *Rev. Mod. Phys.* **1985**, *57*, 617. [CrossRef]

25. He, S.; Sun, K.; Wang, H. Complexity analysis and DSP implementation of the fractional-order Lorenz hyperchaotic system. *Entropy* **2015**, *17*, 8299–8311. [CrossRef]

26. He, S.; Li, C.; Sun, K.; Jafari, S. Multivariate multiscale complexity analysis of self-reproducing chaotic systems. *Entropy* **2018**, *20*, 556. [CrossRef]

27. Munoz-Pacheco, J.M.; Zambrano-Serrano, E.; Volos, C.; Jafari, S.; Kengne, J.; Rajagopal, K. A New Fractional-Order Chaotic System with Different Families of Hidden and Self-Excited Attractors. *Entropy* **2018**, *20*, 564. [CrossRef]

28. Pincus, S. Approximate entropy as a measure of system complexity. *Proc. Natl. Acad. Sci. USA* **1991**, *88*, 2297–2301. [CrossRef]

29. Pincus, S. Approximate entropy (ApEn) as a complexity measure. *Chaos Interdiscipl. J. Nonlinear Sci.* **1995**, *5*, 110–117. [CrossRef]

30. Wang, C.; Ding, Q. A new two-dimensional map with hidden attractors. *Entropy* **2018**, *20*, 322. [CrossRef]

31. Garcia-Martinez, M.; Campos-Canton, E. Pseudo-random bit generator based on multi-modal maps. *Nonlinear Dyn.* **2015**, *82*, 2119–2131. [CrossRef]

32. Francois, M.; Grosges, T.; Barchiesi, D.; Erra, R. Pseudo-random number generator based on mixing of three chaotic maps. *Commun. Nonlinear Sci. Numer. Simul.* **2014**, *19*, 887–895. [CrossRef]

33. Murillo-Escobar, M.A.; Cruz-Hernandez, C.; Cardoza-Avendano, L.; Mendez-Ramirez, R. A novel pseudorandom number generator based on pseudorandomly enhanced logistic map. *Nonlinear Dyn.* **2017**, *87*, 407–425. [CrossRef]

34. Wang, Y.; Liu, Z.; Ma, J.; He, H. A pseudorandom number generator based on piecewise logistic map. *Nonlinear Dyn.* **2016**, *83*, 2373–2391. [CrossRef]

35. Lambic, D. A novel method of S-box design based on chaotic map and composition method. *Chaos Solitons Fractals* **2014**, *58*, 16–21. [CrossRef]

36. Mazloom, S.; Eftekhari-Moghadam, A.M. Color image encryption based on Coupled Nonlinear Chaotic Map. *Chaos Solitons Fractals* **2009**, *42*, 1745–1754. [CrossRef]
37. La Hoz, M.Z.D.; Acho, L.; Vidal, Y. An experimental realization of a chaos-based secure communication using Arduino microcontrollers. *Sci. World J.* **2015**, *2015*, 123080.
38. Acho, L. A discrete-time chaotic oscillator based on the logistic map: A secure communication scheme and a simple experiment using Arduino. *J. Frankl. Inst.* **2015**, *352*, 3113–3121. [CrossRef]
39. Teikari, P.; Najjar, R.P.; Malkki, H.; Knoblauch, K.; Dumortier, D.; Gronfer, C.; Cooper, H.M. An inexpensive Arduino-based LED stimulator system for vision research. *J. Neurosci. Methods* **2012**, *211*, 227–236. [CrossRef] [PubMed]
40. Faugel, H.; Bobkov, V. Open source hard- and software: Using Arduino boards to keep old hardware running. *Fusion Eng. Des.* **2013**, *88*, 1276–1279. [CrossRef]
41. Castaneda, C.E.; Lopez-Mancilla, D.; Chiu, R.; Villafana-Rauda, E.; Orozco-Lopez, O.; Casillas-Rodriguez, F.; Sevilla-Escoboza, R. Discrete-time neural synchronization between an Arduino microcontroller and a Compact Development System using multiscroll chaotic signals. *Chaos Solitons Fractals* **2019**, *119*, 269–275. [CrossRef]
42. Ouannas, A.; Odibat, Z. Generalized synchronization of different dimensional chaotic dynamical systems in discrete-time. *Nonlinear Dyn.* **2015**, *81*, 765–771. [CrossRef]
43. Ouannas, A.; Grassi, G. GA new approach to study co–existence of some synchronization types between chaotic maps with different dimensions. *Nonlinear Dyn.* **2016**, *86*, 1319–1328. [CrossRef]
44. Ouannas, A.; Odibat, Z.; Shawagfeh, N.; Alsaedi, A.; Ahmad, B. Universal chaos synchronization control laws for general quadratic discrete systems. *Appl. Math. Model.* **2017**, *45*, 636–641. [CrossRef]

![entropy logo] *entropy*

MDPI

Article

Dynamics and Entropy Analysis for a New 4-D Hyperchaotic System with Coexisting Hidden Attractors

Licai Liu [1,*] , Chuanhong Du [1], Xiefu Zhang [2], Jian Li [2] and Shuaishuai Shi [3]

[1] School of Electronic and Information Engineering, Anshun University, Anshun 561000, China; duchuanhong@hrbeu.edu.cn

[2] School of Mathematics and Computer Science, Guizhou Education University, Guiyang 550018, China; zhang.xie.fu@163.com (X.Z.); lijian@gznc.edu.cn (J.L.)

[3] School of Information Engineering, Guizhou University of Engineering Science, Bijie 551700, China; gues2017@126.com

* Correspondence: liulicai1981@126.com

Received: 27 February 2019; Accepted: 13 March 2019; Published: 15 March 2019

Abstract: This paper presents a new no-equilibrium 4-D hyperchaotic multistable system with coexisting hidden attractors. One prominent feature is that by varying the system parameter or initial value, the system can generate several nonlinear complex attractors: periodic, quasiperiodic, multiple topology chaotic, and hyperchaotic. The dynamics and complexity of the proposed system were investigated through Lyapunov exponents (LEs), a bifurcation diagram, a Poincaré map, and spectral entropy (SE). The simulation and calculation results show that the proposed multistable system has very rich and complex hidden dynamic characteristics. Additionally, the circuit of the chaotic system is designed to verify the physical realizability of the system. This study provides new insights into uncovering the dynamic characteristics of the coexisting hidden attractors system and provides a new choice for nonlinear control or chaotic secure communication technology.

Keywords: hidden attractor; hyperchaotic system; multistability; entropy analysis

1. Introduction

The chaotic system has great application prospects in the field of image encryption [1–3] and secure communication [4]. For a long time, many chaotic systems composed of certain ordinary differential equations have been explored. This produced a series of classic three-dimensional continuous chaotic systems, including the Lorenz [5–7], Rossler [8], Chua, Chen [9,10], and Liu systems [11]. By adding linear or nonlinear state feedback controllers on 3-D chaotic systems, various 4-D chaotic systems can be constructed [12–14]. Four-dimensional chaotic systems have more complex nonlinear complexity and better randomness than 3-D chaotic systems. These continuous autonomous chaotic systems have common attractors called self-excited attractors because the oscillation is excited from unstable equilibria. At certain initial conditions, traditional self-excited attractors could be tracked from a computational point of view [15].

Recently, the issue of hidden attractors has drawn much attention from the field of nonlinear chaos. The hidden attractors, without equilibrium or with stable equilibrium points, have been found in some continuous chaotic or hyperchaotic systems [16,17]. The basin of attraction for hidden attractors does not intersect with small neighborhoods of any equilibrium point [18–20]. Because the system with hidden attractors has neither homoclinic nor heteroclinic orbits, it has completely different dynamic characteristics from the self-excited attractors [21–23]. In addition, the coexistence of multiple hidden attractors is a strange physical phenomenon called a multistable system [24–26], often encountered in nonlinear dynamic systems. Such a multistable state can greatly improve the complexity of chaotic systems, making these chaotic systems with hidden attractors more suitable for use in chaotic encryption technology. In most cases, however, multistable systems with hidden attractors tend to experience unexpected and potentially disastrous outcomes. Due to the fact that these systems are vulnerable and prone to unpredictably switch to another attractor, multistable systems can cause aircraft crashes [15], drill string failures and breakdowns [27], serious problems during financial crises [28], and catastrophic shifts in ecosystem services [28,29]. Up to now, although predicting a catastrophic bifurcation for multistable systems has been extremely difficult [30], it is still important to uncover and analyze all coexisting attractors in different scientific fields [15].

Motivated by the above research, a new no-equilibrium hyperchaotic system with coexisting hidden attractors is proposed in this work. Up to now, compared with the 3-D hidden attractor system, very little has been published on 4-D hidden attractors, especially for a hyperchaotic system. Therefore, in this paper, a 4-D chaotic system without any equilibrium has been constructed by adding a state variable to a 3-D chaotic system developed by Vaidyanathan and Volos in 2015 [31]. When selecting certain parameters and initial conditions, the Lyapunov exponents (LEs) of the proposed 4-D hyperchaotic system were $LE_1 = 0.0030$, $LE_2 = -0.0079$, $LE_3 = 0.0044$, and $LE_4 = 0$. There were two positive LEs, and the proposed system had hyperchaotic behavior. At this point, the Kaplan–Yorke fractional dimension $D_{KY} = 3.93$.

The existence of chaotic attractors was demonstrated by Lyapunov exponents, a bifurcation diagram, a phase diagram, a time domain diagram, and a power spectral density map. The complexity of the coexisting hidden attractors was also carried out by means of entropy analysis. It was found from the results that the proposed system exhibited extremely complex dynamic characteristics under different initial conditions of the system, such as hidden attractors, quasi-limit cycles, and coexisting attractors with different topological structures. Further, the shape of the hidden attractor was different from the existing attractor. The hidden attractor system was realized by a circuit, for which the experimental results were consistent with the simulation results, further verifying the chaotic characteristics of the system. These qualitative and quantitative studies show that the new system has complex chaotic characteristics.

The rest of this paper is organized as follows. Section 2 describes the mathematical model and chaotic characteristics of the new 4-D system with coexisting hidden attractors. The phenomena exhibited by the system, such as periodicity, multiple coexisting hidden attractors, and quasi-periodic limit cycles, are discussed in Section 3. The information spectral entropy (SE) analysis is given in detail in Section 4. Section 5 presents the circuit of the hyperchaotic system. The inadequacies of the work are discussed in Section 6, and Section 7 summarizes the conclusions.

2. System Description

2.1. Model of the New Chaotic System

A 3-D conservative no-equilibrium chaotic system was developed by Vaidyanathan and Volos [31]. Vaidyanathan's system, having LEs of $LE_1 = -0.0395$, $LE_2 = 0$, and $LE_3 = -0.0395$ and a Kaplan–Yorke dimension of $D_{KY} = 3$, is described as

$$\begin{cases} \dot{x} = ay + xz \\ \dot{y} = -bx + yz \\ \dot{z} = 1 - x^2 - y^2 \end{cases} \quad (1)$$

where x, y, and z are state variables and a and b are parameters of the system. By adding the fourth state variable w, and feeding the third state variable z to the fourth variable w, a new 4-D hyperchaotic system is obtained as

$$\begin{cases} \dot{x} = ay + xz \\ \dot{y} = -bx + yz \\ \dot{z} = 1 - x^2 - y^2 \\ \dot{w} = z(w-1) \end{cases} \tag{2}$$

where x, y, z, and w are state variables and a and b are positive real constant parameters of the system. We know that the equilibrium points of system (2) can be achieved by solving the roots of Equation (3). Simplifying and reorganizing Equation (3), we could get $ay^2 + bx^2 = 0$. Considering that $1 - x^2 - y^2 = 0$ and a and b are positive real numbers, Equation (3) has no solution; that is, there is no equilibrium point. According to the definition of hidden attractors, the system's attractors belong to hidden attractors. Its basin of attraction does not contain neighborhoods of equilibria [32,33].

$$\begin{cases} ay + xz = 0 \\ -bx + yz = 0 \\ 1 - x^2 - y^2 = 0 \\ z(w-1) = 0 \end{cases} \tag{3}$$

2.2. Nonlinear Description of the System

In this subsection, we mainly discuss the nonlinear dynamics of the system with hidden attractors by means of numerical simulation. If there is no special explanation, the simulation step size is 0.01, the ode45 numerical solver is used, and the simulation time is 2000 s in this paper. Figure 1 shows the 3-D attractor projection of system (2), and Figure 2 depicts the 2-D attractor projection of system (2). From the projection phase diagram it can be found that the attractors are different from the scroll or wing shape and belong to a new attractor. Compared with the attractors of system (1), the new system has more abundant attractors, so the dynamic characteristics of the new system are more complicated. Figure 3 is a time series diagram of four state variables of system (2) that indicates that the system is aperiodic, which corresponds with the chaotic feature. The curves in Figure 4, obtained by Fourier transform of the autocorrelation function, show that the variables are continuous and there are no obvious peaks, which accords with the characteristics of chaos.

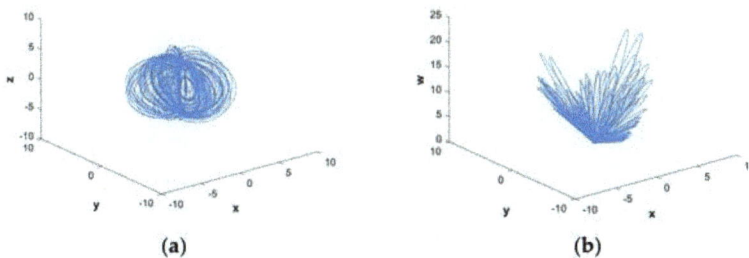

(a)

(b)

Figure 1. *Cont.*

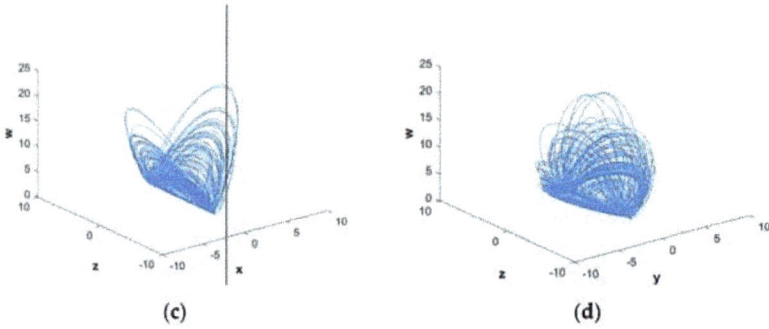

Figure 1. Three-dimensional chaotic attractor of system (2) with parameters $a = 0.05$, $b = 1$ and initial conditions $(x_0, y_0, z_0, w_0) = (-1, -1, 4, 4)$ on the (**a**) (x, y, z) space, (**b**) (x, y, w) space, (**c**) (x, z, w) space, and (**d**) (y, z, w) space.

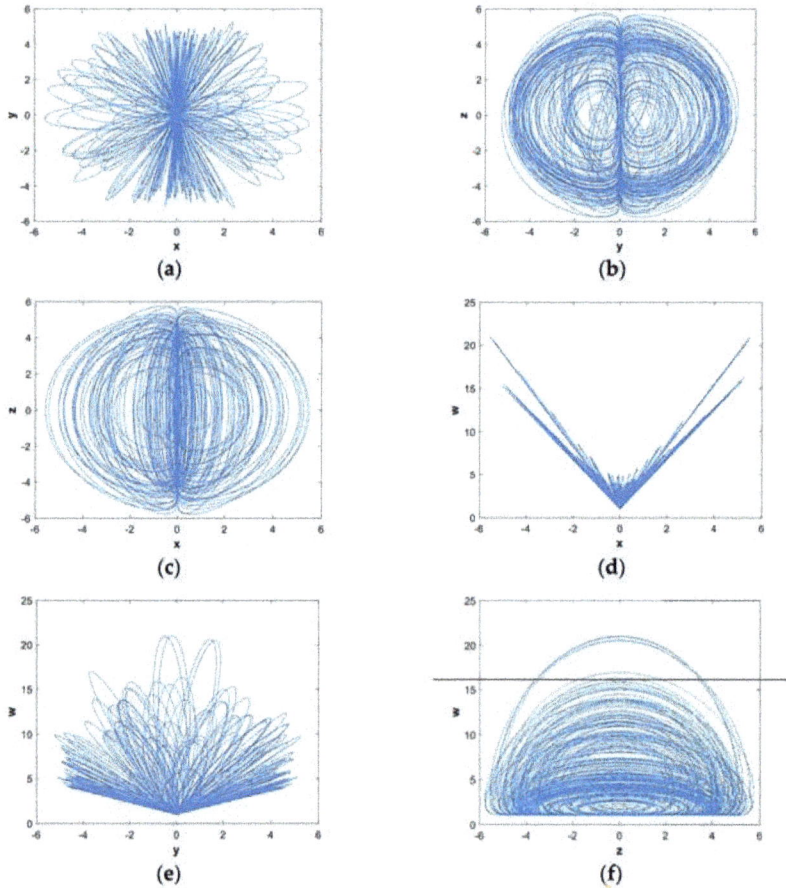

Figure 2. Two-dimensional chaotic attractor of system (2) with parameters $a = 0.05$, $b = 1$ and initial conditions $(x_0, y_0, z_0, w_0) = (-1, -1, 4, 4)$: (**a**) $x - y$ plane; (**b**) $y - z$ plane; (**c**) $x - z$ plane; (**d**) $x - w$ plane; (**e**) $y - w$ plane; and (**f**) $z - w$ plane.

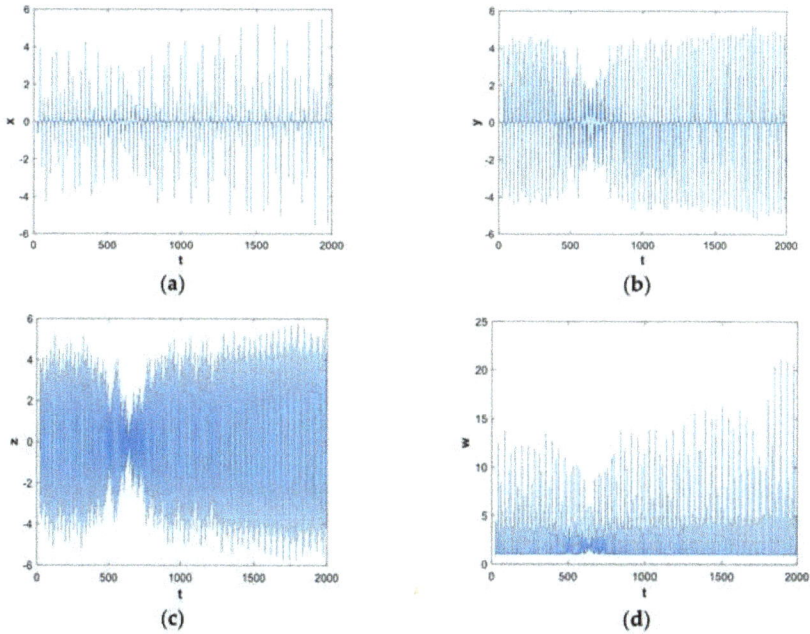

Figure 3. Time series of system (2) with parameters $a = 0.05$, $b = 1$ and initial conditions $(x_0, y_0, z_0, w_0) = (-1, -1, 4, 4)$: (**a**) x variable; (**b**) y variable; (**c**) z variable; and (**d**) w variable.

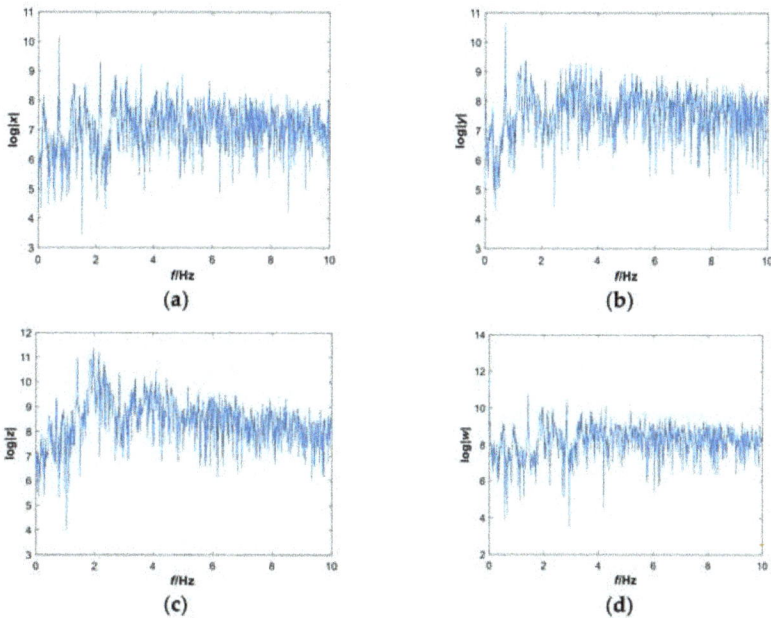

Figure 4. Frequency spectrum of system (2) with parameters $a = 0.05$, $b = 1$ and initial conditions $(x_0, y_0, z_0, w_0) = (-1, -1, 4, 4)$: (**a**) the x variable; (**b**) y variable; (**c**) z variable; and (**d**) w variable.

Using the parameters $a = 0.05$, $b = 1$ and initial conditions $(x_0, y_0, z_0, w_0) = (-1, -1, 4, 4)$, Figure 5 illustrates the LEs of the system computed with the Wolf algorithm [34,35]. They are

$LE_1 = 0.012865$, $LE_2 = -0.0050839$, $LE_3 = -0.0098453$, and $LE_4 = -0.033262$, respectively. One of them is positive and two of them are negative, and the sum of the LEs is negative, so the system is a stable chaotic system with hidden attractors. A further discussion of LEs is given in Section 3.2, which demonstrates that the system is hyperchaotic under certain parameters and proper initial conditions.

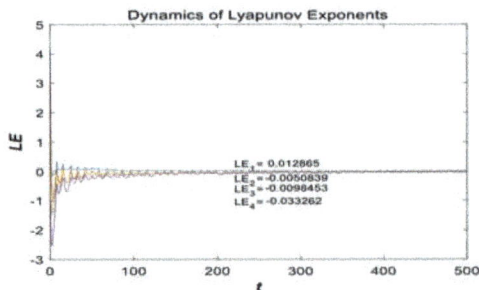

Figure 5. LEs of system (2) in dependence on parameters $a = 0.05$, $b = 1$ and initial value $(x_0, y_0, z_0, w_0) = (-1, -1, 4, 4)$.

To further illustrate that system (2) is chaotic, Figure 6 illustrates a Poincaré map [36–38] with $a = 0.05$, $b = 1$ and initial value $(x_0, y_0, z_0, w_0) = (-1, -1, 4, 4)$. Figure 6(a) and (b) are Poincaré maps in the $x - y$ and $x - w$ planes on a $z = 0$ cross section. As is shown in the cross section of the map, there is a set of points distributed along the line or curve arc with a self-similar fractal structure. Therefore, the Poincaré map of the system also shows the properties of chaos.

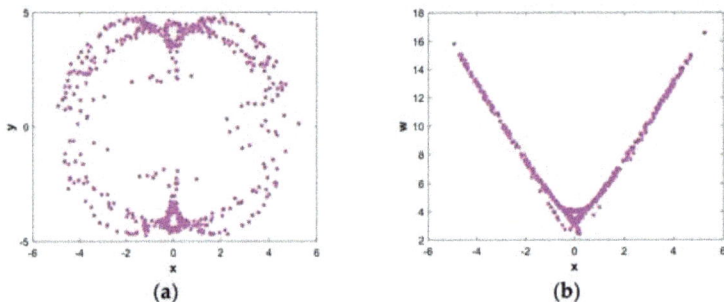

(a)   (b)

Figure 6. Poincaré map of system (2) in dependence on parameters $a = 0.05$, $b = 1$ and initial value $(x_0, y_0, z_0, w_0) = (-1, -1, 4, 4)$ in the (a) $x - y$ plane and (b) $x - \omega$ plane.

3. Nonlinear Dynamics of the System

To date, knowledge of hidden attractors is still insufficient and research results are in their early stages. There is a great need to address the characteristics of nonlinear system with hidden attractors. What we present in this section is an analysis of the chaotic, hyperchaotic, and multistable characteristics of the proposed system in terms of system parameters and initial values of state variables. More specifically, the phenomena of periodicity, coexistence of multiple hidden attractors, and quasi-periodic limit cycles are analyzed and explained in great detail.

3.1. Influence of Parameters on System Dynamic Characteristics

A bifurcation diagram can show the relationship between a system and the variation of parameters and changes in dynamic behaviors, as well as graphically reflect nonlinear behaviors such as chaotic, periodic, and quasi-periodic limit cycles in the system. Therefore, in this section, a bifurcation diagram is used to analyze the influence of different system parameters on the dynamic characteristics of system (2).

The change of dynamic behaviors is shown in the bifurcation diagram in Figure 7(a), from which it can be seen that when fixing $b = 0.05$ and $a \in (0, 5)$ with the initial value $(x_0, y_0, z_0, w_0) = (2, 0, 0, 0)$, the transition to chaos was apparent and the system was in a wide-domain chaotic state when the value of a increased. When $a \in (1.2, 2.7)$, it was obvious that the system was in a chaotic state. Figure 7(b) is the corresponding LE diagram of the system under this simulation parameter. Figure 7(b) shows that two positive Lyapunov exponents also appeared in some ranges, indicating hyperchaos in system (2). When $a \in (2.7, 3.6)$, the maximum Lyapunov exponent of the system was very close to 0, while the remaining LEs were negative or zero, indicating that the system was in the limit circle state or had a hidden periodic attractor. In order to facilitate the analysis of the dynamic behavior, the bifurcation graph and corresponding LE graph when $a \in (0.7, 1)$ and $a \in (4, 5)$ were enlarged locally, as shown in Figures 8 and 9, respectively. During $a \in (0.7, 1)$ and $a \in (4, 5)$, Figures 8 and 9 show that there were complex nonlinear behaviors in these two regions. In addition, the bifurcation boundary line had a certain width of point set, as opposed to being a single point line, which made the system have more abundant dynamic behaviors. Further, because the LE graphs show that the system had two positive LEs, hyperchaotic behavior existed in a large range.

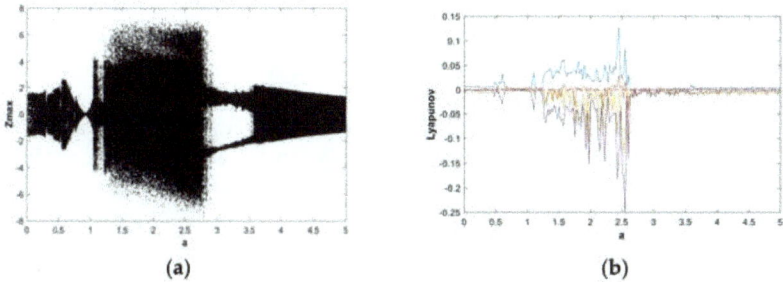

Figure 7. Bifurcation diagram and LEs of system (2) about a, with $b = 0.05$, initial value $(x_0, y_0, z_0, w_0) = (2, 0, 0, 0)$, and $a \in (0, 5)$: (**a**) bifurcation diagram; (**b**) LE graphs.

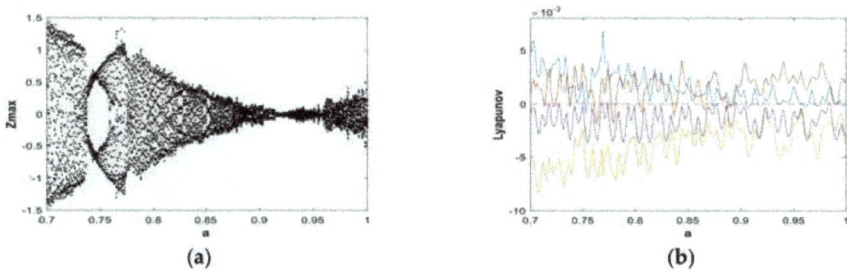

Figure 8. Bifurcation diagram and LEs of system (2) about a, with $b = 0.05$, initial value $(x_0, y_0, z_0, w_0) = (2, 0, 0, 0)$, and $a \in (0.7, 1)$: (**a**) bifurcation diagram; (**b**) LE graphs.

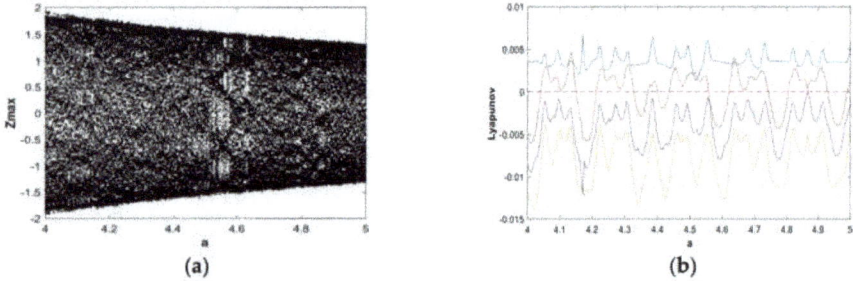

Figure 9. Bifurcation diagram and LEs of system (2) about a, with $b = 0.05$, initial value $(x_0, y_0, z_0, w_0) = (2, 0, 0, 0)$, and $a \in (4, 5)$: (**a**) bifurcation diagram; (**b**) LE graphs.

In order to further analyze the influence of parameters on the hidden attractors system, the system was analyzed using a phase diagram. When $a = 0.88922$, Figure 8(b) shows that the system had a positive LE close to 0 and three negative LEs. The system was in a state of quasi-periodic limit cycle. The 3-D and 2-D projections of the hidden attractor diagram of the system are shown in Figure 10. By increasing the value of parameter a, when $a = 1.0621$, a hidden attractor could be observed (Figure 11). When $a = 1.2$, the LEs were $LE_1 = 0.0030$, $LE_2 = -0.0079$, $LE_3 = 0.0044$, and $LE_4 = 0$, so the system was hyperchaotic. From Figure 12, we can see that there were many strange attractors with different topologies. When $a = 2.982$, the phase diagram of the system was a quasi-periodic limit cycle, as shown in Figure 13, which is consistent with the illustration in Figure 7. It can be seen from Figure 9 that when $a \in (4, 5)$, the bifurcation phenomenon was complicated; hence, the system dynamics behavior was also rich. Figure 9(b) shows that when $a = 4.9$, the system was in a hyperchaotic state, having hidden attractors, as shown in Figure 14. In this scenario, the hidden attractors in the system were different from the attractors when the other parameters were taken. They were novel and unique hidden attractors.

According to the above analysis, there are many kinds of hidden attractors and different topologies in system (2). Therefore, this system has novel attractors in a variety of shapes and has rich dynamic behavior.

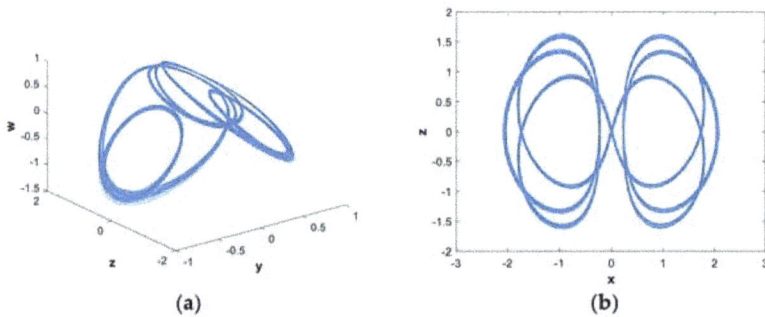

Figure 10. Projections of hidden attractors with parameters $a = 0.88922$, $b = 0.05$ and initial value $(x_0, y_0, z_0, w_0) = (2, 0, 0, 0)$: (**a**) attractor in the $y - z - w$ space; (**b**) attractor in the $x - z$ plane.

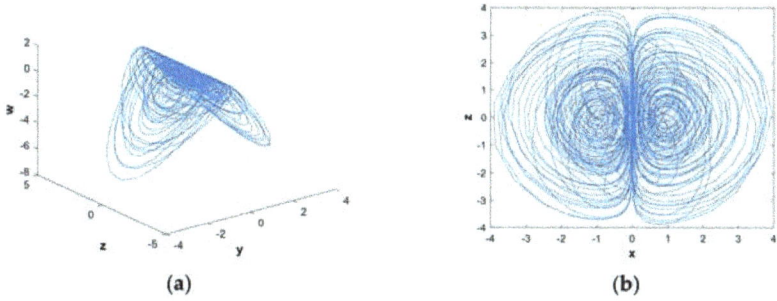

Figure 11. Projections of hidden attractors with parameters $a = 1.0621$, $b = 0.05$ and initial value $(x_0, y_0, z_0, w_0) = (2, 0, 0, 0)$: (**a**) attractor in the $y - z - w$ space; (**b**) attractor in the $x - z$ plane.

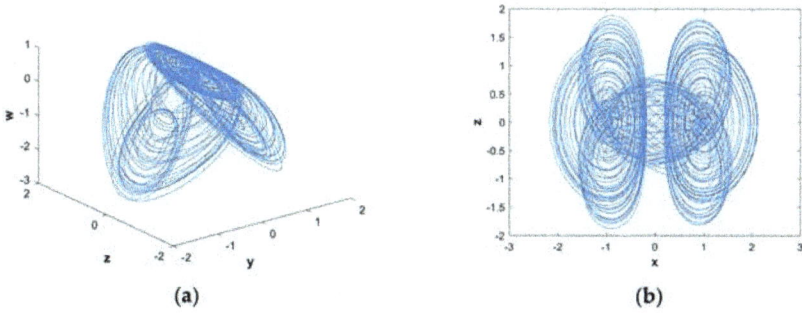

Figure 12. Projections of hidden attractors with parameters $a = 1.2$, $b = 0.05$ and initial value $(x_0, y_0, z_0, w_0) = (2, 0, 0, 0)$: (**a**) attractor in the $y - z - w$ space; (**b**) attractor in the $x - z$ plane.

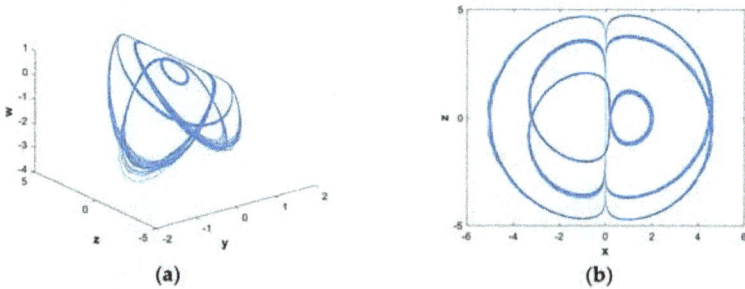

Figure 13. Projections of hidden attractors with parameters $a = 2.982$, $b = 0.05$ and initial value $(x_0, y_0, z_0, w_0) = (2, 0, 0, 0)$: (**a**) attractor in the $y - z - w$ space; (**b**) attractor in the $x - z$ plane.

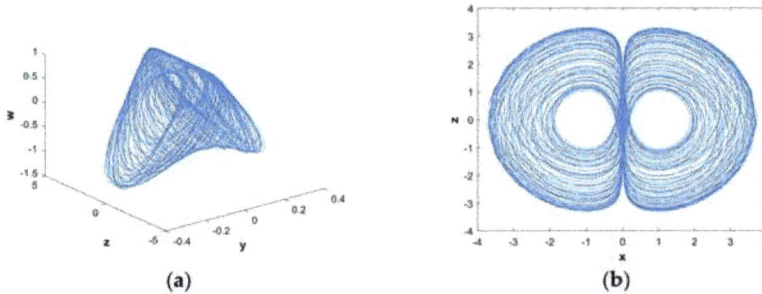

Figure 14. Projections of hidden attractors with parameters $a = 4.9$, $b = 0.05$ and initial value $(x_0, y_0, z_0, w_0) = (2, 0, 0, 0)$: (**a**) attractor in the $y - z - w$ space; (**b**) attractor in the $x - z$ plane.

3.2. Influence of Initials on System Dynamic Characteristics

More recently, the impact of initial values on the dynamic behavior of a system with hidden attractors has been subject to considerable discussion [17,39–41]. Here, we focus on the influence of different initial values on the dynamic characteristics of the hidden attractors in terms of a phase diagram (i.e., projections of attractor) and a bifurcation diagram.

With the parameters of system (2) chosen as $a = 1$, $b = 0.05$, Figure 15 shows the phase diagram of the hidden attractors. In Figure 15, the blue attractors' initial is $(2, 0, 0, 2)$, and the red attractors' initial is $(x_0, y_0, z_0, w_0) = (2, 0, 0, -2)$. It can be seen from the phase diagram that the system space corresponding to these two initial values had a certain symmetrical similarity. Specifically, the size was different and the phase was opposite.

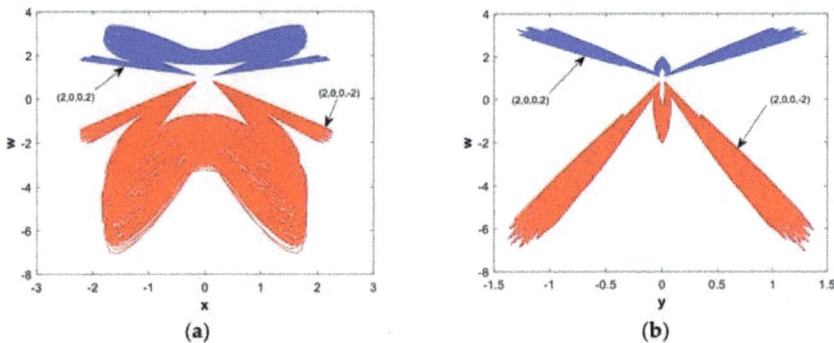

Figure 15. Projections of hidden attractors with different initial conditions. Blue attractors' initial is $(2, 0, 0, 2)$, red attractors' initial is$(2, 0, 0, -2)$, and the parameters are $a = 1$, $b = 0.05$: (**a**) attractor in the $x - w$ plane; (**b**) attractor in the $y - w$ plane.

In order to facilitate the analysis of the influence of different system initial conditions on system dynamics under the same system parameters, the bifurcation diagram and LEs were used again.

There were two initial values of the system $Y0 = (u, u, 0, 0)$ and $Y1 = (u, 0, 0, 0)$, respectively and$u \in [0, 5]$. The parameters chosen were $a = 1$, $b = 0.05$. Figures 16 and 17 present bifurcation diagrams and LEs of system (2) when increasing the value of u. The bifurcation diagrams and LEs shown in Figures 16 and 17 exhibit the process of periodic limit cycles, quasi-periodic limit cycles, and hyperchaos under different initial values. The corresponding LE graphs are also consistent with the bifurcation diagram change. With the $Y0$ point, let $u = 0.01$ and $u = 2.093$, respectively, and for the $Y1$ point, let $u = 1.5$. Thus, we have chosen the initial values $(0.01, 0.01, 0, 0)$,$(2.093, 2.093, 0, 0)$, and$(1.5, 0, 0, 0)$. The colors corresponding to these three initial values are green, blue, and red, respectively. The 2-D numerical simulation phase diagram is shown in Figure 18. Combining Figures 16 and 17, it can be found from the phase diagram of Figure 18 that when the initial value was $(0.01, 0.01, 0, 0)$, the system was in a weak chaotic state; when the initial value was $(2.093, 2.093, 0, 0)$, $LE_1 = 0$, $LE_2 = -0.0076601$, $LE_3 = -0.0067299$, and $LE_4 = -0.0075862$, the system was in a quasi-periodic limit cycle; and when the initial value was $(1.5, 0, 0, 0)$, $LE_1 = 0.0029082$, $LE_2 = -0.0034458$, $LE_3 = 0$, and $LE_4 = -0.0017509$, the system was in a hyperchaotic state. It was also found that the chaotic systems corresponding to different initial values contained hidden attractors with different topological structures.

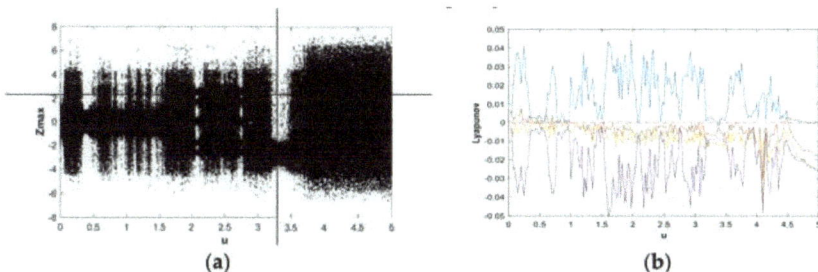

Figure 16. Bifurcation diagram and LEs of system (2) with $a = 1$, $b = 0.05$, initial value $Y0 = (u, u, 0, 0)$, and $u \in [0, 5]$: (**a**) bifurcation diagram; (**b**) LE graphs.

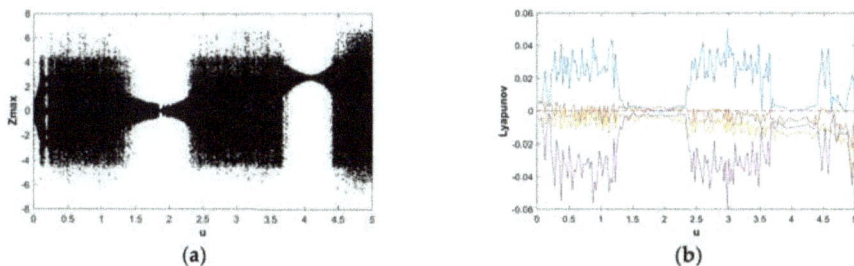

Figure 17. Bifurcation diagram and LEs of system (2) with $a = 1$, $b = 0.05$, initial value $Y1 = (u, 0, 0, 0)$, and $u \in [0, 5]$: (**a**) bifurcation diagram; (**b**) LE graphs.

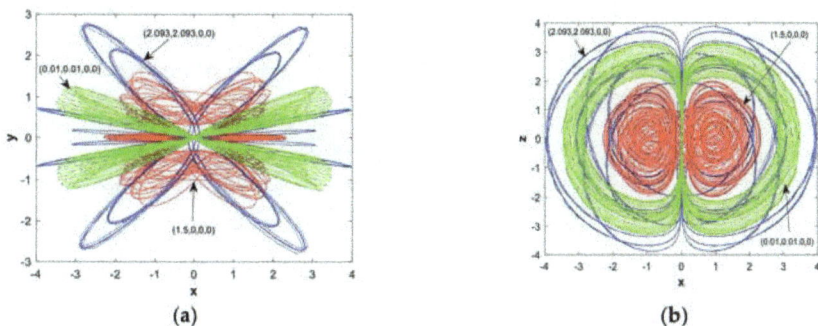

Figure 18. *Cont.*

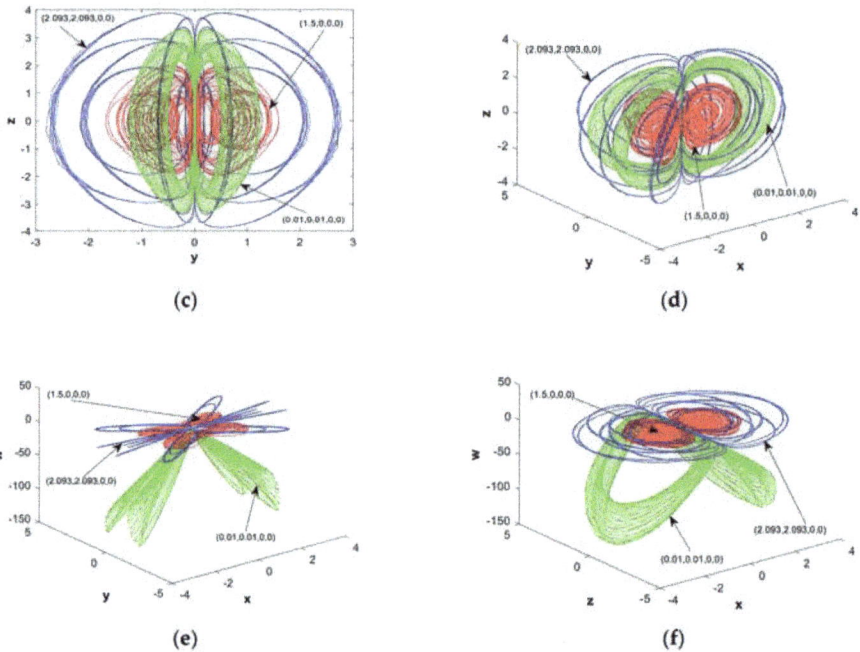

Figure 18. Projections of hidden attractors with different initial conditions. Green attractors' initial is $(0.01, 0.01, 0, 0)$, blue attractors' initial is $(2.093, 2.093, 0, 0)$, red attractors' initial is $(1.5, 0, 0, 0)$, and parameters $a = 1$, $b = 0.05$: (**a**) attractor in the $x - y$ plane; (**b**) attractor in the $x - z$ plane; (**c**) attractor $y - z$ plane; (**d**) in the $x - y - z$ space; (**e**) attractor in the $x - y - w$ space; and (**f**) attractor in the $x - z - w$ space.

Based on the above analysis, it can be concluded that the system is multistable and can produce complex multiple coexisting hidden attractors.

4. Information of Spectral Entropy Analysis

In order to measure the dynamic complexity of chaotic systems (2) with strange hidden attractors, in this section, we discuss system complexity by means of information spectral entropy (SE) analysis [42,43].

We know that another statistical property of dynamical systems is SE, which has a certain relationship with LEs and the Hausdorff dimension. SE is a measure of the chaotic properties of the system. The greater the system complexity value, the stronger the randomness of the system. When such a system is used as a communication key, the security of the information is higher. The complexity of chaotic systems is generally divided into behavioral complexity and structural complexity. At present, there are several algorithms for calculating the complexity of chaotic system behavior, and they are based on the Kolmogorov method and Shannon's entropy. These algorithms are fast and have accurate results. However, the calculation results of high-dimensional chaotic systems will overflow, which may result in the expected results. The structural complexity is the analysis of the energy characteristics in the transform domain. The scope of its action is the entire sequence of the system, not locally, so the results are more global than the behavioral complexity algorithm [18]. In this paper, the SE algorithm of structural complexity was used to analyze the dynamic characteristics of the system.

4.1. Spectral Entropy Complexity Algorithm

By using Fourier transform, the energy distribution was obtained. After that, the corresponding SE value was obtained by combining Shannon entropy. The algorithm requires the following steps.

For the chaotic pseudorandom sequence $\{x(n), n = 0, 1, 2, 3, \cdots, N-1\}$ of length N, the DC part is removed by Equation (4) so that the spectrum more effectively reflects the energy information of the signal:

$$x(n) = x(n) - \overline{x} \tag{4}$$

where $\overline{x} = \frac{1}{N} \sum_{n=0}^{N-1} x(n)$. A discrete Fourier transform (DFT) was then performed on the $x(n)$ to obtain Equation (5):

$$X(k) = \sum_{n=0}^{N-1} x(n) e^{-j\frac{2\pi}{N}nk} = \sum_{n=0}^{N-1} x(n) W_N{}^{nk} \tag{5}$$

where $k = 0, 1, 2, 3, \cdots, N-1$. The relative power spectrum was calculated for the transformed sequence $X(k)$ by taking the first half of the sequence for calculation. According to the Parseval theorem, the power spectrum value of a certain frequency point is determined by Equation (6):

$$S(k) = \frac{|X(k)|^2}{N} \tag{6}$$

where $k = 0, 1, 2, 3, \cdots, \frac{N}{2} - 1$. Then, the total power of the sequence can be defined as Equation (7):

$$S_{total} = \frac{\sum_{k=0}^{N/2-1} |X(k)|^2}{N} \tag{7}$$

The probability of the relative power spectrum for the sequence is shown in Equation (8):

$$p_k = \frac{S(k)}{S_{total}} = \frac{\frac{1}{N}|X(k)|^2}{\frac{1}{N} \sum_{k=0}^{N/2-1} |X(k)|^2} = \frac{|X(k)|^2}{\sum_{k=0}^{N/2-1} |X(k)|^2} \tag{8}$$

From statistical knowledge, we know that $\sum_{k=0}^{N/2-1} p_k = 1$. Combined with the Shannon entropy solving method, the SE of the signal could be obtained using Equation (9):

$$SE = \sum_{k=0}^{N/2-1} p_k \ln \frac{1}{p_k} \tag{9}$$

In Equation (9), if $p_k = 0$, then $p_k \ln p_k = 0$. The equation converges to $\ln(\frac{N}{2})$, and for the convenience of comparative analysis, the SE is normalized. Then, the normalized SE calculation formula could be obtained with Equation (10) [42]:

$$SE(N) = \frac{SE}{\ln(N/2)} \tag{10}$$

It can be seen from the above algorithm process that the higher the degree of imbalance for the sequence power spectrum distribution, the simpler the structure of the sequence spectrum, resulting in a stronger law of oscillation for the signal. Correspondingly, a smaller SE stands for smaller complexity; otherwise, the complexity is greater.

4.2. Influence of Parameters on Entropy

It can be seen from the discussion in Section 3.1 that the change of system parameters has a great influence on the nonlinear dynamic behavior of the system, which affects the complexity of the system. Therefore, studying the influence of system parameters on SE is necessary. Figure 19 shows how parameters a and b impact the SE, where parameter $a = 1$ or parameter $b = 0.05$, and the initial

$(x, y, z, \omega) = (2, 0, 0, 0)$. With the great fluctuations around $a = 1$, the SE attenuated to 0.1. This is because the system at this time was in the quasi-periodic limit-cycle state with minimal complexity, as shown in Figure 10. Figure 19 indicates that when $SE \in (0, 0.5)$, the regions of parameter a were greater than parameter b. So, in this region, parameter b was more sensitive than parameter a in terms of change. In the range of $b \in (2, 5)$, the SE value was high, and the fluctuation was not large. It is worth noting that when $a, b \in (2, 2.8)$, there was a similar complexity value.

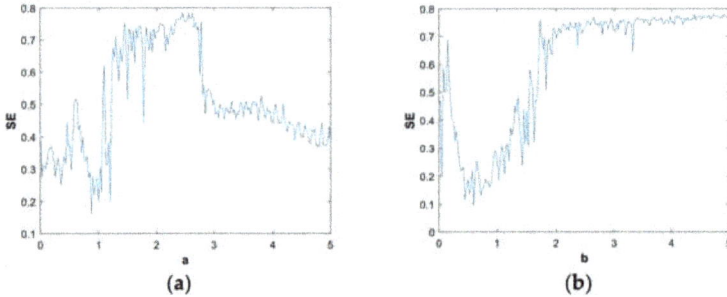

Figure 19. SE vs. parameters of the system, initial is $(x, y, z, \omega) = (2, 0, 0, 0)$: (a) $b = 1$, $a \in (0, 5)$; (b) $a = 0.05$, $b \in (0, 5)$.

4.3. Influence of Initials on Entropy

The initial value condition is a major factor affecting the dynamic behavior of the system, which was introduced in Section 3.2. In order to study the influence of the initial value of the system on the nonlinear behavior of the system, the degree of influence of the initial value on the nonlinear behavior of the system was further measured from the perspective of entropy. In the SE graphs of Figure 20, we used the parameters $a = 1$, $b = 0.05$. Suppose there are three types of initial conditions: $IN0 = (u, u, 0, 0)$, $IN1 = (u, 0, 0, 0)$, and $IN2 = (0.01u, 0, 0, -2u)$, respectively; u is a variable, and $u \in (0, 5)$. The relationship between the variables u and SE is shown in Figure 20. As can be seen from the figure, for the system complexity, the SE of $IN0$, $IN1$, and $IN2$ were alternate variations when $u \in (0, 0.3022)$; when $u \in (0.3022, 0.6179)$, the SE was $IN1 > IN2 > IN0$; when $u \in (0.6179, 0.83)$, $IN1$ was always greater than the $IN2$ and $IN0$; when $u \in (0.83, 1.0076)$, $IN1 > IN2 > IN0$; when $u \in (1.0076, 1.5503)$, $IN1$ was in the transition from chaotic to nonchaotic; when $u \in (1.5503, 2.3397)$, the SE was $IN0 > IN2 > IN1$; the values of $IN1$ and $IN0$ alternated but were always greater than the $IN2$ when $u \in (2.3397, 3.985)$; when $u \in (3.985, 4.2637)$, it was approximately $IN0 > IN2 > IN1$; when $u \in (4.2637, 5)$, the value of SE was $IN0 > IN1 > IN2$. As is shown in Figure 20, the change of $IN2$ was relatively flat, the corresponding system had less complexity, and the dynamic behavior of the system was much less than that of the other two initial value systems. The $IN0$ curve fluctuations were varied and intense, so the corresponding system was rich in dynamic behavior. The $IN1$ curve changes were more orderly, and the corresponding system also contained rich nonlinear characteristics.

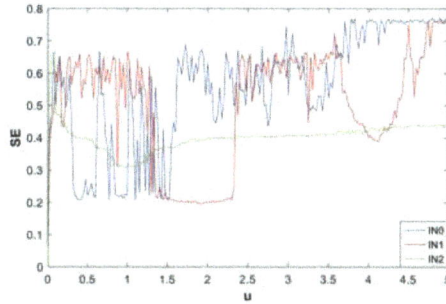

Figure 20. SE vs. initials of the system, where $IN0 = (u, u, 0, 0)$ (Blue), $IN1 = (u, 0, 0, 0)$ (Red), and $IN2 = (0.01u, 0, 0, -2u)$ (Green); $u \in (0, 5)$.

The above analysis only deals with three simple forms of initial value problems. Since there are numerous initial values for the system, the complexity of the system varies with the initial value and corresponds to infinite variety, which further illustrates that the system has a very rich dynamic behavior.

4.4. Characteristic Analysis of Chaotic Diagram of System Entropy

The previous sections described the relationship between system parameters, initial system values, and system nonlinear dynamics complexity. In the following section, the chaotic characteristic distribution of the system complexity is discussed from the perspective of the interaction of system parameters a and b. In order to observe the distribution of chaotic SE more clearly and graphically, a contour map with different color schemes is used to show the chaotic property of SE, using flipud (hot) mode. A contour map was obtained of the complexity of the chaotic system vs. system parameters, as shown in Figure 21, where $a \in (0, 5)$, $b \in (0, 5)$, and the initial $(x_0, y_0, z_0, w_0) = (2, 0, 0, 0)$. The figure shows that under the same initial value, the adjacent different color boundary lines are the contour lines; that is, the adjacent contour lines are filled with the same color, and the color used for filling was done by flipud (hot) mode. Due to the variety of colors in the figure, there are many values for the corresponding system entropy, which is consistent with the results discussed in Section 4.2. The main color distribution in the graph is red and black, or similar, and the corresponding system has a larger entropy value, whereas the other colored areas are smaller. This shows that the system was in a hyperchaotic or chaotic state in a wide range when varying parameters a or b, which is consistent with the conclusion from the system bifurcation diagram. Therefore, a more detailed color distribution contour map can be obtained by way of subdividing system parameters and changing the system initial value.

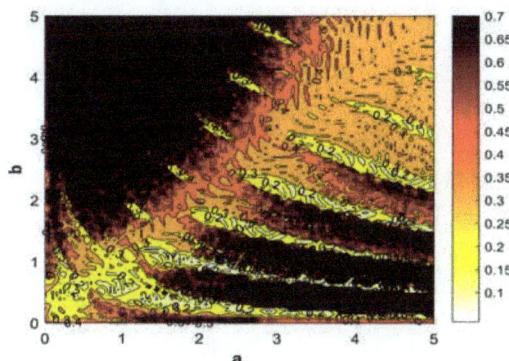

Figure 21. Chaotic characteristics of SE vs. the parameters of the system, with $a \in (0, 5)$, $b \in (0, 5)$, and the initial $(x_0, y_0, z_0, w_0) = (2, 0, 0, 0)$.

5. Circuit Design

The nonlinear dynamic behavior of system (2) is discussed here in detail based on numerical simulation, which verified that the system had abundant dynamic behavior. This section presents a novel circuit implementation for a hidden attractor system.

5.1. Improved Modular Circuit Design

Electronic circuit realization is of great physical significance for the application of chaos theory [14,44]. Chaotic circuit design mainly includes three methods: individualized design, modular design, and improved modular design. Although an individualized design needs fewer circuit components, it does require more prior knowledge. The modular design approach is based on a dimensionless state equation that is universal and versatile but requires more components. The improved modular design method can combine the advantages of the former two methods very well. By comparing the state differential equation with the actual circuit's state differential equation, the total coefficients of the system can be determined, which can minimize the number of components. Therefore, we adopted the improved modular design method to design the system circuit.

We used TL082 operational amplifiers, AD633 analog multipliers, some linear resistors, and capacitors to form this 4-D signal generator circuit system. Since the power supply voltage was $\pm 15V$, in order to make the signal of the system not exceed the linear dynamic range of the operational amplifier with $\pm 13.5V$, the state variable was rescaled using Equation (11). Introduce new state variables u_x, u_y, u_z, and u_w. Let

$$\begin{cases} x = u_x \\ y = u_y \\ z = u_z \\ w = 2u_w \end{cases} \tag{11}$$

Substituting Equation (11) in Equation (2), we obtain Equation (12):

$$\begin{cases} \frac{du_x}{dt} = au_y + u_x u_z \\ \frac{du_y}{dt} = -bu_x + u_y u_z \\ \frac{du_z}{dt} = 1 - u_x{}^2 - u_y{}^2 \\ \frac{du_w}{dt} = 0.5u_z(2u_w - 1) \end{cases} \tag{12}$$

In order to make the circuit parameters better match the system, set the time scale to τ_0, and $\tau_0 = 10^4$; then, perform time transformation. The new time variable is τ, and $t = \tau_0 \tau$; then, $dt = \tau_0 d\tau$. Hence, Equation (12) can be written as

$$\begin{cases} \frac{du_x}{d\tau} = \tau_0(au_y + u_x u_z) \\ \frac{du_y}{d\tau} = \tau_0(-bu_x + u_y u_z) \\ \frac{du_z}{d\tau} = \tau_0(1 - u_x{}^2 - u_y{}^2) \\ \frac{du_w}{d\tau} = 0.5\tau_0 u_z(2u_w - 1) \end{cases} \tag{13}$$

From the constraint relationship of Equation (13), the corresponding circuit equation can be designed as (14):

$$\begin{cases} \frac{du_x}{d\tau} = \frac{1}{C_1}\left(\frac{u_y}{R_1} + \frac{g_1 u_x u_z}{R_2}\right) \\ \frac{du_y}{d\tau} = \frac{1}{C_2}\left(-\frac{u_x}{R_3} + \frac{g_2 u_y u_z}{R_4}\right) \\ \frac{du_z}{d\tau} = \frac{1}{C_3}\left(\frac{1}{R_7} - \frac{g_3 u_x{}^2}{R_5} - \frac{g_4 u_y{}^2}{R_6}\right) \\ \frac{du_w}{d\tau} = \frac{1}{C_4}\left(\frac{g_5 u_w}{R_9} - \frac{1}{R_8}\right)u_z \end{cases} \tag{14}$$

where C_1, C_2, C_3, and C_4 are the integral capacitance; g_1, g_2, g_3, and g_4 are the gain of five multipliers; u_x, u_y, u_z, and u_w are the output variables of the integrator, which correspond to the system-state variable of Equation (13); and $R_i(i = 1, 2, 3, \cdots 9)$ is the corresponding resistance. In order to get the parameters of the circuit, we compared Equations (13) and (14), and then obtained Equation (15). Thus, the specific resistance value is shown in Equation (16):

$$
\begin{cases}
\frac{1}{C_1 R_1} = a\tau_0 \\
\frac{g_1}{C_1 R_2} = \tau_0 \\
\frac{1}{C_2 R_3} = b\tau_0 \\
\frac{g_2}{C_2 R_4} = \tau_0 \\
\frac{g_3}{C_3 R_5} = \tau_0 \\
\frac{g_4}{C_3 R_6} = \tau_0 \\
\frac{1}{C_3 R_7} = \tau_0 \\
\frac{1}{C_4 R_8} = 0.5\tau_0 \\
\frac{g_5}{C_4 R_9} = \tau_0
\end{cases}
\tag{15}
$$

$$
\begin{cases}
R_1 = \frac{1}{a\tau_0 C_1} \\
R_2 = \frac{g_1}{\tau_0 C_1} \\
R_3 = \frac{1}{b\tau_0 C_2} \\
R_4 = \frac{g_2}{\tau_0 C_2} \\
R_5 = \frac{g_3}{\tau_0 C_3} \\
R_6 = \frac{g_4}{\tau_0 C_3} \\
R_7 = \frac{1}{\tau_0 C_3} \\
R_8 = \frac{1}{0.5\tau_0 C_4} \\
R_9 = \frac{g_5}{\tau_0 C_4}
\end{cases}
\tag{16}
$$

In order to calculate and guarantee the unity of the circuit parameters, let the five multipliers and the four capacitors have equal gains, namely, $g_i = 0.1V(i = 1, 2, 3, \cdots 5)$ and $C_i = 10nF(i = 1, 2, 3, 4)$. When $b = 0.05$ is substituted into Equation (16), we can obtain $R_1 = \frac{10}{a}k\Omega$, $R_3 = \frac{10}{b}k\Omega = 200k\Omega$, $R_2 = R_4 = R_5 = R_6 = R_9 = 1k\Omega$, $R_7 = 10k\Omega$, and $R_8 = 20k\Omega$. The circuit schematic, designed using Kirchhoff's laws, is shown in Figure 22, where ux, uy, uz, and uw marked in the figure correspond to the output variables u_x, u_y, u_z, and u_w, respectively, and the resistance parameter of the inverter is $R_{10} = R_{11} = R_{12} = R_{13} = R_{14} = R_{15} = 10k\Omega$. To observe the phase diagrams of circuits with hidden attractors, a sliding rheostat $R_1 = 20k\Omega$ was used (Figure 22).

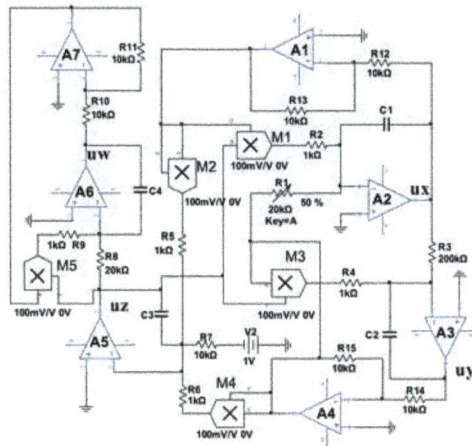

Figure 22. Circuit exhibiting hidden attractors without equilibrium.

5.2. Multisim Results

We used Multisim14.0 software to build the circuit shown in Figure 22. It is well known that the components used in Multisim software are highly compatible with the actual components. Therefore, the specific circuit scheme can well reflect actual circuit performance. Here, by adjusting the tap of the sliding rheostat to change the value of R_1, the projections of hidden attractors observed from the oscilloscope are displayed in Figure 23, which agree well with the phase diagrams of Equation (2) shown in Section 3.1. This also confirmed that the proposed system is physically achievable.

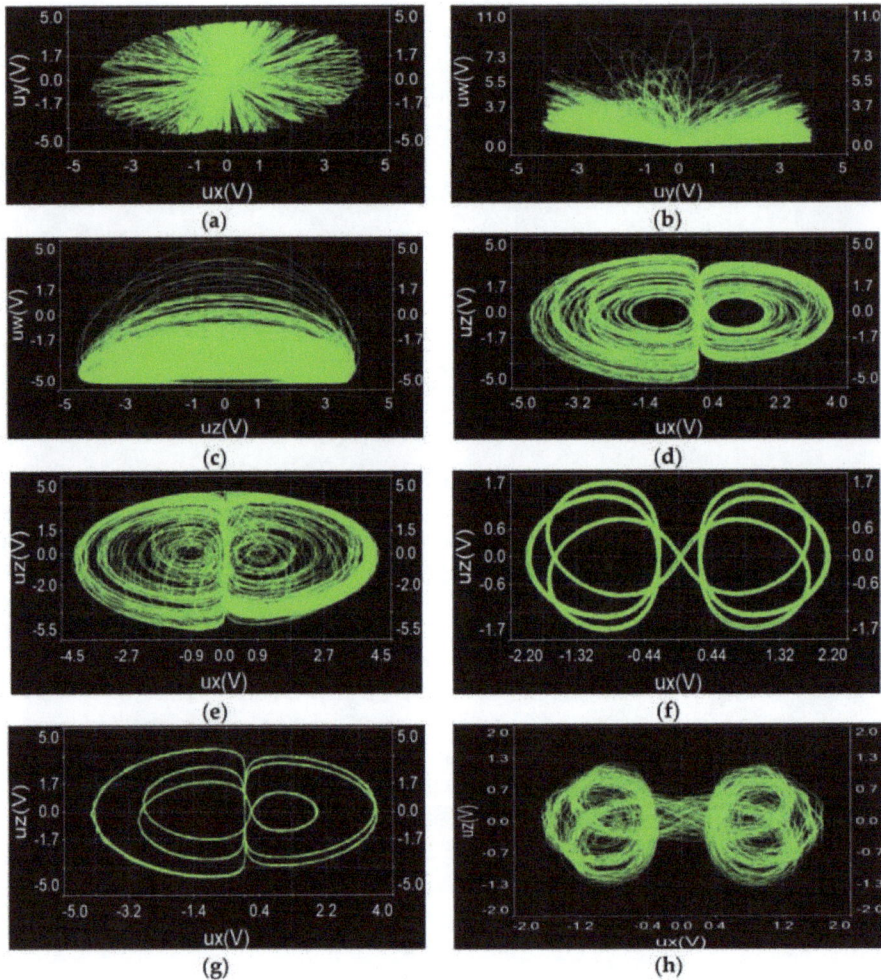

Figure 23. Hidden chaotic attractors of circuit (14) : (**a**) $a = 1$, $R_1 = 10k\Omega$, $u_x - u_y$ plane; (**b**) $a = 1$, $R_1 = 10k\Omega$, $u_y - u_w$ plane; (**c**) $a = 1$, $R_1 = 10k\Omega$, $u_z - u_w$ plane; (**d**) $a = 4.9$, $R_1 = 2.041k\Omega$, $u_x - u_z$ plane; (**e**) $a = 1.0621$, $R_1 = 9.415k\Omega$, $u_x - u_z$ plane; (**f**) $a = 0.88922$, $R_1 = 11.246k\Omega$, $u_x - u_z$ plane; (**g**) $a = 2.982$, $R_1 = 3.353k\Omega$, $u_x - u_z$ plane; (**h**) $a = 1.2$, $R_1 = 8.333k\Omega$, $u_x - u_z$ plane.

6. Discussion

The multisim results confirmed that the proposed hyperchaotic system with hidden attractors does not have transitional chaos or transient behavior. The slightly changing values of electronic components can greatly affect the state of a system with hidden attractors. Although the nonlinear characteristics of this system have been carefully studied, the impact of accuracy and simulation time on the system remains to be further studied. This aspect also requires scholars to do deep theoretical research work.

7. Conclusions

In this study, a chaotic mathematical model with hidden attractors was constructed. Firstly, system parameters and initial values were found to affect system dynamics and system complexity. From the analysis of the bifurcation characteristics of the system parameters, it was found that there were complex hidden dynamic behaviors, such as periodicity, quasi-periodic, chaotic, and hyperchaotic. In particular, under different initial conditions, different topologies of chaotic attractors or quasi-periodic limit cycles coexisted with chaotic attractors, and quasi-periodic limit cycles coexisted with chaotic attractors of various topologies. This shows that the proposed system has multistable characteristics. Moreover, the entropy of the system was analyzed from several aspects, such as the entropy of changing parameters, the entropy of different initial values, and the entropy of the chaotic characteristics of the system parameters, proving that the proposed system has very rich dynamic characteristics. Finally, the behaviors of hidden attractors were observed in electronic circuits by the method of improved modular design.

The numerical simulations and circuit implementation presented in this paper prove that the proposed system is a multistable system. Because it is very sensitive to the initial value and has a rich topological structure, the system is suitable for encryption applications.

Author Contributions: Conceptualization, L.C. and C.D.; Methodology, L.C. and S.S.; Software, X.Z.; Validation, L.C., C.D., J.L., and X.Z.; Formal Analysis, L.C.; Writing—Original Draft Preparation, L.C.; Writing—Review & Editing, L.C., C.D., and X.Z.; Funding Acquisition, L.C. and C.D.; All authors have read and approved the final manuscript.

Funding: This research was funded by Natural Science Research Youth Project of the Department of Education of Guizhou Province of China, grant number [KY [2015] 465, and KY [2015] 470], Tripartite Joint Funds for Science and Technology Department of Guizhou Province of China , grant number [LH [2015] 7698, and LH [2015] 7697].

Acknowledgments: The authors would like to thank the three anonymous reviewers for their constructive comments and insightful suggestions.

Conflicts of Interest: The authors declare no conflict of interests.

References

1. Zhang, Y.; Wang, X. A new image encryption algorithm based on non-adjacent coupled map lattices. *Appl. Soft. Comput.* **2015**, *26*, 10–20. [CrossRef]
2. Zhang, Y.; Wang, X. Spatiotemporal chaos in mixed linear–nonlinear coupled logistic map lattice. *Physica A* **2014**, *402*, 104–118. [CrossRef]
3. Zhang, Y.; Wang, X. A symmetric image encryption algorithm based on mixed linear–nonlinear coupled map lattice. *Inf. Sci.* **2014**, *273*, 329–351. [CrossRef]
4. Luo, C.; Wang, X. Chaos Generated from the Fractional-Order Complex Chen System and Its Application to Digital Secure Communication. *Int. J. Mod. Phys. C* **2013**, *24*, 1350025. [CrossRef]
5. Lorenz, E.N. Deterministic nonperiodic flow. *J. Atmos. Sci.* **1963**, *20*, 130–141. [CrossRef]
6. Wang, X.; Wang, M. A hyperchaos generated from Lorenz system. *Physica A* **2008**, *387*, 3751–3758. [CrossRef]
7. Luo, C.; Wang, X. Chaos in the fractional-order complex Lorenz system and its synchronization. *Nonlinear Dyn.* **2012**, *71*, 241–257. [CrossRef]
8. Rössler, O.E. An equation for continuous chaos. *Phys. Lett. A* **1976**, *57*, 397–398. [CrossRef]
9. Leonov, G.A.; Kuznetsov, N.V.; Vagaitsev, V.I. Localization of hidden Chua's attractors. *Phys. Lett. A* **2011**, *375*, 2230–2233. [CrossRef]
10. Wang, X.; He, Y. Projective synchronization of fractional order chaotic system based on linear separation. *Phys. Lett. A* **2008**, *372*, 435–441. [CrossRef]

11. Wang, X.; Wang, M. Dynamic analysis of the fractional-order Liu system and its synchronization. *Chaos* **2007**, *17*, 033106. [CrossRef] [PubMed]

12. Volos, C.; Maaita, J.-O.; Vaidyanathan, S.; Pham, V.-T.; Stouboulos, I.; Kyprianidis, I. A Novel Four-Dimensional Hyperchaotic Four-Wing System With a Saddle–Focus Equilibrium. *IEEE Trans. Circuits Syst. II Express Briefs* **2017**, *64*, 339–343. [CrossRef]

13. Bo, Y.U.; Guosi, H. Constructing multiwing hyperchaotic attractors. *Int. J. Bifurc. Chaos* **2010**, *20*, 727–734.

14. Yujun, N.; Xingyuan, W.; Mingjun, W.; Huaguang, Z. A new hyperchaotic system and its circuit implementation. *Commun. Nonlinear Sci. Numer. Simul.* **2010**, *15*, 3518–3524. [CrossRef]

15. Dudkowski, D.; Jafari, S.; Kapitaniak, T.; Kuznetsov, N.V.; Leonov, G.A.; Prasad, A. Hidden attractors in dynamical systems. *Phys. Rep.* **2016**, *637*, 1–50. [CrossRef]

16. Singh, J.P.; Roy, B.K. Hidden attractors in a new complex generalised Lorenz hyperchaotic system, its synchronisation using adaptive contraction theory, circuit validation and application. *Nonlinear Dyn.* **2018**, *92*, 373–394. [CrossRef]

17. Bao, B.; Jiang, T.; Wang, G.; Jin, P.; Bao, H.; Chen, M. Two-memristor-based Chua's hyperchaotic circuit with plane equilibrium and its extreme multistability. *Nonlinear Dyn.* **2017**, *89*, 1157–1171. [CrossRef]

18. Munoz-Pacheco, J.; Zambrano-Serrano, E.; Volos, C.; Jafari, S.; Kengne, J.; Rajagopal, K. A New Fractional-Order Chaotic System with Different Families of Hidden and Self-Excited Attractors. *Entropy* **2018**, *20*, 564. [CrossRef]

19. Wang, C.; Ding, Q. A New Two-Dimensional Map with Hidden Attractors. *Entropy* **2018**, *20*, 322. [CrossRef]

20. Xu, G.; Shekofteh, Y.; Akgül, A.; Li, C.; Panahi, S. A New Chaotic System with a Self-Excited Attractor: Entropy Measurement, Signal Encryption, and Parameter Estimation. *Entropy* **2018**, *20*, 86. [CrossRef]

21. Jafari, S.; Sprott, J.C.; Nazarimehr, F. Recent new examples of hidden attractors. *Eur. Phys. J. Spec. Top.* **2015**, *224*, 1469–1476. [CrossRef]

22. Feng, Y.; Wei, Z. Delayed feedback control and bifurcation analysis of the generalized Sprott B system with hidden attractors. *Eur. Phys. J. Spec. Top.* **2015**, *224*, 1619–1636. [CrossRef]

23. Leonov, G.A.; Kuznetsov, N.V.; Mokaev, T.N. Hidden attractor and homoclinic orbit in Lorenz-like system describing convective fluid motion in rotating cavity. *Commun. Nonlinear Sci. Numer. Simul.* **2015**, *28*, 166–174. [CrossRef]

24. Jaros, P.; Perlikowski, P.; Kapitaniak, T. Synchronization and multistability in the ring of modified Rössler oscillators. *Eur. Phys. J. Spec. Top.* **2015**, *224*, 1541–1552. [CrossRef]

25. Li, C.; Hu, W.; Sprott, J.C.; Wang, X. Multistability in symmetric chaotic systems. *Eur. Phys. J. Spec. Top.* **2015**, *224*, 1493–1506. [CrossRef]

26. Sprott, J.C. Strange attractors with various equilibrium types. *Eur. Phys. J. Spec. Top.* **2015**, *224*, 1409–1419. [CrossRef]

27. Leonov, G.A.; Kuznetsov, N.V.; Kiseleva, M.A.; Solovyeva, E.P.; Zaretskiy, A.M. Hidden oscillations in mathematical model of drilling system actuated by induction motor with a wound rotor. *Nonlinear Dyn.* **2014**, *77*, 277–288. [CrossRef]

28. Scheffer, M.; Carpenter, S.; Foley, J.A.; Folke, C.; Walker, B. Catastrophic shifts in ecosystems. *Nature* **2001**, *413*, 591–596. [CrossRef] [PubMed]

29. Rietkerk, M.; Dekker, S.C.; de Ruiter, P.C.; van de Koppel, J. Self-organized patchiness and catastrophic shifts in ecosystems. *Science* **2004**, *305*, 1926–1929. [CrossRef] [PubMed]

30. Scheffer, M.; Bascompte, J.; Brock, W.A.; Brovkin, V.; Carpenter, S.R.; Dakos, V.; Held, H.; van Nes, E.H.; Rietkerk, M.; Sugihara, G. Early-warning signals for critical transitions. *Nature* **2009**, *461*, 53–59. [CrossRef] [PubMed]

31. Vaidyanathan, S.; Volos, C. Analysis and adaptive control of a novel 3-D conservative no-equilibrium chaotic system. *Arch. Control Sci.* **2015**, *25*, 333–353. [CrossRef]

32. Hu, X.; Liu, C.; Liu, L.; Ni, J.; Li, S. Multi-scroll hidden attractors in improved Sprott A system. *Nonlinear Dyn.* **2016**, *86*, 1725–1734. [CrossRef]

33. Chen, G.; Kuznetsov, N.V.; Leonov, G.A.; Mokaev, T.N. Hidden Attractors on One Path: Glukhovsky–Dolzhansky, Lorenz, and Rabinovich Systems. *Int. J. Bifurc. Chaos* **2017**, *27*, 1750115. [CrossRef]

34. Wolf, A.; Swift, J.B.; Swinney, H.L.; Vastano, J.A. Determining Lyapunov exponents from a time series. *Physica D* **1985**, *16*, 285–317. [CrossRef]

35. Wolf, A. Lyapunov exponent estimation from a time series. Documentation added. *Acta Biochim. Pol.* **2013**, *60*, 345–349.

36. Holmes, P. Poincaré, celestial mechanics, dynamical-systems theory and "chaos". *Phys. Rep.* **1990**, *193*, 137–163. [CrossRef]
37. Lauritzen, B. Semiclassical Poincare map for integrable systems. *Chaos* **1992**, *2*, 409–412. [CrossRef] [PubMed]
38. Kuznetsov, A.P.; Kuznetsov, S.P.; Mosekilde, E.; Stankevich, N.V. Co-existing hidden attractors in a radio-physical oscillator system. *J. Phys. A Math. Theor.* **2015**, *48*, 125101. [CrossRef]
39. Bao, H.; Wang, N.; Bao, B.; Chen, M.; Jin, P.; Wang, G. Initial condition-dependent dynamics and transient period in memristor-based hypogenetic jerk system with four line equilibria. *Commun. Nonlinear Sci. Numer. Simul.* **2018**, *57*, 264–275. [CrossRef]
40. Kingni, S.T.; Pham, V.-T.; Jafari, S.; Woafo, P. A chaotic system with an infinite number of equilibrium points located on a line and on a hyperbola and its fractional-order form. *Chaos Solitons Fractals* **2017**, *99*, 209–218. [CrossRef]
41. Kapitaniak, T.; Mohammadi, S.; Mekhilef, S.; Alsaadi, F.; Hayat, T.; Pham, V.-T. A New Chaotic System with Stable Equilibrium: Entropy Analysis, Parameter Estimation, and Circuit Design. *Entropy* **2018**, *20*, 670. [CrossRef]
42. He, S.; Sun, K.; Wang, H. Complexity Analysis and DSP Implementation of the Fractional-Order Lorenz Hyperchaotic System. *Entropy* **2015**, *17*, 8299–8311. [CrossRef]
43. Abedi, M.; Moghaddam, M.M.; Fallah, D. A Poincare map based analysis of stroke patients' walking after a rehabilitation by a robot. *Math. Biosci.* **2018**, *299*, 73–84. [CrossRef] [PubMed]
44. Shahzad, M.; Pham, V.T.; Ahmad, M.A.; Jafari, S.; Hadaeghi, F. Synchronization and circuit design of a chaotic system with coexisting hidden attractors. *Eur. Phys. J. Spec. Top.* **2015**, *224*, 1637–1652. [CrossRef]

entropy

MDPI

Article

Bogdanov Map for Modelling a Phase-Conjugated Ring Resonator

Vicente Aboites [1],*[iD], David Liceaga [2], Rider Jaimes-Reátegui [3][iD] and Juan Hugo García-López [3][iD]

1   Centro de Investigaciones en Óptica, Loma del Bosque 115, 37150 León, Mexico
2   División de Ciencias e Ingenierías, Universidad de Guanajuato, Loma del Bosque 107, 37150 León, Mexico; d.l.s.0612@gmail.com
3   Centro Universitario de los Lagos, Universidad de Guadalajara, Enrique Diaz de León 1144, Paseos de la Montaña, Lagos de Moreno, 47460 Jalisco, Mexico; rider.jaimes@gmail.com (R.J.-R.); jhgarcial@gmail.com (J.H.G.-L.)
*   Correspondence: aboites@cio.mx; Tel.: +52-4774414200

Received: 24 October 2018; Accepted: 4 December 2018; Published: 10 April 2019

Abstract: In this paper, we propose using paraxial matrix optics to describe a ring-phase conjugated resonator that includes an intracavity chaos-generating element; this allows the system to behave in phase space as a Bogdanov Map. Explicit expressions for the intracavity chaos-generating matrix elements were obtained. Furthermore, computer calculations for several parameter configurations were made; rich dynamic behavior among periodic orbits high periodicity and chaos were observed through bifurcation diagrams. These results confirm the direct dependence between the parameters present in the intracavity chaos-generating element.

Keywords: spatial dynamics; Bogdanov Map; chaos; laser; resonator

---

1. Introduction

Matrix description of optical systems through $ABCD$ matrices (Equation (8)) naturally produces iterative maps with rather complex dynamics. Several publications have dealt with the $ABCD$ law and the iterative maps it produces. Belanger [1] has generalized the $ABCD$ propagation law for optical systems Onciul [2], using the Kirchhoff integral, derives a generalized $ABCD$ propagation law for general astigmatic Gaussian beams through misaligned optical systems, Bastiaans [3] shows under what condition the well-known $ABCD$ law that can be applied to describe the propagation of one-dimensional Gaussian light through first-order optical systems (or $ABCD$ systems) can be extended to more than one dimension; in the two-dimensional (or higher-dimensional) case, an $ABCD$ law only holds for partially coherent Gaussian light for which the matrix of second-order moments of the Wigner distribution function is proportional to a symplectic matrix. Tian [4] presents an iterative method for simulating beam propagation in nonlinear media using Hamiltonian ray tracing in which the Wigner distribution function of the input beam is computed at the entrance plane, used as the initial condition for solving the Hamiltonian equation; he gives examples for the study of periodic self-focusing, spatial solitons and the Gaussian–Schell model in Kerr-effect media. Finally, Siegman [5] and Tarasov [6] shown how to describe a laser resonator with iterative matrix optics by ray propagation through cascaded optical elements. This kind of map has been successfully applied before to the description of laser beams within optical resonators. This treatment has been explored for several other maps, obtaining several chaos-generating intracavity elements that are based on the dynamical behavior from widely diverse maps, such as the Ikeda map [7], Standard map [8], Tinkerbell map [9–11], Duffing map [11,12], logistic map [13] and the Henón map [11,14]. Throughout this article the Bogdanov Map will be used to describe a ring-phase conjugated resonator, while the resultant iterative matrix system is analyzed. In the following Section 2, a quick derivation of the Bogdanov map is sketched following reference [15], then will convert our *two*-dimensional mapping into a theoretical optical element that will produce the same complex dynamical behavior as the Bogdanov map within a phase-conjugated ring resonator. To accomplish this, we introduce the $ABCD$ matrix formalism that is commonly used in paraxial optics [16], allowing us to represent each optical component as a 2×2 matrix. Moving forward with the previously obtained results, finding what we call *Bogdanov beams*; these are beams that propagate within the resonator following dynamics of the Bogdanov map. In Section 3, we discuss the results obtained from numerical calculations displaying the rich dynamics of the system, as it is shown in the bifurcation diagrams as a function of the intracavity chaos-generating element parameters. Finally, Section 4 presents the conclusions.

2. Material and Methods

2.1. Bogdanov Map

This map was originally conceived by Bogdanov while studying the universal unfolding of the double-zero-eigenvalue singularity [17] (also called *Bogdanov–Takens* or *cusp*), which is the equivalent of a vector field invariant under a rotation of the plane by 2π. The Bogdanov map can be obtained by means of discretization using the Euler method on the Bogdanov vector field. Next, to be thorough and closely follow reference [15], we proceed to sketch a quick derivation of the Bogdanov Map.

$$\begin{aligned} \dot{y} &= \theta, \\ \dot{\theta} &= 0 \end{aligned} \tag{1}$$

This vector field has a codimension-two fixed point at the origin, known as a double-zero-eigenvalue singularity; the normal form of this can be written as follows:

$$\begin{aligned} \dot{y} &= \theta + \lambda y^2, \\ \dot{\theta} &= \eta y^2 \end{aligned} \tag{2}$$

where $\lambda \neq 0$, $\eta \neq 0$. A two-parameter *versal* unfolding for this normal form, which contains all possible qualitative dynamical behavior near Equation (2), can be given:

$$\begin{aligned} \dot{y} &= \theta + v_2 y + \lambda y^2, \\ \dot{\theta} &= v_1 + \eta y^2 \end{aligned} \tag{3}$$

The unfolding given above is not unique and a versal unfolding or deformation such as Equation (3) contains all possible qualitative dynamical behavior that can occur near the singularity. By restricting our attention to the region away from the saddle-node bifurcations, the Hamiltonian system of ordinary equations first considered by Bogdanov can be obtained,

$$\begin{aligned} \dot{y} &= \theta \\ \dot{\theta} &= y(y-1) \end{aligned} \tag{4}$$

once again, a two-parameter versal unfolding is obtained for Equation (4),

$$\begin{aligned} \dot{y} &= \theta \\ \dot{\theta} &= u_1\theta + y(y-1) + u_2 y\theta \Xi(y, u_1, u_2) + u_2^2 \theta^2 \Phi(y, \theta, u_1, u_2) \end{aligned} \tag{5}$$

By taking the vector field from Equation (5), and applying the backward Euler discretization method to the first equation ($\dot{y}$) and the forward Euler method to the second equation ($\dot{\theta}$), both with step length h, we obtain

$$
\begin{aligned}
y_{n+1} &= y_n + h\theta_{n+1} \\
\theta_{n+1} &= \theta_n + hu_1\theta_n + hy_n(y_n - 1) + hu_2 y_n \theta_n \Xi(y_n, u_1, u_2) + hu_2^2 \theta_n^2 \Phi(y_n, \theta_n, u_1, u_2)
\end{aligned}
\tag{6}
$$

now making $\Xi(y_n, u_1, u_2) = 1$, $\Phi(y_n, \theta, u_1, u_2) = 0$ and multiplying the second equation by h. Finally, making the change of variables $u_1 = \varepsilon/h$, $u_2 = \mu/h$, $h\theta = \tilde{\theta}$, $h^2 = k$ and dropping the tilde from θ, we get the Bogdanov Map.

$$
\begin{aligned}
y_{n+1} &= y_n + \theta_{n+1} \\
\theta_{n+1} &= \theta_n + \varepsilon\theta_n + ky_n(y_n - 1) + \mu y_n \theta_n
\end{aligned}
\tag{7}
$$

The Bogdanov map is a planar quadratic map, conjugate to the Hénon-area-preserving map in its conservative limit ($\varepsilon = \mu = 0$). Here, ε and μ are related to the Bogdanov vector field, while k plays the role of step length in the discretization, such that for a small k, the map behavior will resemble the original vector field. The dissipative Hopf parameter ε determines the birth and growth from the origin for the primary Hopf invariant circle; the stability of this circle is determined by μ, while the Hamiltonian discretization parameter k determines the birth and growth of the island chains.

2.2. Paraxial Matrix Analysis

The description of ray or Gaussian optics with matrices turns both the analysis and composition of optical systems into a simple and straightforward task, since this technique allows us to represent the behavior of any optical element as a 2×2 matrix. Cylindrical symmetry is used around the optical axis, so that for any given position z both the perpendicular distance of any ray to the optical axis (y) and its angle with the same axis (θ) can be defined; thus, any optical system can be represented by an $[ABCD]$ matrix,

$$
\begin{pmatrix} y_{n+1} \\ \theta_{n+1} \end{pmatrix} = \begin{pmatrix} A & B \\ C & D \end{pmatrix} \begin{pmatrix} y_n \\ \theta_n \end{pmatrix}
\tag{8}
$$

In passive optical elements (mirrors, lenses, interfaces between two media, etc.), elements A, B, C, D are constant; nevertheless, for nonlinear optical elements, they are not necessarily constant, but may be functions of different parameters; The description of an optical system described by a Bogdanov Map requires (from Equation (7)) that the coefficients A, B, C, D be:

$$
\begin{pmatrix} A & B \\ C & D \end{pmatrix} = \begin{pmatrix} 1 & \dfrac{\theta_{n+1}}{\theta_n} \\ k(y_n - 1) & 1 + \varepsilon + \mu y_n \end{pmatrix}
\tag{9}
$$

where the value $\dfrac{\theta_{n+1}}{\theta_n}$ can be written as

$$
\frac{\theta_{n+1}}{\theta_n} \equiv 1 + \varepsilon + y_n\left[\frac{k}{\theta_n}(y_n - 1) + \mu\right]
$$

In Figure 1, we sketch the diagram of our optical system, where the $[a, b, c, e]$ matrix is the unknown map generating device, located between the plain mirrors M_1 and M_2 at a distance $d/2$, while M_3 is a phase-conjugated mirror. For this system, the total transformation $[ABCD]$ matrix for a complete round trip is written as follows:

$$\begin{pmatrix} A & B \\ C & D \end{pmatrix} = \begin{pmatrix} 1 & 0 \\ 0 & -1 \end{pmatrix} \begin{pmatrix} 1 & d \\ 0 & 1 \end{pmatrix} \begin{pmatrix} 1 & 0 \\ 0 & 1 \end{pmatrix} \begin{pmatrix} 1 & d/2 \\ 0 & 1 \end{pmatrix}$$
$$\times \begin{pmatrix} a & b \\ c & e \end{pmatrix} \begin{pmatrix} 1 & d/2 \\ 0 & 1 \end{pmatrix} \begin{pmatrix} 1 & 0 \\ 0 & 1 \end{pmatrix} \begin{pmatrix} 1 & d \\ 0 & 1 \end{pmatrix} \tag{10}$$

which gives

$$= \begin{pmatrix} a + \frac{3cd}{2} & b + \frac{3d}{4}(2a + 3cd + 2e) \\ -c & -\frac{3cd}{2} - e \end{pmatrix} \tag{11}$$

$$A = a + \frac{3cd}{2}$$
$$B = b + \frac{3d}{4}(2a + 3cd + 2e)$$
$$C = -c$$
$$D = -\frac{3cd}{2} - e$$

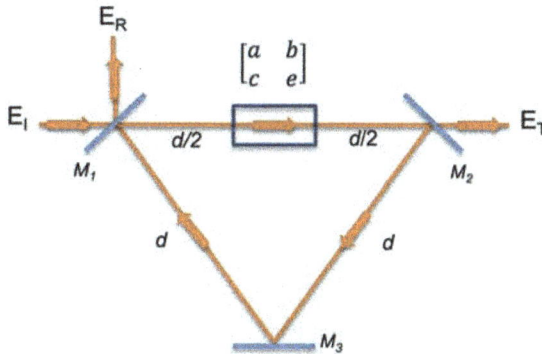

Figure 1. Phase-conjugated ring resonator with an intracavity chaos-generating element.

To reproduce the behavior of the Bogdanov map by means of a ray within the optical ring resonator, each round trip described by (y_n, θ_n) must be considered as an iteration of the Bogdanov map. Next, we take the previously obtained $[ABCD]$ matrix elements of the Bogdanov map, Equation (9), and equate them to the total $[ABCD]$ matrix of the resonator, Equation (11); this in order to generate the round-trip map dynamics for (y_{n+1}, θ_{n+1}). Note here that the results obtained are only valid for a small b value, $(b \approx 0)$: this is because before and after the matrix element $[a, b, c, e]$, there is a propagation of $(d - b)/2$. Meanwhile, for a general case, Equation (11) ought to be replaced by the following:

$$\begin{pmatrix} A & B \\ C & D \end{pmatrix} = \begin{pmatrix} 1 & 0 \\ 0 & -1 \end{pmatrix} \begin{pmatrix} 1 & d \\ 0 & 1 \end{pmatrix} \begin{pmatrix} 1 & 0 \\ 0 & 1 \end{pmatrix} \begin{pmatrix} 1 & \frac{d-b}{2} \\ 0 & 1 \end{pmatrix}$$
$$\times \begin{pmatrix} a & b \\ c & e \end{pmatrix} \begin{pmatrix} 1 & \frac{d-b}{2} \\ 0 & 1 \end{pmatrix} \begin{pmatrix} 1 & 0 \\ 0 & 1 \end{pmatrix} \begin{pmatrix} 1 & d \\ 0 & 1 \end{pmatrix} \tag{12}$$

which gives

$$\begin{pmatrix} a - \frac{c}{2}(b-3d) & \frac{1}{4}\left[b^2c - 2b(-2+a+3cd+e) + 3d(2a+3cd+2e)\right] \\ -c & \frac{1}{2}(bc - 3cd - 2e) \end{pmatrix} \tag{13}$$

$$A = a - \frac{c}{2}(b - 3d)$$

$$B = \frac{1}{4}\left[b^2c - 2b(-2 + a + 3cd + e) + 3d(2a + 3cd + 2e)\right]$$

$$C = -c$$

$$D = \frac{1}{2}(bc - 3cd - 2e)$$

This is the total round-trip transformation matrix for the general case.

2.3. Bogdanov Beams

We define 'Bogdanov beams' as beams that behave on the y_n and θ_n optical ray parameters according to the Bogdanov Map given by Equation (7), i.e., Beams produced in the above optical resonator that undergo the Bogdanov map dynamics will be called 'Bogdanov beams'. To obtain the Bogdanov beams, the matrix elements of Equation (9) must be equaled to the elements of Equation (11), thus giving the system.

$$a + \frac{3cd}{2} = 1$$
$$b + \frac{3d}{4}(2a + 3cd + 2e) = 1 + \varepsilon + y_n\left[\frac{k}{\theta_n}(y_n - 1) + \mu\right]$$
$$-c = k(y_n - 1)$$
$$-\frac{3cd}{2} - e = 1 + \varepsilon + \mu y_n \tag{14}$$

This system is solved to obtain the $[a, b, c, e]$ matrix elements. Therefore, the intracavity matrix that produces Bogdanov Beams is

$$\begin{pmatrix} a & b \\ c & e \end{pmatrix} = \begin{pmatrix} [1 + \frac{3}{2}kd(y_n - 1)] & \frac{\theta_{n+1}}{\theta_n} - \frac{3}{2}d\left\{\frac{3}{2}kd(y_n - 1) - \varepsilon - \mu y_n\right\} \\ -k(y_n - 1) & -[1 + \varepsilon + \mu y_n + \frac{3}{2}kd(y_n - 1)] \end{pmatrix} \tag{15}$$

2.4. General Case for Bogdanov Beams

Taking the elements of matrix Equation (9) and equating them to the ones of matrix Equation (13), we get the following system, which is analogous to Equation (14):

$$a - \frac{c}{2}(b - 3d) = 1$$

$$\frac{1}{4}\left[b^2 c - 2b\alpha + 3d\beta\right] = 1 + \varepsilon + y_n\left[\frac{k}{\theta_n}(y_n - 1) + \mu\right]$$

$$-c = k(y_n - 1)$$

$$\frac{1}{2}(bc - 3cd - 2e) = 1 + \varepsilon + \mu y_n$$

(16)

Here $\alpha = (-2 + a + 3cd + e)$ and $\beta = (2a + 3cd + 2e)$.

Solving the system found in Equation (16), we find two new $[a, b, c, e]$ matrices, Equations (17) and (18). These matrices contain all the dynamic information of the Bogdanov map taking into account the thickness b of the intracavity element,

$$\begin{pmatrix} a & b \\ c & e \end{pmatrix} = \begin{pmatrix} \frac{1}{60\theta_n}(\vartheta_n - \gamma_n) & \frac{1}{3k\theta_n(y_n - 1)}(\varphi_n + \gamma_n) \\ k(1 - y_n) & \frac{1}{60\theta_n}(\varrho_n - \gamma_n) \end{pmatrix}$$

(17)

$$\begin{pmatrix} a & b \\ c & e \end{pmatrix} = \begin{pmatrix} \frac{1}{60\theta_n}(\vartheta_n + \gamma_n) & \frac{1}{3k\theta_n(y_n - 1)}(\varphi_n - \gamma_n) \\ k(1 - y_n) & \frac{1}{60\theta_n}(\varrho_n + \gamma_n) \end{pmatrix}$$

(18)

were $\gamma, \vartheta, \varphi, \varrho$ are defined as:

$$\gamma_n \equiv \{\theta_n[-12k^2(y_n - 1)^2 y_n + \theta_n[36k^2 d^2(y_n - 1)^2 + (2 + \varepsilon + \mu y_n)^2 \\ -12k(y_n - 1)(1 + \varepsilon + \mu y_n + d(-1 + \varepsilon + \mu y_n))]]\}^{1/2}$$

$$\vartheta_n \equiv \theta_n(8 + \varepsilon + 12kd(y_n - 1) + \mu y_n)$$

$$\varphi_n \equiv -\theta_n(2 + \varepsilon + 3kd(y_n - 1) + \mu y_n)$$

$$\varrho_n \equiv \theta_n(-4 - 5\varepsilon + 12kd(y_n - 1) - 5\mu y_n)$$

the intracavity chaos-generating matrix, whose b_n element is given as follows;

$$b_n \equiv \frac{1}{3k\theta_n(y_n - 1)}(\varphi_n - \gamma_n)$$

(19)

3. Results

3.1. Computer Calculations

The dynamic behavior of the phase-conjugated resonator in phase space was studied through numerical iteration of the obtained matrices, Equations (17) and (18). To find valid trajectories on the phase plane values for y_n, θ_n must be real numbers at every iteration, diverging trajectories are only mathematical possibilities since they cannot be related to any physical reality given that they do not meet the stability requirements to stay within the resonator. Also, the b_n intracavity element from the matrices must be greater than zero at every iteration, while being smaller than the mirror resonator separation distance. These conditions ensure that the trajectories are on the real phase plane and within a stable trajectory, greater than zero at every iteration, given that the b_n element is related to the total round-trip distance traveled by the Bogdanov beam within the resonator. The last condition ensures that no 'negative distances' are traveled.

Iterations were carried out using Equation (18) for values of the control parameter d, where the iterations (y_n, θ_n) have physical meaning. The system displays high periodicity for $0.91 < d < 1$, Figure 2a. Also, a short region of low periodicity appears within a high periodicity range where $d = 0.99$ Figure 2b. For $d > 0.9925$, the Bogdanov beam resonator exhibits a period-doubling route to chaos, Figure 2c.

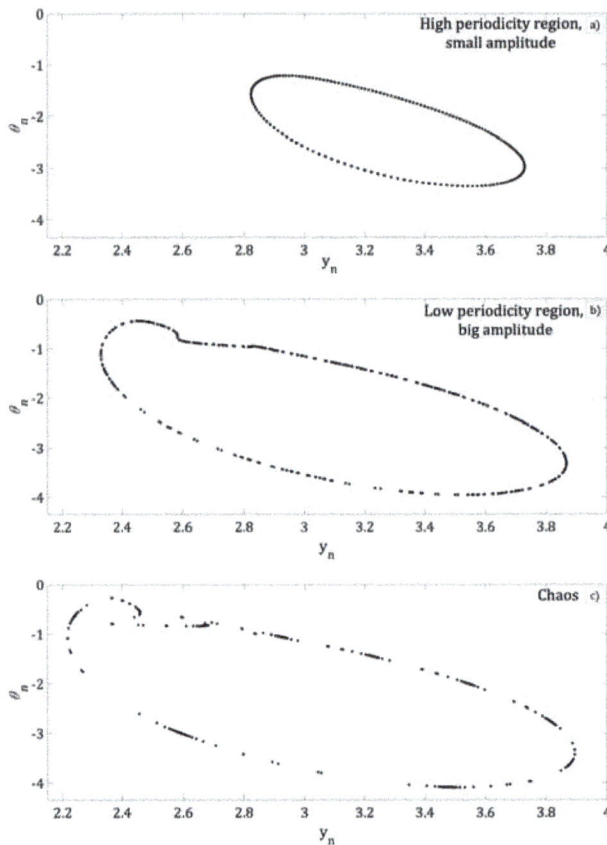

Figure 2. Phase space (y_n, θ_n), equivalent to a round trip inside the resonator for (**a**) $d = 0.95$, (**b**) $d = 0.99$ and (**c**) $d = 0.998$; in all cases $k = 0.295$, $\varepsilon = 0.01$ and $\mu = -0.1$.

The bifurcation diagram of b_n with was obtained to understand the dependence of the intra cavity nonlinear element b_n with respect to parameters; d, k and ε of the Bogdanov map. Advantages of this bifurcation diagram is that it gives a global view of the dynamic element b_n as one or several parameters are changed.

Figure 3a shows the bifurcation diagram of local max of b_n as a function of parameter d. In this figure, high periodicity is interrupted by regions of low periodicity windows and a route to chaos by period-doubling is shown. The same result is also shown while plotting the temporal Inter Peak Intervals (IPI) of b_n as the parameter d is varied, Figure 3b. Comparing Figure 3a and Figure 3b, it is shown that Figure 3b clearly illustrates a rich dynamics that shows high periodicity for $0.91 < d < 1$, Figure 2a, interrupted by low periodicity windows of for $d = 0.99$, Figure 2b. The bifurcation diagrams show a route to chaos due to period-doubling, Figure 2c.

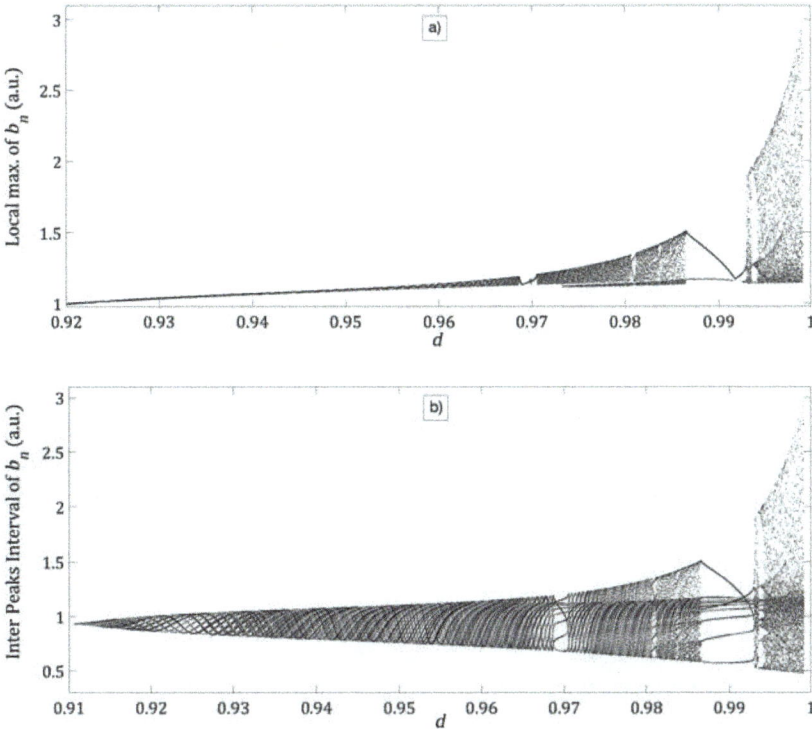

Figure 3. (**a**) Bifurcation diagram of local max of b_n as a function of parameter d. (**b**) Temporal inter peak interval (IPI) of b_n as a function of parameter d; in both plots, the following fixed values were used: $k = 0.295$, $\varepsilon = 0.01$ and $\mu = -0.1$.

As can be seen, the dependence of the intra cavity nonlinear element b_n to parameter d of the phase-conjugated ring resonator has been shown. In the following figures, dependence of b_n on the k and ε parameters of the Bogdanov map will be displayed. Figure 4 shows the bifurcation diagram of the local max of b_n as a function of parameter k. Although Figure 4 is qualitatively similar to Figure 3a clear difference is noted when the low periodicity windows are considered. It can be observed that when the parameter k is increased, the region of high periodicity is interrupted by windows with low values of periodicity, i.e., for $k = 0.2964$, and for $k = 0.30735$ exhibits a route to chaos by period-doubling. The phase space (y_n, θ_n) for particularly cases of high and low periodicity and chaos is shown in Figure 5a–c respectively.

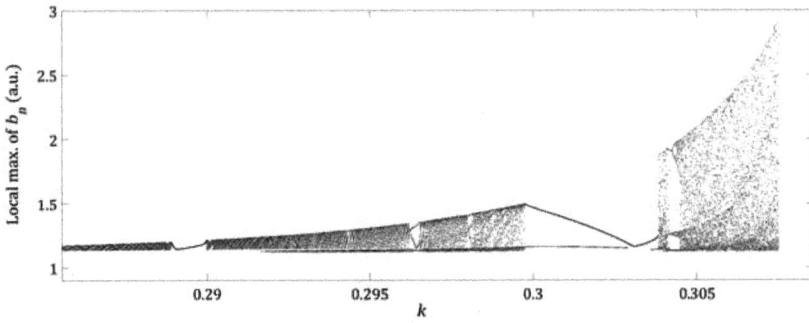

Figure 4. Bifurcation diagram of local max of b_n as a function of parameter k, for $d = 0.9837$, $\varepsilon = 0.01$ and $\mu = -0.1$.

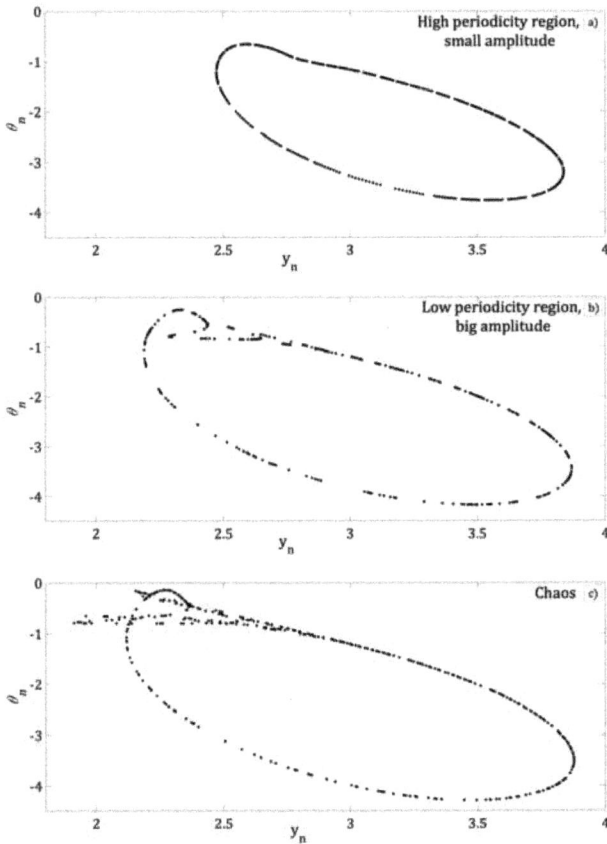

Figure 5. Phase space (y_n, θ_n). High periodicity for (**a**) $k = 0.2925$, low periodicity for (**b**) $k = 0.30434$ and chaos for (**c**) $k = 30735$, in all cases $d = 9837$, $\varepsilon = 0.01$ and $\mu = -0.1$.

In addition, the phase space (y_n, θ_n) for different values of k with d fixed in chaotic region are plotted in Figure 5, while the bifurcation diagram of local max b_n as a function of ε for same values of k and d, are plotted in Figure 6. In this figure, we can see that for $k = 0.2894$, Figure 5a, the bifurcation diagram of local max b_n presents a high periodicity for all the range of control parameter ε; see Figure 6a. With further increase of parameter k to values of $k = 0.30434$, Figure 5b, the bifurcation diagram Figure 6b show a short interval of ε where the local max of b_n exhibits low periodicity windows that interrupts a region of high periodicity. Finally, for $k = 0.30735$ (chaotic region of Figure 5c), Figure 6c shows regions of high periodicity interrupted by low periodicity windows and a large region of route to chaos by period-doubling as control parameter ε is increased.

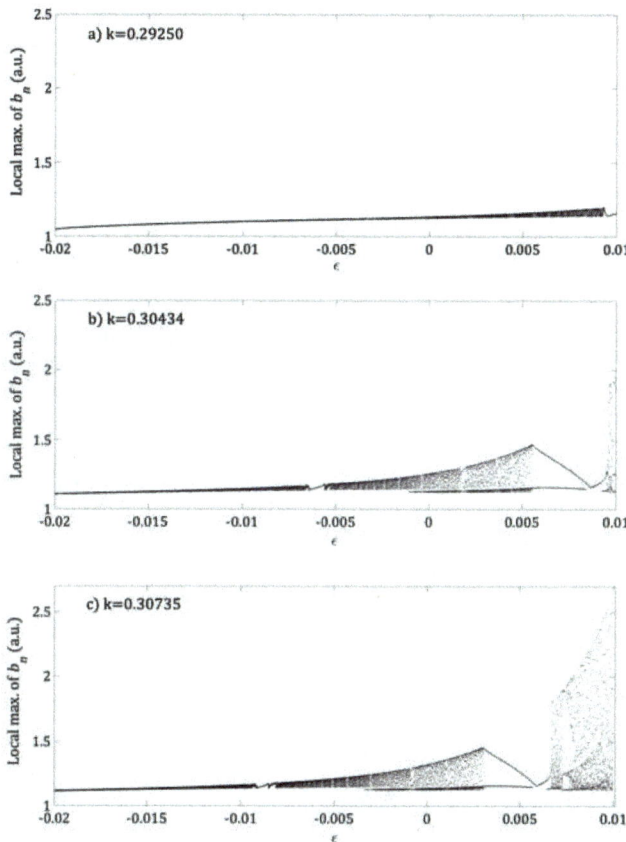

Figure 6. The bifurcation diagram of local max b_n as a function of ε for three different values of k with d fixed in the chaotic region, and $\mu = -0.1$. (**a**) $k = 0.2925$, (**b**) $k = 0.30434$ and (**c**) $k = 0.30735$.

4. Conclusions

In this paper, a matrix transformation over the Bogdanov map is proposed to obtain an intra cavity element that can yield the same rich, dynamical behavior within a phase-conjugated ring resonator. We began our study by obtaining the Bogdanov Map through the use of Euler method for discretization over the Bogdanov Vector Field; then, we introduced the paraxial matrix analysis (or $ABCD$ propagation law): this was done in order to simplify the analysis for the complete resonator system, enabling us to express this system as a simple dynamical matrix Equation (8). Once these central concepts had been introduced, we proceed to obtain what we call "Bogdanov Beams", which are beams produced in an optical resonator undergoing the Bogdanov map dynamics. Then, we studied a simple case of 'Bogdanov Beams' where the thickness of the intra cavity element is considered to be negligible. Next, we moved on to the general case, where the thickness of the intracavity element is greater than zero. While it may seem a trivial difference, this general case introduces a new parameter d in our final matrix transformation, which adds up to the three initial parameters from the Bogdanov Map (k, ε, μ), therefore increasing the dimension of the problem and contributing to the non-linearity of the map. Once the explicit expressions for the general case were obtained, Equations (17) and (18), computer programs were made that allowed us to search the 4-dimensional parameter space for combinations that yield stable trajectories; this is no easy task, since the stability of the trajectories is also dependent on the initial values (y_0, θ_0), due to this, often the trajectories will not have physical meaning; it is important to remark that we analyzed valid intervals of the parameters $(k, \varepsilon, \mu$ and $d)$. We have found that the intracavity element, b_n, Equation (19), is responsible for the different dynamic behavior of the optical resonator. The response of b_n to the parameters $(k, \varepsilon, \mu$ and $d)$ by bifurcation diagrams of local max and IPI of time series of b_n has been accomplished.

The dependence of b_n with respect to d, which is the distance between plain mirrors of the phase-conjugated ring resonator showed low, high periodicity and route to the chaos by period-doubling behavior, see Figure 3. Similar behavior was observed when the dependence of b_n was analyzed with respect to the parameters k, ε while μ and d were fixed, see Figure 4. Interesting results were found for the dependence of b_n on the parameter ε for different fixed values of k. For a small value of $k = 0.2925$, the bifurcation diagram shows high periodicity of low amplitude, see Figure 6a. With an increment of $k = 0.30434$, we have low periodicity windows within high periodicity regimens, see Figure 3b. Finally, at $k = 0.30735$, the bifurcation diagram of local max of b_n, shows rich dynamics, with low and high periodicity regions and a route to chaos by period-doubling, see Figure 6c.

Based on the behavior observed, we conclude that the matrix transformation used was successful in generating a dynamical system that preserves the main structures found in the Bogdanov map. The practical implementation of an intracavity element is a complex technical challenge far beyond the aim of this work. Interested readers on this matter may consult reference [9].

Author Contributions: V.A. conceived and designed the work; D.L., J.H.G.-L. and R.J.-R. performed the simulations; V.A. and J.H.G.-L. analyzed the data; R.J.-R. contributed with analysis tools; V.A. wrote the paper.

Acknowledgments: J.H.G.-L. and R.J.-R. acknowledge to the University of Guadalajara for financial support under the projects R-0138/2016, Agreement RG/019/2016 and RC/075/2018, Agreement RG/006/2018, UdeG, Mexico. V.A. acknowledges support and useful conversations with Ernst Wintner from TU-Wien and Matei Tene from TU-Delft. The authors acknowledge the professional English proof reading service provided by Mario Ruiz Berganza.

Conflicts of Interest: The authors declare no conflict of interest.

References

1. Bélanger, P.A. Beam propagation and the ABCD ray matrices. *Opt. Lett.* **1991**, *16*, 196–198. [CrossRef] [PubMed]
2. Onciul, D. ABCD propagation law for misaligned general astigmatic Gaussian beams. *J. Opt.* **1992**, *23*, 163. [CrossRef]
3. Bastiaans, M. ABCD law for partially coherent Gaussian light, propagating through first-order optical systems. *Opt. Quantum Electron.* **1992**, *24*, S1011–S1019. [CrossRef]
4. Tian, L. Iterative nonlinear beam propagation using Hamiltonian ray tracing and Wigner distribution function. *Opt. Lett.* **2010**, *35*, 4148–4150.
5. Siegman, A.E. *Lasers*; University Science Books: Herndon, VA, USA, 1986.
6. Tarasov, L. *Physique Des Processus Dans Les Generaeurs De Rayonnement Optique Coherent*; (trans. from Russian); Mir, Moscu: Gent, Belgium, 1981.

7. Aboites, V.; Liceaga, D.; Kir'yanov, A.; Wilson, M. Ikeda Map and Phase Conjugated Ring Resonator Chaotic Dynamics. *Appl. Math. Inf. Sci.* **2016**, *10*, 1–6. [CrossRef]
8. Aboites, V.; Wilson, M.; Lomeli, K. Standard Map Spatial Dynamics in a Ring-Phase Conjugated Resonator. *Appl. Math. Inf. Sci.* **2015**, *9*, 2823–2827.
9. Aboites, V.; Barmenkov, Y.; Kiryanov, A.; Wilson, M. Tinkerbell beams in a non-linear ring resonator. *Results Phys.* **2012**, *2*, 216–220. [CrossRef]
10. Aboites, V.; Wilson, M. Tinkerbell chaos in a ring phase-conjugated resonator. *Int. J. Pure Appl. Math.* **2009**, *54*, 429–435.
11. Aboites, V.; Barmenkov, Y.; Kiryanov, A.; Wilson, M. Bidimensional dynamics maps in optical resonators. *Rev. Mex. Fis.* **2014**, *60*, 13–23.
12. Dignowity, D.; Wilson, M.; Rangel-Fonseca, P.; Aboites, V. Duffing spatial dynamics induced in a double phase-conjugated resonator. *Laser Phys.* **2013**, *23*, 076002. [CrossRef]
13. Aboites, V. Dynamics of a LASER Resonator. *Int. J. Pure Appl. Math.* **2007**, *36*, 345–352.
14. Aboites, V.; Huicochea, M. Henon beams. *Int. J. Pure Appl. Math.* **2010**, *65*, 129–136.
15. Arrowsmith, D.K.; Cartwright, J.H.; Lansbury, A.N.; Place, C.M. The Bogdanov Map: Bifurcations, Mode Locking, and Chaos in a Dissipative System. *Int. J. Bifurc. Chaos* **1993**, *3*, 803–842. [CrossRef]
16. Hallbach, K. Matrix Representation of Gaussian Optics. *Am. J. Phys.* **1964**, *32*, 90. [CrossRef]
17. Bogdanov, R.I. Versal deformations of a singularity of a vector field on the plane in the case of zero eigenvalues. *Sel. Math. Sov.* **1981**, *1*, 373–388.

MDPI

St. Alban-Anlage 66

4052 Basel

Switzerland

Tel. +41 61 683 77 34

Fax +41 61 302 89 18

www.mdpi.com

Entropy Editorial Office

E-mail: entropy@mdpi.com

www.mdpi.com/journal/entropy